DROSS CITY

urban metabolism resilience and recycling projects for the drosscapes

etited by
carlo gasparrini
anna terracciano

re-cycle italy

NAPOLI

2013 - 2016

team

Carlo Gasparrini [Principal Investigator] **Vito Cappiello** **Massimo Fagnano** **Fabrizia Ippolito** **Rocco Lafratta** **Marina Rigillo** **Michelangelo Russo** **Roberto Serino**

lab

Fabrizia Ippolito [Local Coordinator] **Anna Terracciano** [Operational Coordinator] **Libera Amenta** **Susanna Castiello** **Daniele Cannatella** **Emanuela De Marco** **Cecilia Di Marco** **Davide di Martino** **Francesca Fasanino** **Enrico Formato** **Paola Galante** **Massimo Lanzi** **Francesco Stefano Sammarco** **Antonella Senatore** **Ciro Sepe** **Sabrina Sposito** **Danilo Vinaccia**

partners

Saxion Universities of Applied Sciences [The Netherlands] **Universitat Politécnica de Catalunya** [Spain] **Master de Arquitectura del Paisaje** [Spain] **Autorità di Bacino Campania Centrale** **Comune di Napoli** [Department of Urban Planning] **CNR: IRC Istituto Ricerche sulla combustione** **Progetto LIFE Ecoremed**

index

166 PART 4 **RECYCLING PROJECTS, LANDSCAPE PROJECTS, CITY PLANNING**

6

Introduction

CRISTIN
NAPOLI
IMO 9455985

DrossCity. About the City of waste

Carlo Gasparrini

The city of waste and garbage - the "Drosscity" - is not only an efficient naming practice referred to the numerous aspects of the "dark side of change"[1]. It expresses a deeper demand, that of bringing forward a different narration of the City, able to interpret and represent the dissipative dynamics connected to its metabolism and to the challenges of urban phenomena, its functioning and pathologies. This narration of the City begins from the landscape and ecological consequences produced by those processes, thus subverting the traditional approach that goes from the general to the specific, from a systematic interpretation of the whole to detailed analyses of specific fallouts on the urban space. The reasons for this inversion are multiple. First of all, the persuasion that the communicative and argumentative strength of dross and waste mappings can significantly contribute to move the focus of attention towards a better knowledge of the dimensions and intensity of these dissipative phenomena, of which we still have a modest and surely underestimated perception. This rise of awareness is crucial for the consolidation - in the diversified reality of public and private stakeholders and the widespread culture of inhabitants, manufacturers, users and consumers of the City - of an ecological culture oriented towards urban regeneration. The communicative strength of a shared narrative, able to connect the "here" and "now" through wider and systematic diagrammatic interpretations on urban scale, fuels relevant widespread cognitive processes from both the tactical and strategic points of view.
A second reason, closely connected to the first, is innate in the inseparable link between dissipative processes and the life cycle depletion of certain natural and anthropic resources. The traditional functioning, renovation and crisis of certain production sectors, such as the agricultural, extractive, construction, industrial and SMEs, but also of the distribution, trade, storage and logistics, as well as the real estate sector and the managerial crises of big networks and metabolic "urban machines" particularly linked to the waste and water

1. See Lynch K., Wasting away, Sierra Club Books, 1990 (translated in Italian by Southwork M. and Andriello V., "Deperire", Cuen, 1992).

cycles, produce consequences on the life cycles of the City's structuring resources. Water and soil in particular but also certain building fabrics and infrastructures become part of the "dross" geography, often resulting from a diffused critical ecological and social condition that urges for more attention and action to defend and enhance the "common goods".
The germs of an alternative, growth-oriented, socially inclusive and economically innovative urban metabolism stem from this set of values and the business initiatives attributable to it. A metabolism that fills the open spaces and the built environment, fostering new life cycles of the resources and diversified forms of management anchored to discrete-sized territories controlled by local communities. This discretization can shape and elicit broader systemic achievements based on alternative scenarios in relation with incremental dynamics, with a low or zero rate of dissipation and under high social control. Such metabolism no longer claims the centralization of modernity. Instead, it advocates a constellation of "hybrid machines" corresponding to forms of aggregation of local scale, capable of ensuring effective forms of management of water, soil, energy, waste and soft mobility and of creating a network for the necessary multi-level decisions.

An updated notion of the meaning and geography of drosscapes

"Dross" and "waste", but more generally urban metabolism and environmental problems, are no news in the urban debate and in our current experience of the city. These issues have been largely discussed in modern times and, more recently, were considered exciting and potentially fruitful opportunities for the future of the contemporary city[2]. Often concealed, they started developing within the nineteenth-century hygienist culture of the new technological networks and urban remediation, and lasted across the past two centuries. They justified some paradigmatic choices of important plans in the first half of the twentieth century, such as those of Cologne and

2. C. Gasparrini, "Editoriale. Oltre la sostenibilità", Crios n. 9, 2015.

Copenhagen by Fritz Schumacher and Steen Eiler Rasmussen in the '20s and '40s respectively. They fed the need to replace an "outdated" "urban planning" with the introduction of an "urban biology" in the vision of José Luis Sert in the '40s. They helped introducing the pioneering concept of "Metabolism of cities" by Abel Wolman[3] in the mid-60s, often resumed and abandoned over the last fifty years. Those issues have then found adequate space and a central position within the broader environmental issue that, in recent years, has promptly appeared on the agendas of all Western countries, driven by a major concern for the "disease" that had spread from the cities to the entire planet. Among the most significant contributions of the last decades, we can quote the micro stories of "Wasting away" by Kevin Lynch[4] and the extra-large stories of the great and photogenic North American "drosscapes" narrated by Alan Berger[5]. But the most explicit attempt to propose an urban-scale interpretative dimension for the territories of waste and dross remains that of Lars Lerup[6]. The dynamic coexistence, the contamination and the conflict between places in the development phase ("stim") and refused places ("dross") is still relevant as a reference for the narratives of urban metabolism. Nevertheless, this field of work is virtually absent in the national urban plans and policies, with the mere exception of timid sectoral programs set aside from the city and its actors. Terms like "drosscape", "wasteland" and "brownfield" have been used for years in the stories of these territories in the cities of the planet. But their underlying meanings express very different situations. The term "drosscape" contains the concept of "dross" as scrap[7], which has a wider latitude than "waste" - if considered in a limiting connotation of "junk" - and also a wider field of observation strictly confined to contaminated, abandoned territories of industrial disposal. At the same time it contains the suffix of the term landscape ("scape") that extends its gaze far beyond the mere description of the phenomena of alteration of the affected lands and soils ("land" or "field"), opening up to a multidimensional interpretation and project. The extension of the field inaugurated by Berger in the North American experience led to a configuration of an articulated taxonomy[8] that insists on the association between drosscapes and the deindustrialization processes, particularly pervasive and widespread in the United States[9].

In the "Recycle-Italy" research project, the universities that investigated this field of study[10] have outlined a methodological repositioning in the European and Italian context through both a narrative strategy of urban scale, and a reformulation of taxonomy based on the many environmental, economic, functional and social dynamics that determine the exhaustion of the life cycles of the urban and suburban areas and their progressive metamorphosis into drosscapes, sometimes characterized by the emergence of new life cycles of different nature and role. In this sense, they have proposed a systemic view of such changes that is closely related to the dissipative

3. A. Wolman, "The metabolism of cities", Scientific American, n.213, 1965.

4. See K. Lynch, "Wasting away", quote.

5. See A. Berger, "Drosscape. Wasting Land in Urban America", Princeton Architectural Press, 2007. and A. Berger, "Designing the Reclaimed Landscape", Taylor and Francis, 2008 and A. Berger, "Systemic Design can change the world", SUN, 2009.

6. See L. Lerup, "Stim & Dross: Rethinking the Metropolis", in "Assemblage" n°25, MIT Press, 1995

7. See A. Berger, "Drosscape. Wasting Land in Urban America", quote.

8. Residual voids in the fabric ("landscapes of dwelling"); temporary depository ("landscapes of transition"); interstitial spaces ("landscapes of infrastructure"); landfills ("landscapes of obsolescence"); abandoned shopping malls ("landscapes of exchange"); military bases and other brownfields ("landscapes of contamination"). See A. Berger, "Drosscape. Wasting Land in Urban America", quote.

9. "Drosscapes accumulate in the wake of the socio - and spatio - economic processes of deindustrialization, post-Fordism and technological innovation. [They] are located in the declining, neglected and deindustrializing areas of cities", A. Berger, "Drosscape. Wasting Land in Urban America", quote.

10. University of Rome "La Sapienza", University of Naples "Federico II", University of Chieti-Pescara "G. D'Annunzio" e Polytechnic of Bari. A first resume of the outcomes of this activity can be found and analyzed in R. Pavia, R. Secchi, C. Gasparrini (edited by), "Il territorio degli scarti e dei rifiuti", Aracne, 2014, that combines the theoretical considerations, the interpretations and the early project implementations of the first two years of research. See C. Gasparrini, Recycling drosscapes in Europe, in R. Bocchi, L. Fabian, S. Munarin (edited by), "Atlante Recycle Italy", Letter n. 22, 2016

Thomas Hirschhorn, "Flamme éternelle", Palais de Tokio, Paris

Zürich, "Out to Sea_The Plastic Garbage Project", Museum für Gestaltung, 2012

Federico Figa Talamanca, Hillary Powell's The Games, 2007

processes of the urban metabolism, including some extreme and illegal forms of its dynamics.
These waste areas constitute a critical porosity of the town. Here, the environmental outcomes caused by the exhaustion and the lack of control over the production, agricultural and commercial cycle, and the distribution and logistics not only produce derelict areas and abandoned buildings, brownfields and residual and interstitial spaces. In addition, as already mentioned, they affect and even accelerate the exhaustion of the life cycle of soils as well as the cycle of surface and ground waters by critically compromising the traditional urban metabolism, which is also linked to the energy and waste cycles[11]. The demand expressed by abandoned soils and buildings due to the exhaustion of the expansive construction cycle is another aspect to be considered. Starting from the tourist segment, it produces increasingly damaged patterns, long affected by social and economic replacement processes with new categories of users and inhabitants, temporary or permanent, generated by the migration of "wasted human beings" escaping from "wasted lives"[12]. Such a process intercepts the variegated and unstable nebula of the urban sprawl, of the large monofunctional urban boundaries and of the most vulnerable fabrics. It undermines the mosaic of agricultural landscapes and infiltrates in the interstices of the infrastructure network and its satellite activities - including the precarious logistics and the scrapping of used vehicles - fed by the often deepening gaps in the civic and industrial waste cycle in which the segment of toxic debris has acquired, over time, a huge role[13]. The macro and micro porosity of the city are crossed by flows and expectations of different consistency and origin, hindered by specialized or uncertain and transient uses and the leftovers of quickly disappeared economies, and by approaches to ecology that fail to work due to fragmented decisions and an irrational use of resources. Behind this approach is the awareness that we are in a phase of radicalization of the ecological, economic and social processes related to the multiplicity of environmental risks especially in urban areas, also as a result of the occurring climate change, that amplifies those risks. This requires an update of the concept of urban metabolism, since it has flinched its way through the last two centuries deliberately neglected in its planning and management by a major segment of the planning culture. This update encourages policies, plans and programs that are strongly characterized by resilient objectives, projects and practices, able to fertilize the quality and adaptability of the urban landscapes, recovering and raising some narrative traditions of the city in the recent past on the basis of new environmental paradigms .
We can no longer rely on a linear metabolism that configures the city as a great urban machine, consumer of unlimited resources and producer of waste to be disposed of, according to a functionalist "input-output" dynamic that is strongly pervasive due to the resulting dissipative processes.

11. See C. Gasparrini, "Riciclare drosscapes a Napoli", in S. Marini e V. Santangelo (edited by), "Nuovi cicli di vita per architetture e infrastrutture della città e del paesaggio", Re-Cycle Itay n.01, Aracne, 2013

12. Z. Bauman, "Wasted lives. Modernity and its Outcasts", Polity Press, 2003.

13. See P. Bélanger, "Landscape as Infrastructure", in "Landscape Journal" n°28, 2009; P. Bèlanger, "Landscape infrastructure: urbanism beyond engineering", in S. N. Pollalis, A. Georgoulias, S. J. Ramos e D. Shodek, "Infrastructure sustainability and design", Routledge, 2011. On the same topic, with reference to the situation in Neaples, see C. Gasparrini, "Riciclare drosscapes a Napoli", quote, and C. Gasparrini, "The waste side of change. Drosscape and reverse city", in "Crios" n°8, 2014

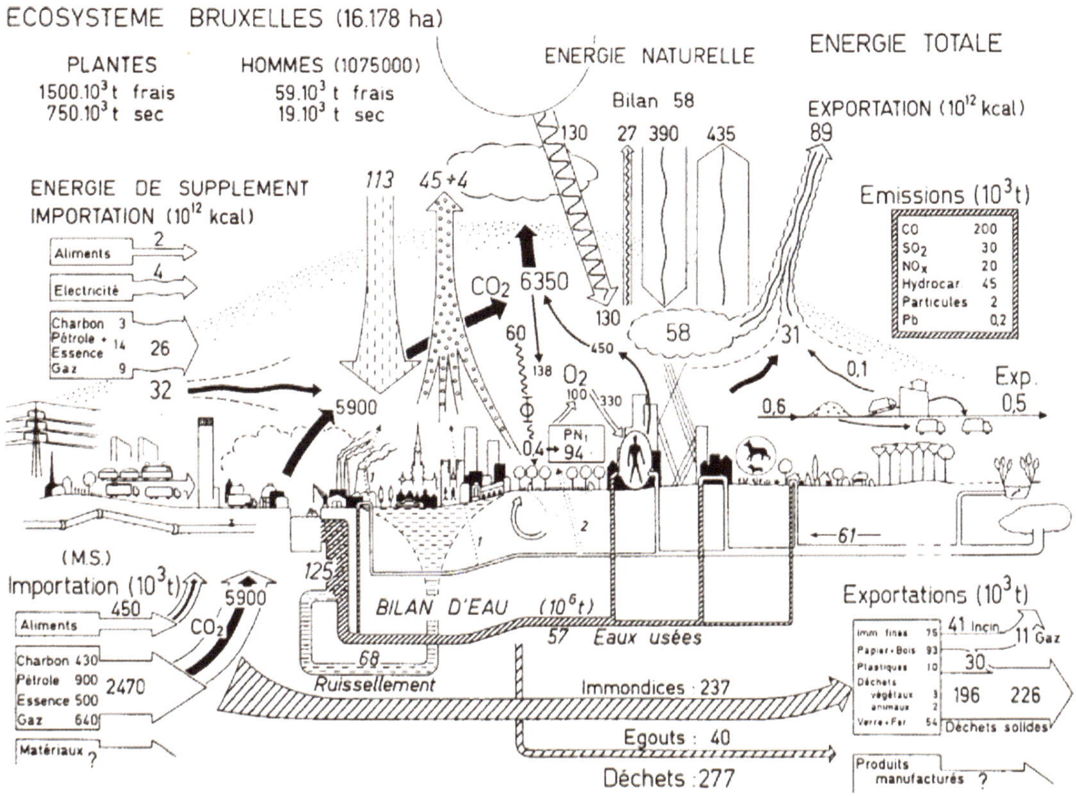
ECOSYSTEME BRUXELLES (16.178 ha)
PLANTES
1500.10³t frais
750.10³t sec
HOMMES (1075000)
59.10³ t frais
19.10³ t sec
ENERGIE NATURELLE
Bilan 58
ENERGIE TOTALE
EXPORTATION (10¹² kcal)
89
130
27 390 435
113 45+4
ENERGIE DE SUPPLEMENT
IMPORTATION (10¹² kcal)
Aliments 2
Electricité 4
Charbon 3
Pétrole + Essence 14
Gaz 9
26
32
CO2 6350
60
138
O2
100
330
450
130
58
31
PN1 94
0,4
5900
Emissions (10³t)
CO 200
SO2 30
NOx 20
Hydrocar 45
Particules 2
Pb 0,2
0,1
Exp. 0,5
0,6
61
(M.S.)
Importation (10³t)
450
Aliments
CO2
5900
125
Charbon 430
Pétrole 900
Essence 500
Gaz 640
2470
Matériaux ?
BILAN D'EAU (10⁶t)
68
Ruissellement
57 Eaux usées
Immondices : 237
Egouts : 40
Déchets : 277
Exportations (10³t)
Imm. fines 75
Papier + Bois 93
Plastiques 10
Déchets végétaux 3
animaux 2
Verre + Fer 54
41 Incin.
11 Gaz
30
196 226
Déchets solides
Produits manufacturés ?

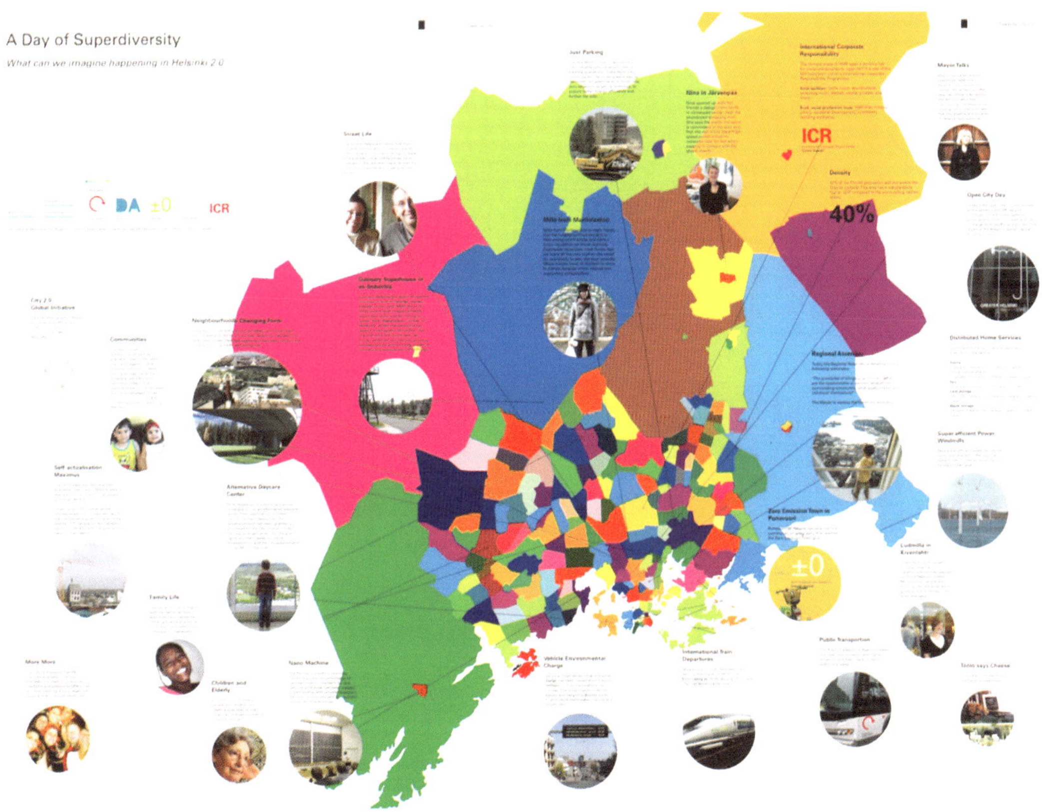
A Day of Superdiversity
What can we imagine happening in Helsinki 2.0
DA ±0
ICR
40%
±0

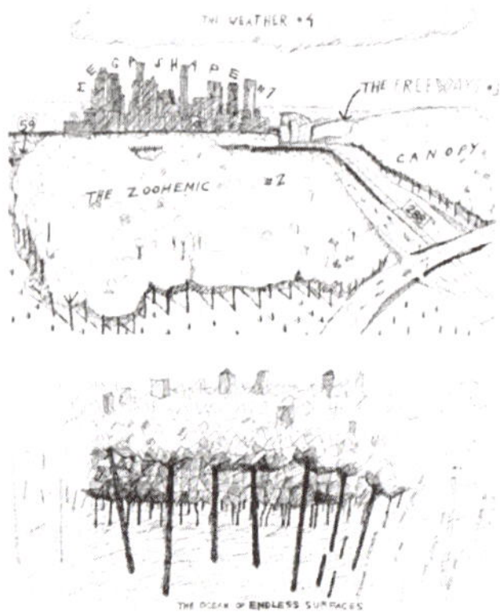

This image is taken from L. Lerup, "Stim & Dross: Rethinking the Metropolis", in "Assemblage" n. 25, MIT Press, 1995.

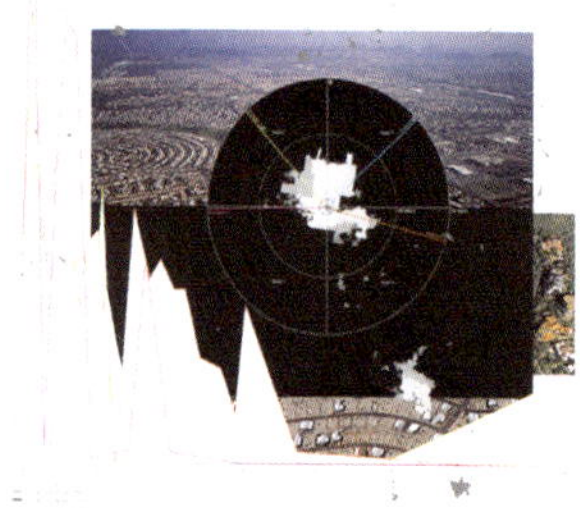

This image is taken from A. Berger, "Drosscape. Wasting Land in Urban America", Princeton Architectural Press, 2007.

On the opposite page: DrossDataScape in Italy, by Anna Terracciano.

It is therefore necessary to achieve a more complex metabolism, decentralized and equitable, based on recycling and on a more articulated synergy between an essential systemic dimension, rationalized in multilevel interactions, and widespread / incremental forms of management. This synergy must be able to respond simultaneously, with resilient projects and practices, to the scarcity of non-renewable resources, the irregularities and latency of the risks generated by urban development in recent decades, and the changing climatic conditions which exacerbate these dynamics[14].

The return of "monsters" in urban planning

In recent decades, the pervasive rhetoric of the landscape and the space-time simplifications of cartographic representations have not been able to describe the contextual dynamics of miniaturization / bigness and overlay / spacing of urban materials in the contemporary city. Neither did they provide an account of the dynamics of the flows passing through them, making it impossible to read their relationship with the places and with the irregular, wavy patterns of their uses by varying combinations of self-referential social actors.

Therefore, those rhetorics and simplifications have also intentionally and overtly opposed the re-emergence of "monsters"[15] linked to the crazy dynamics of urban metabolism, those places filled with junk and dross that are often invisible to the distant and mechanic eyes of aerophotogrammetric detections and standardized topographic codifications, both over and under the "skin of the planet."[16]

Today, however, the "drosscapes/monsters" are once again seeking adequate space in our interpretive descriptions and this requires a radical rethinking of the traditional tools to shift the priority -established by modernity - from the scales to the values. Or, perhaps, it is even possible to reach a virtuous balance between these two urban narration options, yet so distant and conflicting.[17]

"The city as we know it hides a multitude of unknown activity beyond its memory and topography". This makes up "the dark side of our allegedly sterilized reality" in which the line of separation between the desirable and the unwanted is drawn[18]. When the dross and the waste come into contact with "the respectable and the prized", the pervasive outcomes of this loss of control in the process undermine the city and the rationality of the separation between "what is valuable and what is not".

The construction of appropriate modes of interpretation of the "dark side of change" - to use again the aforementioned definition by Kevin Lynch - is absolutely necessary to handle those dilated spaces caused by urban explosion, in which the "monsters" of urban metabolism live, happily ignored by the current cartographic productions.

This approach does not, in fact, belong to the traditional - zenithal or diagonal - mode of representation. Over the past few decades, it has rather been the search methodology of

14. C. Gasparrini, "Politiche e progetti urbani di fronte alle domande ambientali della città diffusa", in F. D. Moccia, M. Sepe (edited by), "Una politica per le città italiane", Inu Publishers, "Accademia", 2015

15. The term comes from a popular essay by F. Farinelli, "I mostri, l'Atlante e il mondo", in Id., "I segni del mondo. Immagine cartografica e discorso geografico in età moderna", La Nuova Italia, 1992

16. This definition originated the title of W. B. Logan's, "La pelle del pianeta. Storia della terra che calpestiamo", Bollati Boringhieri, 2011

17. See C. Gasparrini, "Waste, Drosscape and Project in the Reverse City", in R. Pavia, R. Secchi, C. Gasparrini (edited by), "Il territorio degli scarti e dei rifiuti", Aracne, 2014. On the same topic, see P. O. Rossi, "Dross mapping. Le terre dei mostri" in the same text.

18. J. Scanlan, "On garbage", Reaktion Books, 2005

f_t3|dr_05
f_t4|dr_05
f_t4|dr_04
2|dr_01
f_t5|dr_15
f_t3|dr_07
f_t5|dr_07
f_t5|d
U
R
V

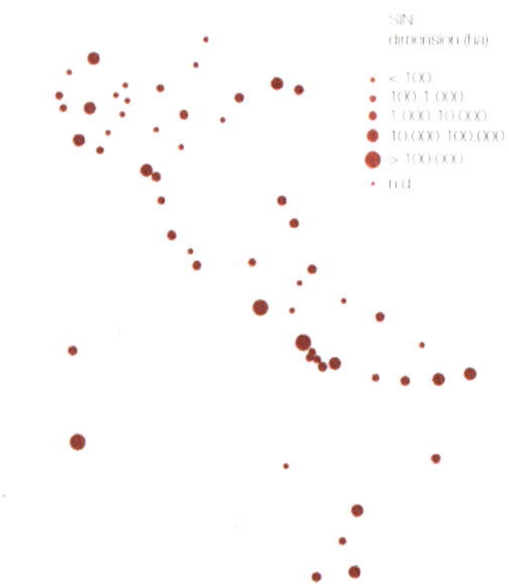

Sites of National (SIN) or Regional (SIR) Interest. Source: Decree of the Minister for Environment of 11/01/2018.

Industrial districts and sites at high risk of accident (RIR). Source: ISPRA, Rapporto 2013.

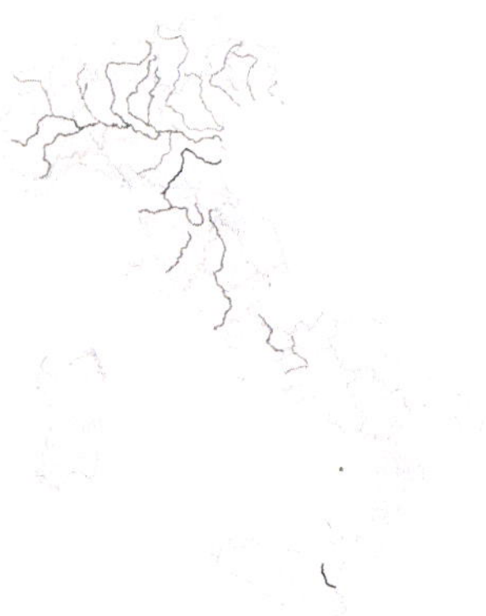

Quality of surface waters. Source: ISPRA, Rapporto 2013; Legambiente, Rapporto 2016.

site-specific photographers, writers and artists. More rarely, it was evident in the images of the city in the plans and projects of architects and urban planners.
On the other hand, this is not about adding a further "layer" to GIS (Geographic Information Systems), but rather to substantially modify the forms and techniques of representation through an interpretive gaze supported by an explicit planning intent, reaffirming the centrality of topology with respect to topography . In fact, the dynamic and three-dimensional representation of the soil and water life cycle - in crisis due to the pervasive processes of production of city waste - puts under tension both the traditional descriptive categories of space and the abstract and discontinuous temporal measurement of the surveys. Thefore, the need to intercept and properly describe those processes tends to produce a "deformation" of the space / time in order to render their intensity and variability.
The augmented reality we experience and explore everyday through Google Earth and its derivatives suggests a different way of interpretive experimentation and representation, able to meet some needs that we now perceive as ordinary and familiar. We demand non-linear narratives that can enable two-directional scale crossing through a continuous switch from the general overview to the multidimensional collage of images, data, testimonies and stories of urban space and its constantly evolving practices. In this sense, the interpretative description maintains a fertile tension between the selective and communicative design of the city "in between" - deformed by its "monsters" and his projects - and the continuous incursions in the variety of its materials, places and flows. After gaining some experience in this field of research, some of the Universities involved proposed forms of drossmapping able to explain the stratigraphic / relational dimension of the drosscape life cycles and of some of the supply chains that generated them, in space and time and in relation to the new content of the recycle project. The maps that represent the geography of dross do not only tell us the dissipative processes caused by the cycles of production, management, commerce, logistics and real estate - legal and illegal, seen and unseen. Neither do they tell us only the effects that those processes produce on the consumption and exhaustion of the life cycles of limited resources such as water and soil and the resulting impoverishment of the ecosystem. They also tell us about the crisis of a decision-making and management model of localization, operation, treatment and disposal of the waste generated by the economic activities and the machines of urban metabolism, generally opaque and poorly controlled by local communities. A model that resulted in a widespread distrust towards public action and strong actors, sometimes unjustified and paradoxically paralyzing. Traditional maps do not reveal these dynamics, nor the processes that support them, the integrated dimension of the risks they produce , their interactions with the quality of living. The "monsters" have

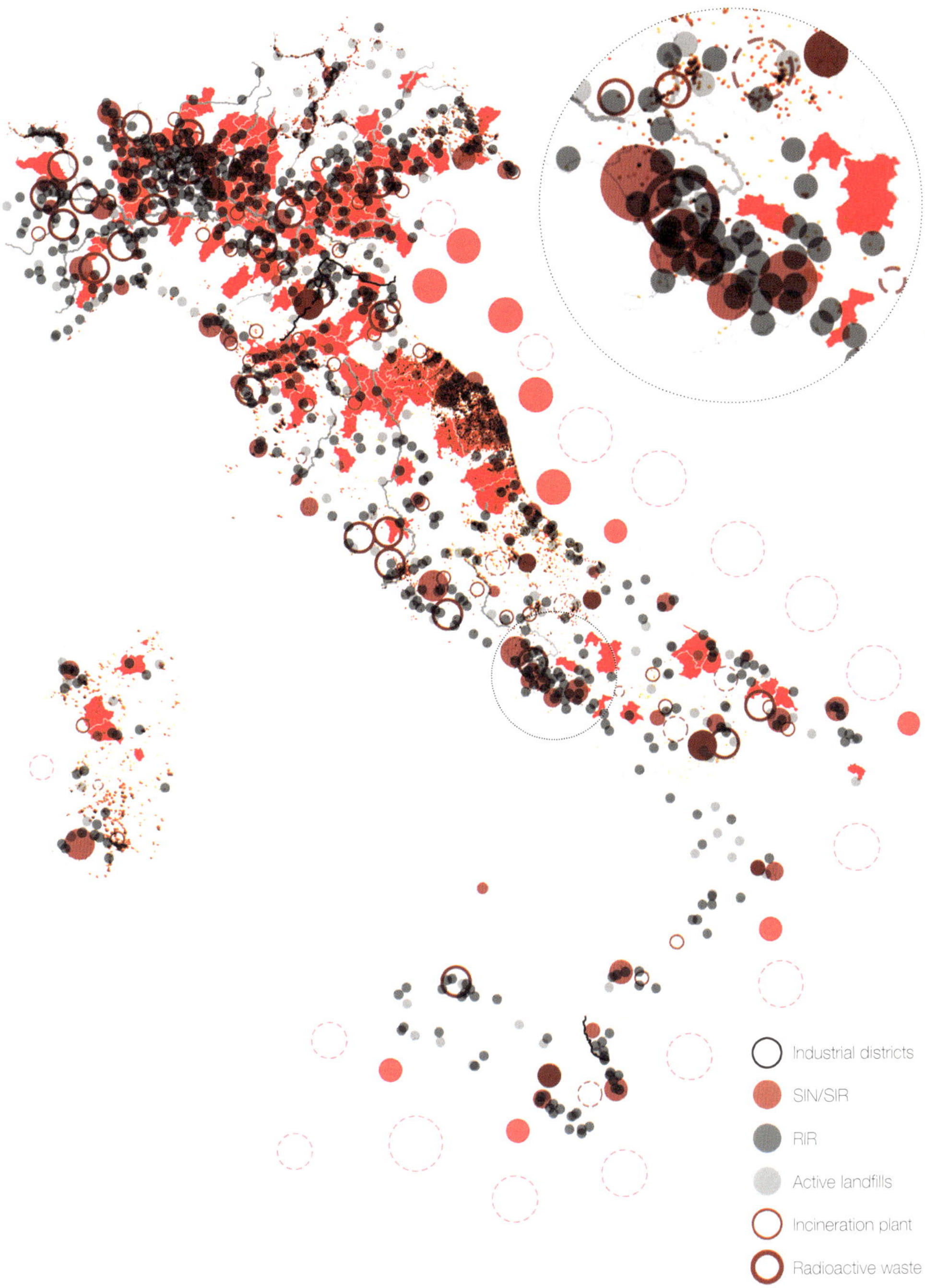

Areas for the depletion of main life cycles, resources and industrial supply chains. Data processing includes several sources and databases, such as:
www.distretti.org/
www.minambiente.it/
www.isprambiente.gov.it/it
www.sviluppoeconomico.gov.it/
www.wwf.it/

Legambiente, "Reclamation of Polluted Sites, Chimaera or reality?", Rome, 2014; Legambiente, "Rivers and legality. Monitoring of illegality and the status of italian rivers", 2016; Parlamentary Commission of Inquiry on illicit waste trafficking based on regional data, 2012; MATTM and INAIL, Piano Nazionale Amianto: stato di attuazione e prospettive future.

Asbestos

Class 0
Class 1
Class 2
Class 3
Class 4
Class 5

Hydrocarbon

Operating platforms
Platforms on hold
Wells on the ground
Wells on hold

disappeared from the current planning representations. Best case scenario, a constellation of blacks holes to be targeted as specialists or landscape decorators of spatial and ecological phenomena that cannot be put in question. Therefore, the narrative proposed by the "Recycle - Italy" research project and presented in this book has a significant value for the future of the city because it can question, together with the existing urban metabolism, the same ways of living and producing (in) the city, to prioritize, and to approach questions, lifestyles, and social practices that urban planning is struggling to identify.

Images are taken from "Detroit Strategic Framework Plan", 2012.

The new forms of the 'recycling drosscapes' project

The narrative of the "dross" and of the metabolism that originates this phenomenon values and promotes new, different and updated insights and views, stimuli and hypotheses that, in most cases, were already present in the roots of the recent and previous planning. In particular, this narrative identifies other urban stories of recent decades, resulting in a process of mutual fertilization. Let's start with the story of "reverse city"[19], the overthrowing in perspective that, over the last 30 years, encouraged us to focus on the open spaces and to promote the "soil project"[20]. A different way of looking at the city planning that has changed over time, acquiring a central role that is not only morphological, also but geo-strategic, social and connected to the landscape. It is about a new, quality public town built in the net-fabric of the urban explosion and its metamorphoses. A second story, set in the large context of an increasingly pervasive environmental issue, builds up on the concerns related to the multiple risks produced by climate change and their amplifications, focusing on resilience policies and adaptive strategies. In this sense, it identifies the natural and man-made resources currently at risk and urges the application of the necessary resilience strategies and tactics, working on urban tales that are strongly characterized by a systemic rationality and, at the same time, by a social awareness whose incremental frame of blue and green infrastructure is perhaps the more robust and meaningful field of work to regain possession of compromised resources and re-think the city[21].

In fact, the triangulation of these three stories - open spaces and soil project; resilience and blue and green infrastructures; metabolism and urban recycling - come together in a common process to deeply renovate the interpretative instruments and forms of the project that can ward off the ever-present temptation of using the easy way-out: sectorial segmentation.

The territorial and landscape dimension of drosscapes requires multi-scale recycling strategies, to cover anything from single aspects to large urban and territorial relations. Strategies that are able to interpret the interaction between the critical aspects of the environment, the infrastructure and the settlements and the opportunities to build entirely new landscapes, alternative economic models and sustainable energy cycles, within

19. For an overview on the different positions and experiences on this topic though history see the chapter "Reverse City" in P. Viganò, "La città elementare", Skira 1999

20. B. Secchi, "Progetto di suolo", in "Casabella", n. 520-521, 1986

21. See C. Gasparrini, "Un cambio di paradigma per l'urbanistica delle città resilienti", in "Urbanistica", n. 154, 2016

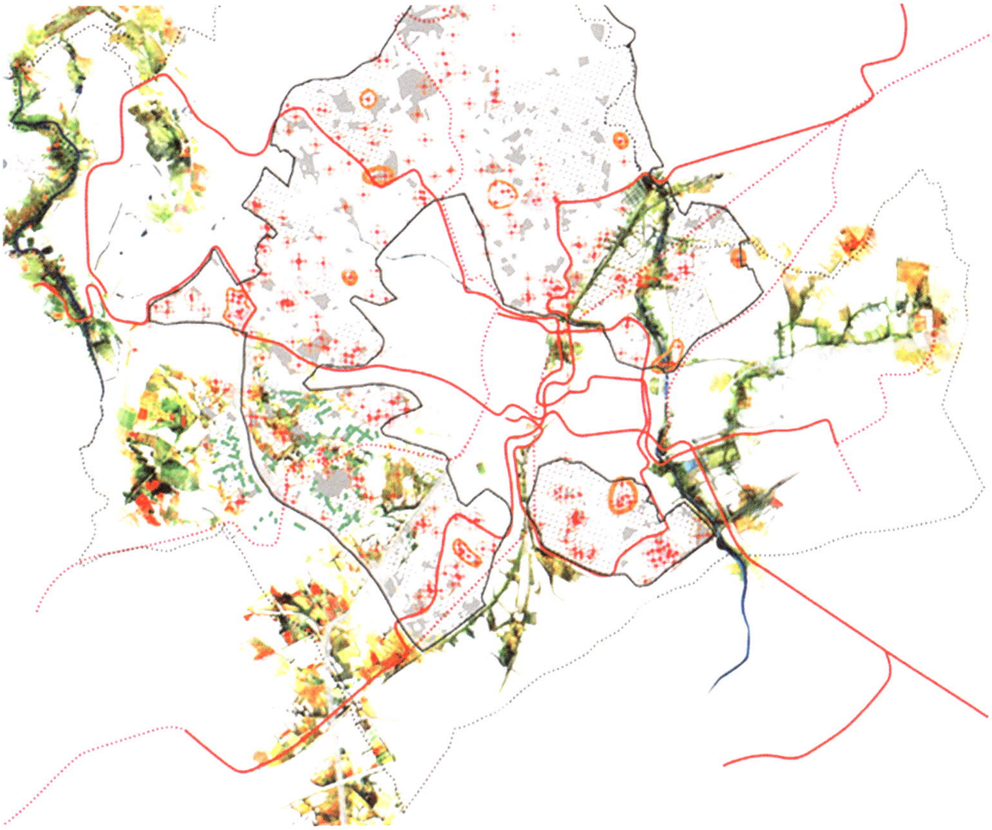

Images are taken from "Etude stratégique urbaine et paysagère de gestion des délaissés", Colocò, 2009-2012.

Images are taken from "Detroit Strategic Framework Plan", 2012.

scenarios envisaging ecological regeneration and a spatial reconfiguration of the contemporary city[22].

The ecological and urban impact of drosscapes, as well as their effects on production and meaning, extend well beyond the compromised sites, involving a multiplicity of spaces that are not only brown but also grey and green, often deeply affected by the contamination of water, soil and air, with a net-shaped domino effect on substantial parts of the ecosystems and on the urban fabric.

Around the concept of drosscape revolve a growing number of projects, policies, resources, widespread actions of recycling and non-traditional practices for - sometimes temporary - uses, in a structural phase of scarce public funds. This mainly takes place in areas of dense urban sprawl and interstitial spaces, abandoned and contaminated, that the public sector cannot handle by itself . The paradigms of the multi-scale recycling project stress the need to hold together in time and space, and through forms of multilevelled governance, the local dimension of the tactics - fragmentary and diffuse - with the centrality of adaptive and resilient systemic strategies. They also urge the search for rules, addresses and updated forms of representation for the innovation of plans and projects, with reference to compatible uses, temporary or long-term, diagrammatic simulations, parameters to "measure" the effectiveness of project proposals, and the activation modes of direct and indirect resources.

This growing attention is part of a trajectory that combines the objective to envisage new life cycles of 'discarded' buildings and open spaces in the various forms of urban settlement with a resurfacing of the geostrategic characters of the city in urban planning, after a long period of expansion and land use. The most significant strategic perspective of the recycling of drosscapes is in fact to participate in a wider and more radical shift of attention and priority towards two of the major directions in planning. On the one hand, the central role of blue, green and slow infrastructure as an innovative landscaping model of the ecologic dimension of the city and its metabolism. Another key factor is the 'public' offer of high-value spaces and a renewed productive background based on green economy[23].

On the other hand, the need to stimulate the urban-environmental patterns of existing settlements currently in crisis to enable the recycling processes of these resources and to rethink incrementally the same quality of the built environment. Such patterns consist of water and green networks, road infrastructures and energy, soils and the areas of scrap and waste. Research in the field has explored, in the urban area of Naples, the planning conditions and trajectories in terms of multi-scale hypercycle. They achieve a different urban metabolism[24], produce quality landscapes, define virtuous combinations of uses, come into harmony with the construction of new infrastructure networks (slow mobility, micrologistics, digital networks, energy networks) and aim at stimulating

22. Cfr. M. Mostafavi, G. Doherty, "Ecological urbanism", Lars Muller Publishers, 2010

23. See C. Gasparrini, "Città da riconoscere e reti eco-paesaggistiche", in "PianoProgettoCittà-PPC", 2011 and R. Pavia, "Il passo della città", Donzelli, 2015 (in particular, the chapter "Reti infrastrutturali e ambientali")

24. See C. Gasparrini, "Editoriale. Oltre la sostenibilità", quote.

2020

2040

Images are taken from DELVA Landscape Architects, Metabolic & Studioninedots, "Circular cities. Designing post-industrial Amsterdam, the case of Buiksloterham", 2016.

Images from "Detroit Strategic Framework Plan", 2012.

On the opposite page, images are taken from DELVA Landscape Architects, Metabolic & Studioninedots, "Circular cities. Designing post-industrial Amsterdam, the case of Buiksloterham", 2016.

economic activities that are in tune with the demands originated by multiple chains of recycling.
The project must therefore rely primarily on the simulation of non-deterministic scenarios and on a concatenation of nonlinear actions as part of a broader, reticular landscape vision. Thus, the potential for ecological regeneration and a three-dimensional urban redesigning of the "new soils", defined as" synthetic surfaces "of the contemporary city, becomes a major aspect within this approach[25]. This kind of quality planning requires a strategic and adaptive approach to the regeneration of drosscapes. It also starts from a cultural, technical and procedural redefinition of the "reclamation project" to get over the more traditional sectoral practices, identifying it as the indispensable substratum of a more comprehensive, ecologically oriented landscape plan. It is a conceptual and constructive process of variable geometry which borrows, in its dynamic movement between scales, many of the tools and practices of the landscape disciplines. As already mentioned, it suggests a project idea which is simultaneously stratigraphic / relational in space and resilient / adaptive over time.
Relational interactions of the dynamics of use and behavior of the "reverse city" - with its space and its real and formal uses, with the flows that feed the dross and waste areas and their rehabilitation, with the need for transformation expressed through plans and projects, with existing practices of appropriation of these spaces denied by a multitude of public and private actors - are closely related to the stratigraphic interactions moving in the three dimensional set made up by soil / subsoil / topsoil. Hence, the proposal for inductive and experiential narratives to start new and credible life cycles. Therefore, the project required by drosscapes needs to combine the speed, duration and possible progression in the life cycles applied to materials, products, spaces, buildings and parts of the city - up, and sub hypercycle - while managing the relational implications in terms of overall quality over time. This project proposal expresses important innovations in urban planning. The centrality of the major issues of urban metabolism for all of us as inhabitants, producers and consumers is a key driver in giving impetus to a different dimension of values that goes beyond housing, workplaces and ontrade premises. It provides vitality and concreteness to the inevitable - yet often abstract - need for a much advocated multiscale approach. The centrality of places and microscale goes beyond the traditional motives of participation because it is powered by the demands of a circular society and economy[26], confronted with the direct management of the recycling processes without intermediaries and with increased business initiatives in green and smart areas. From communities to active citizenship, the regionalization of the consciousness of the fluxes through the consciousness of the places is also promoted by the discretization of the networks within material and immaterial clusters of variable

25. P. Bélanger, "Synthetic surfaces", In C. Waldheim, "The Landscape Urbanism Reader", Princeton Architectural Press, 2006

26. See A. Bonomi, F. Della Puppa, R. Masiero, "La società circolare. Fordismo, capitalismo molecolare, sharing economy", Comunità concrete 2016

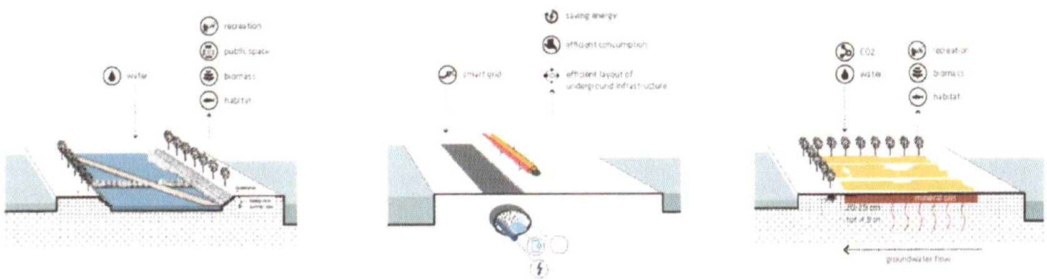
recreation
public space
biomass
habitat
water
saving energy
efficient consumption
smart grid
efficient layout of underground infrastructure
CO2
water
recreation
biomass
habitat
groundwater flow

23

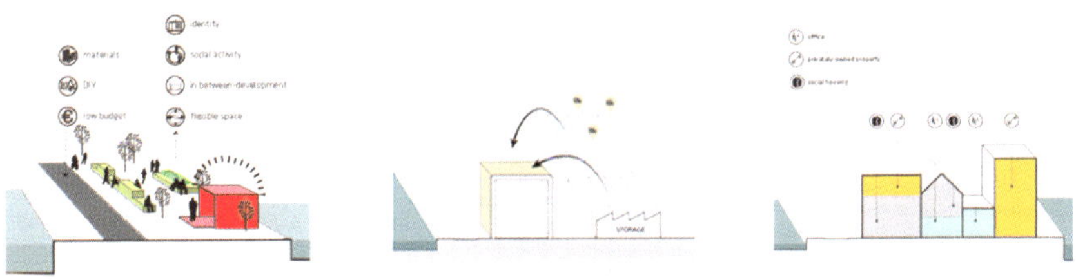
identity
materials
social activity
DIY
in between development
low budget
flexible space
STORAGE

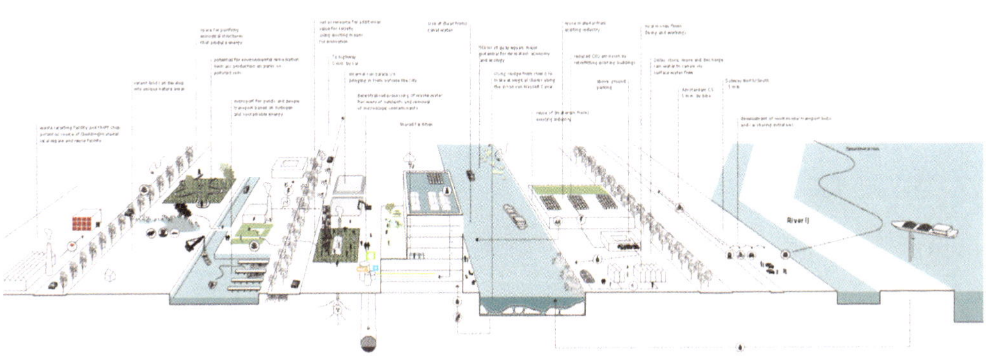
River IJ

Masterplan, "Ambito 13"

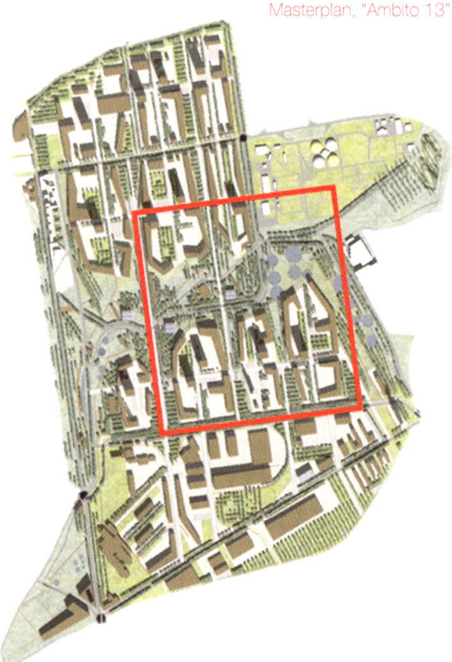

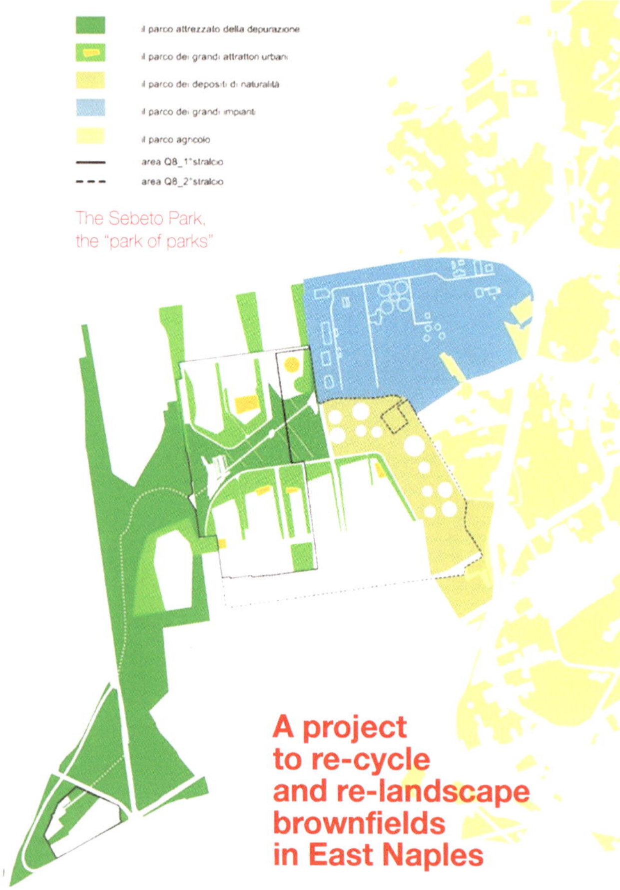

The Sebeto Park, the "park of parks"

A project to re-cycle and re-landscape brownfields in East Naples

Territorial analysis: air view of the territory before the demolitions took place.

Planivolumetric of the PUA Q8 project and overall view of the public and private green areas.

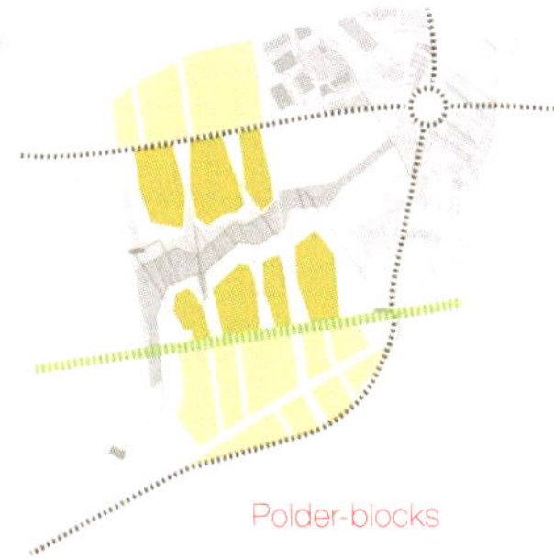

Polder-blocks

Images in these pages are taken from the Q8 areas of East Naples for the PUA (Implementing Urban Plan), by the studio Gasparrini.

Recycling of soils

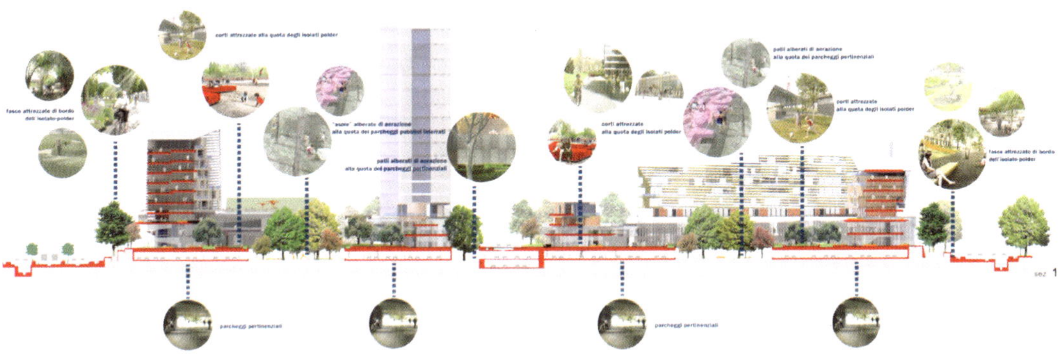

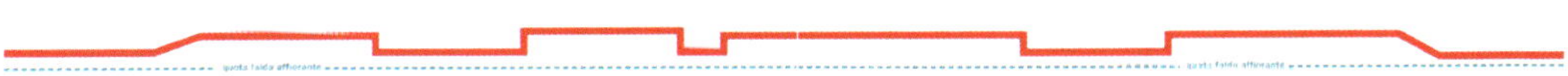

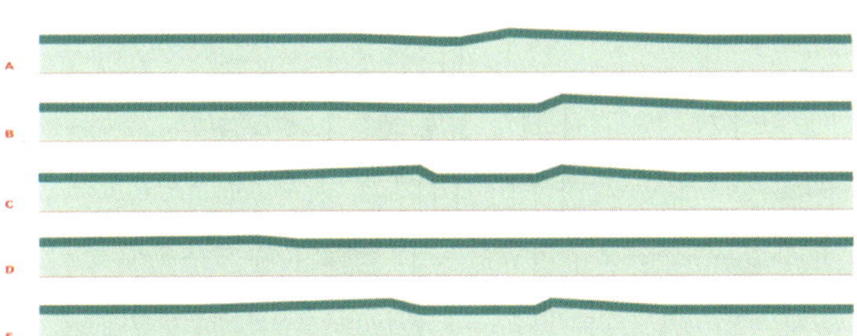

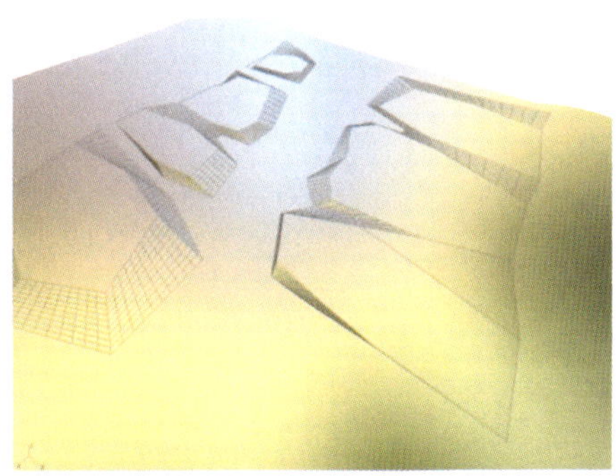

The new soil under the polder-blocks allowing surface water to flow through the aquifer

The "depuration park", the big tank and canals for the hybrid phytoremediation systems"

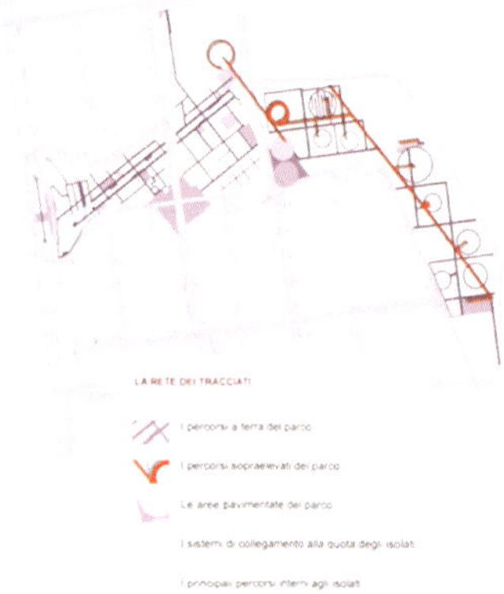

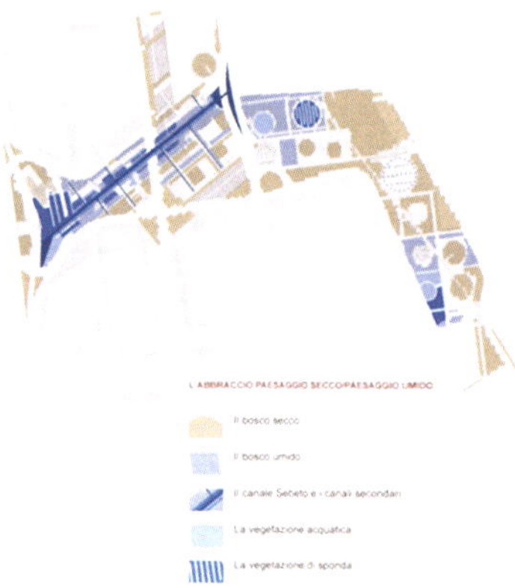

The park's themes

Phytoremediation park: section
of the Sebeto canal

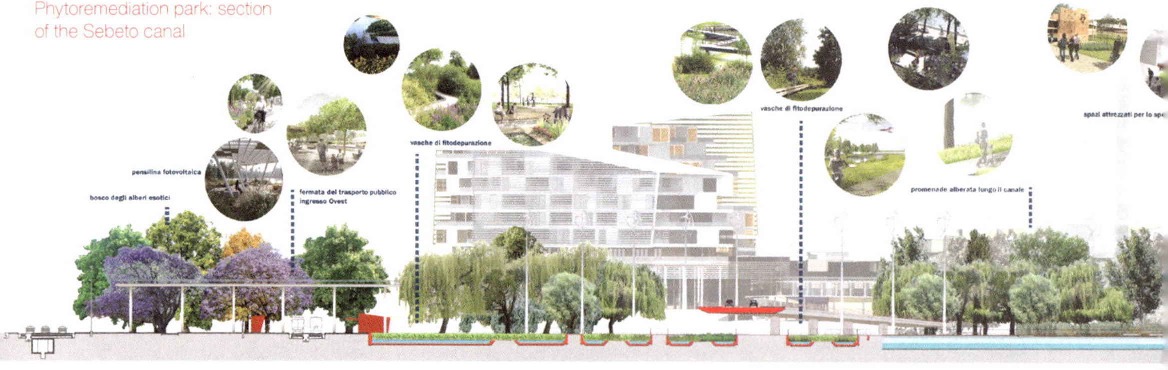

Recycling of water

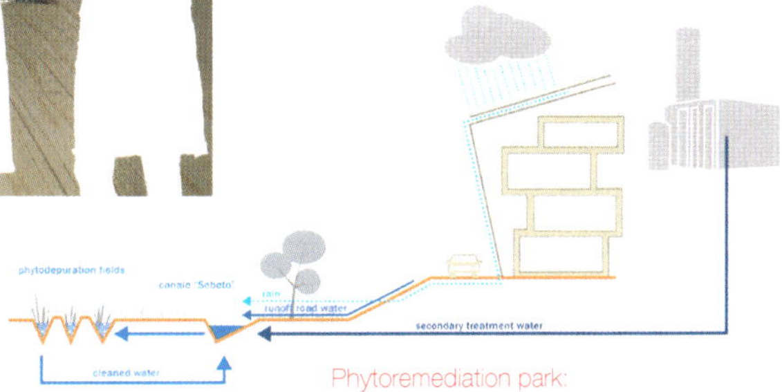

Phytoremediation park: water cycle functioning

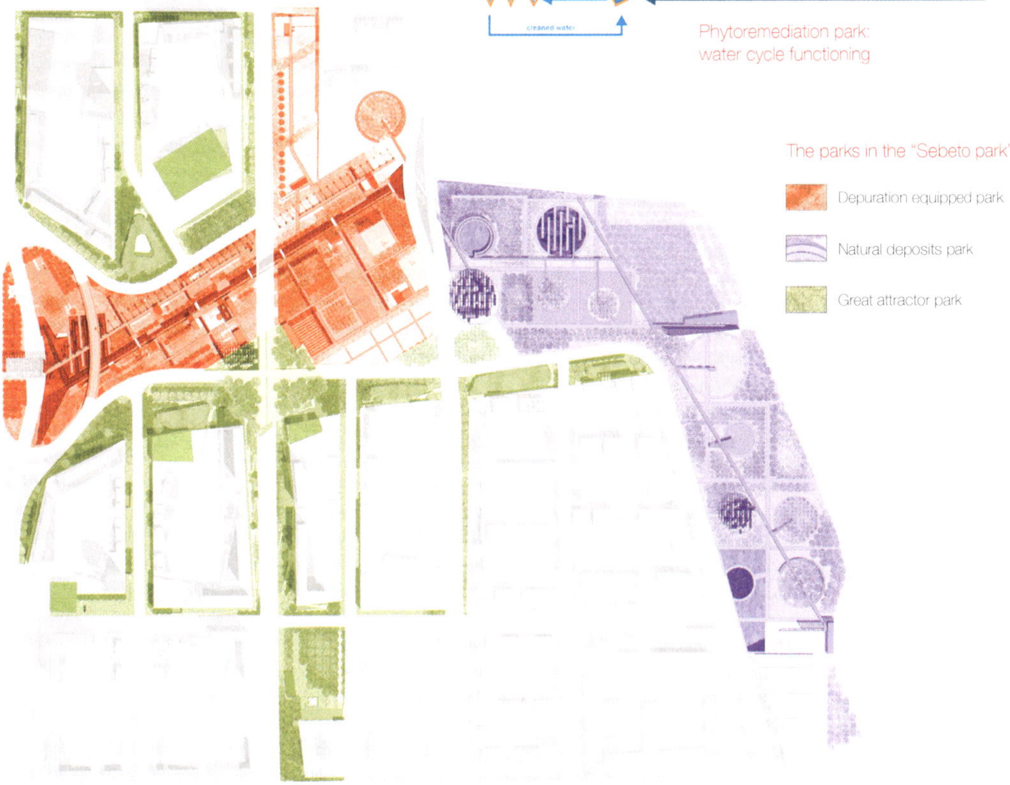

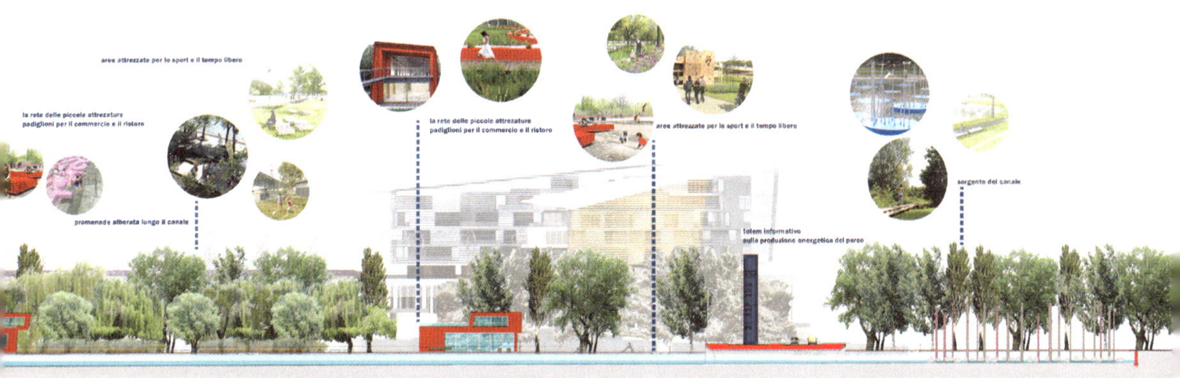

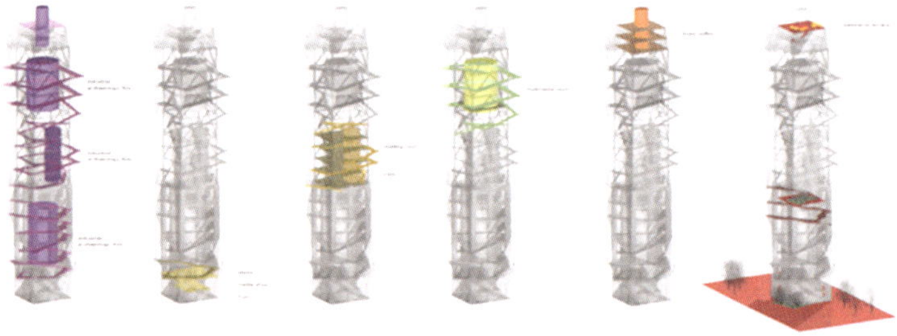

Recycling of the TOC tower from the old refinery to the city's "outlook tower".

Recycling of industrial areas

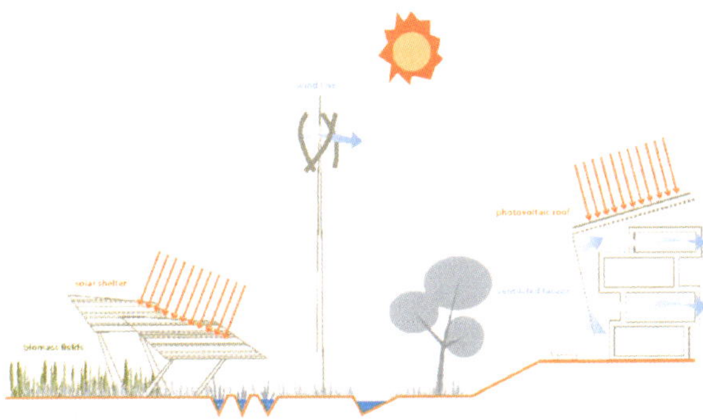

Recycling of energy

Phytoremediation park: process of energy production from renewable sources.

AGROCITÉ
INTERNAL CYCLES

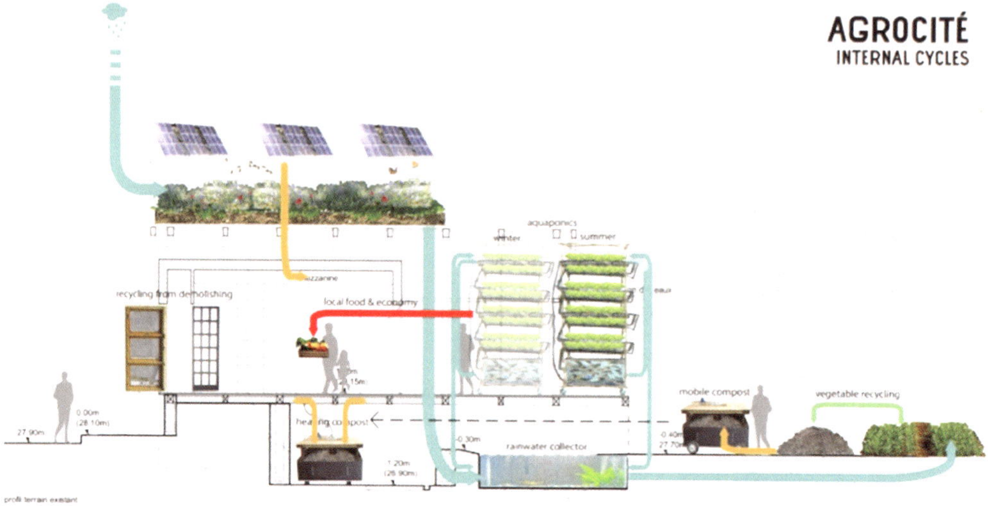

principles

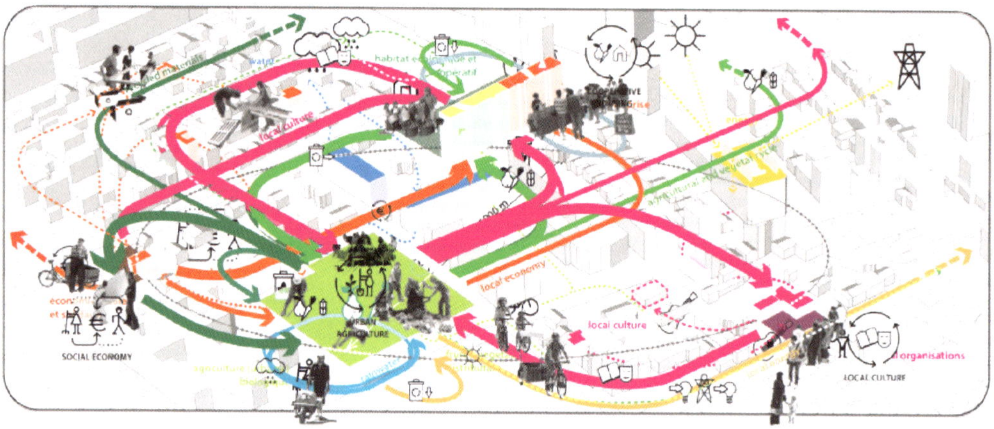

geometry. The practices of self-organization, multi-actor partnerships, as well as the new forms of active leadership and covenantal management of resources originated dynamics that are unusual even for the urban metabolism, and go beyond the traditional decision-making, management and financial models. This brings back the debate on the inevitability of fragmentation in the contemporary city, on its spatial and social self-referentialism, without actually falling into the crystallizing trap of neighborhood unity or conforming dimension. Instead, it prefers ever-evolving community and social structures, and cultures of living that cannot be defined once and for all. The places involved in this regeneration process will certainly be the subjects of traditional and legitimate attention from the surrounding communities. Nevertheless, they will increasingly be crossed by many organized groups and a wider network of people that base their relationship on mutual exchange, who will be able to place these areas within more complex fluxes in matter of value, finance and fruition, of urban scale and beyond, and to determine alliances and perspectives of different kinds with the local communities. The aim is to bring the consciousness of the place, the quality of "warm connections" of the new public spaces and the recycling and regeneration tactics within richer and denser relational perspectives. In conclusion, such processes become key factors for a "place-based" strategic awareness and thus for the set up of integrated and shared projects for more welcoming and increasingly resilient cities.

Image taken from A. Branzi, D. Donegani, A. Petrillo, C. Raimondo and T. Ben David, "Agronica", Domus Accademy for Philips, 1995.

Image taken from the "Vall d'en Joan" project by Battle i Roig.

Friendly Floatees, 2015

Images from R-URBAN, "AgroCité", 2012 http://r-urban.net/en/projects/agrocite/

Part 1

Life cycles of drosscapes and urban metabolism. A redefinition of the field

There is "an inseparable link between dissipative processes and the life cycle depletion of certain natural and anthropic resources. The traditional functioning, restructuring and crisis of manufacturing sectors (...) but also of distribution and trade, warehousing and logistics, of the real estate market itself, as well as all the managerial crises of the great networks or of the metabolism's urban machines, linked to the waste and water cycles, are reflected into the life cycle of the City's structural resources. (...) The City's macro and microporosities are crossed by fluxes and expectations of different consistencies and nature, cluttered by specialised, temporary or undefined, uses, by the wrecks of quickly perished economies and by critical environments, product of fragmented decision-making and of the irrational use of resources. (...) Water and soil but also infrastructures and several urban fabrics are part of the dross geography, often followed by a diffuse condition of deep environmental and social challenges that call on action for the protection and enhancement of common goods. From these dynamics and several business initiatives linked to the values mentioned above an alternative and incremental urban metabolism takes shape, based on social inclusiveness and an innovative economy. A kind of metabolism that penetrates open and built spaces while urging new life cycles for the resources and geometrically variable methods for the management of modest size territories which can be controlled by local communities."

Carlo Gasparrini

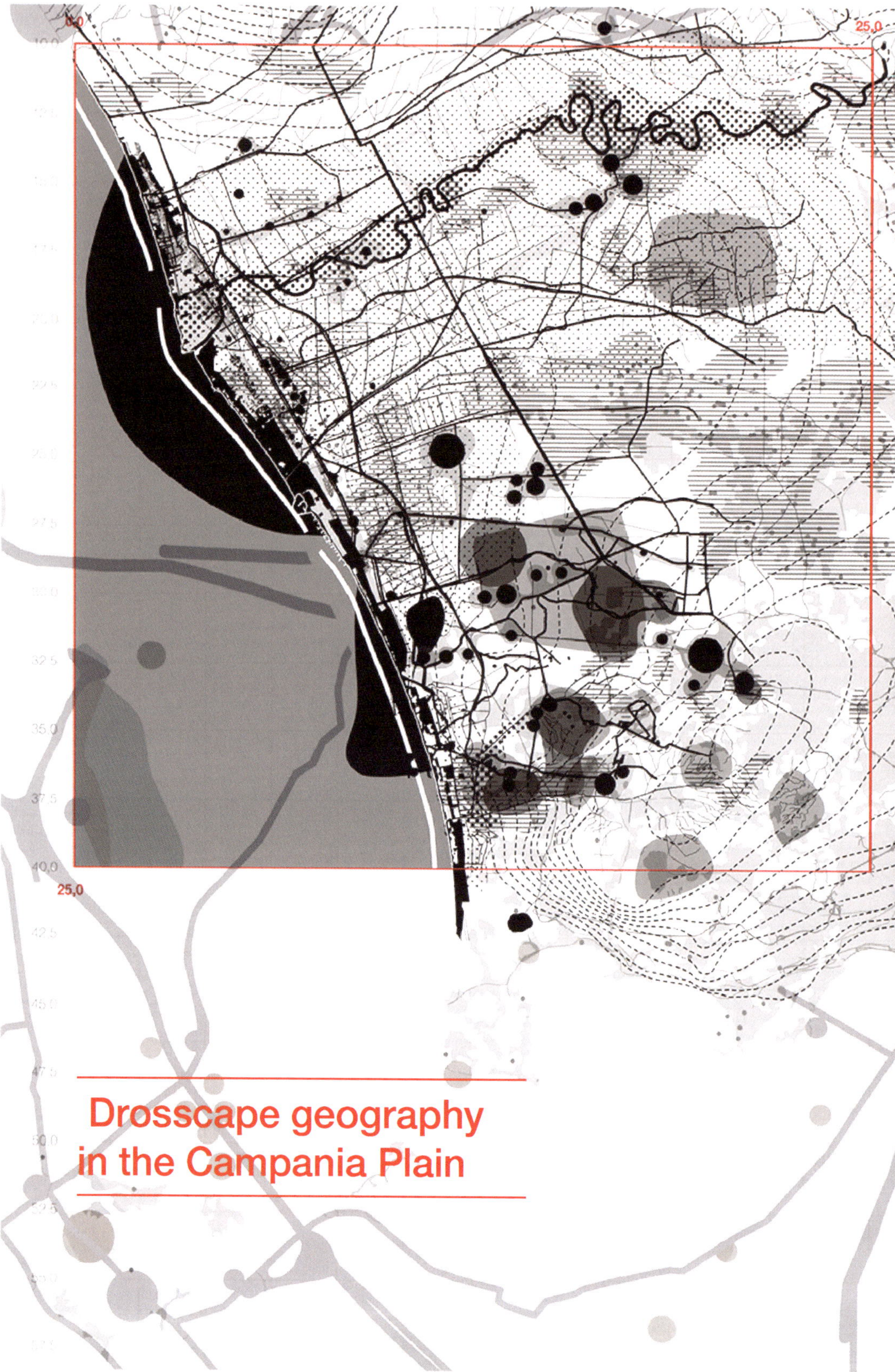

Drosscape geography in the Campania Plain

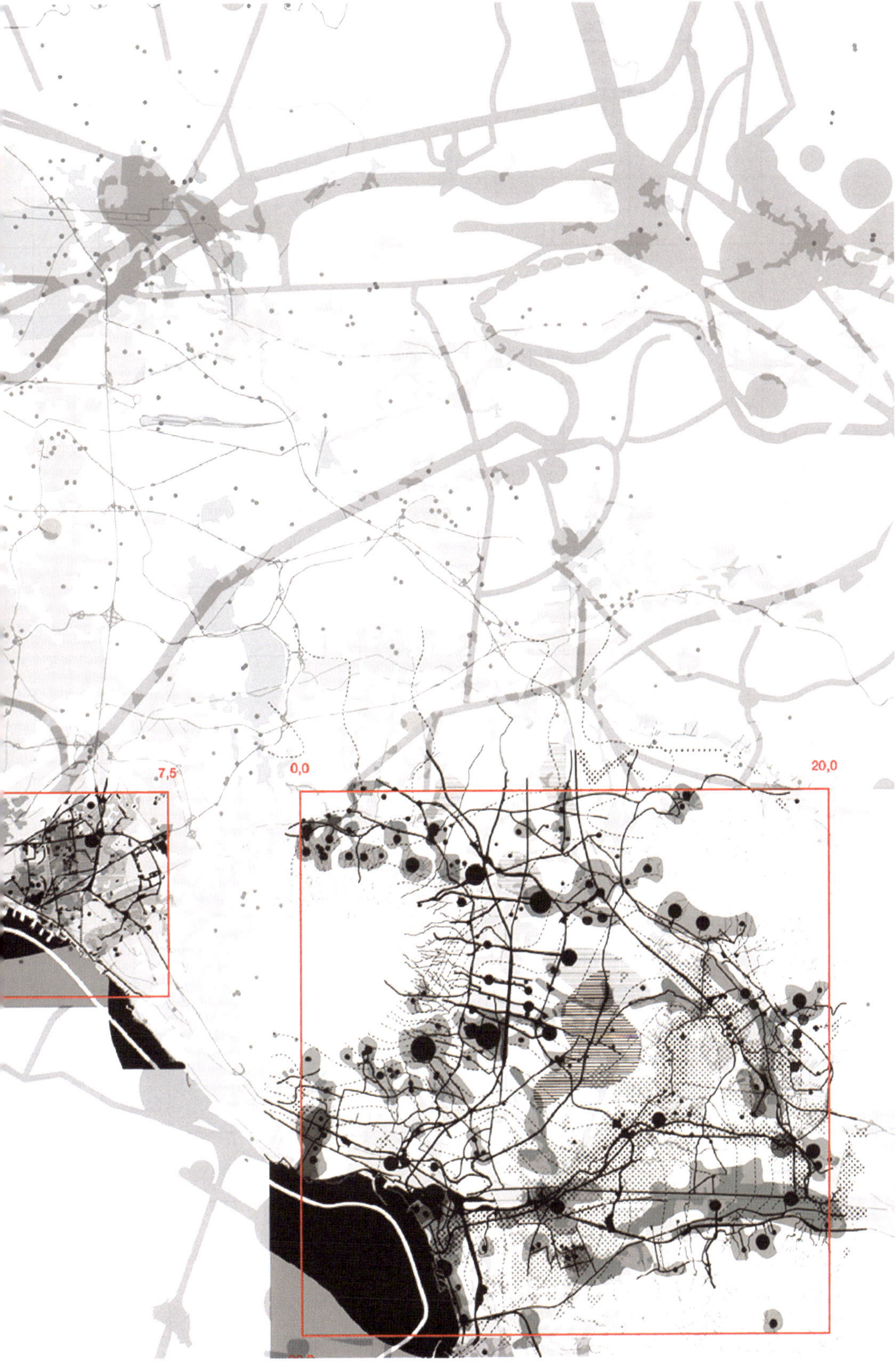
7,5
0,0
20,0

Recycling and metabolism - how to rethink the project

Michelangelo Russo

Contemporary urban planning cannot be expected to be merely a tool to increase the land value, aimed at producing economic growth: the idea that the city is a space where the building industry can generate monetary value is now inadequate. At the same time, the expansive growth that has inevitably characterized the urban planning of the twentieth century can no longer be considered a social and collective goal nowadays.

Over time the urban, economic, and productive growth has shown its inability to produce a social value of the space. In urban planning, growth - as an indiscriminate expansion of the urbanized land and occupation of agricultural soils, and the tendency of the city to take over the countryside - clearly showed the limits of the expansive model of the late twentieth century, which requires a radical rethinking.

The contemporary territory is increasingly fragile and its equilibria are increasingly threatened by the lack of attention to the values of the environment and the ecosystem and the necessary forms of compatibility that their treatment requires; a fragility that generates risks, constantly undermining the conditions of its habitability. The territory is like an organism, with its pathologies and its life to be understood as alternating cycles, as a progressive and unstable transformation of its metabolism.

"Metabolism" is a "metaphor" that alludes to the functioning of the territory as a network of ecosystems, as a context in which the laws that link the structures of urban environments are rooted, together with the reasons for their growth and decline, as a combination of effects produced by the different cycles of life that cross the cities and change its function and shape. The city's metabolism is based on the input and output flows (of raw materials that feed the production and / or transformation processes, as in the case of the building cycle; of the production of resources and energy flows; of ecological components, such as water or soils). Input flows are transformed through the metabolic processes of the

territory and become different substances, channeled into other outgoing flows.
The idea that the territory is endowed with its own metabolism adds a new sensibility for the environment and its ecology - a sensitivity that finds progressive acceptance and consent in the contemporary planning policies and projects. The concept of metabolism is associated with a view of the city as the complex integration of more ecosystems rather than a single ecosystem. They regulate energy exchanges, and the exchange of waste materials that take place between the city and its ecological environment.
Urban structures are undergo metabolic processes that define life cycles: the end of life cycles represents the phase in which the processing and supply chains produce flows of non-usable materials: waste materials that are transformable in terms of "sub-cycles", sometimes no longer usable, destined for disposal.
The meaning of urban metabolism, however, goes beyond metaphor: its political value is all the more important as its communicative power - that can be perceived by common knowledge with a language accessible to all - is pervasive.
Dealing with such metabolism in the above-mentioned terms requires a set of technical and planning skills that affect the cognitive, analytical and interpretative dimension of the territory and its spectrum, as well as that of its government and its plan.
In fact, the territory has to be studied and described as a sequence of life cycles through a non-zenithal and univocal look, thus a selective approach that defines intentional geographies of widespread, sometimes latent values, produced by the various metabolic processes of the city. Such values shall be recognized, read, interpreted.
Nowadays, the geographies of change can be redefined by recognizing the waste produced in the Fordist factory's lifecycle, during the twentieth-century economic boom in construction, by carbon-centered energy-based structures.

Other waste-producing systems include traditional infrastructures, on iron and rubber; the public city, in the residential neighborhoods built by the centralist post-war governments. They are mapping a transforming territory and draw attention on one or more fields that interweave in their consistency and relational value.
There are systems of dismantled industrial buildings, urban agglomerations subject to technological obsolescence, in need of profound renovation; logistic platforms and spaces that are abandoned or underused, over-saturated landfills which can no longer be used or exceeded any environmental compatibility threshold. Large quantities of water are dispersed in the environment, incorrectly channeled, or delivered in a promiscuous way; there are permanently polluted watercourses. In addition: roads and viaducts; old railroads; old galleries in disuse; shopping malls and large distribution spaces, made obsolete during a global 'de-malling' process that affects large urban and metropolitan areas. And above all, a massive amount of ordinary building that occupied the open spaces on the Italian territory, which clogged the interstices of the consolidated city and expanded into the wreaths of the urban sprawl, without a design or a project, without architectural quality or technological appropriateness, despite any law on housing security and without respect for the landscape and the territory. We are talking about networks of objects, building components and infrastructures, industrial and residential buildings that have changed the shape of the Italian territory, altering the millennial ecological relationship between the communities and their territories in Italy, from the post-war period to the present, and which enter - in terms of security, reliability, ecological relationships and resilience - a terminal phase without return to their life cycles. It is a potential patrimony to start a great project for the Italian territory and its landscape. Indeed, many life cycles end without allowing any material recovery of the residues, while other life cycles can be profoundly and radically modified, or even reversed. For example, in the large residential neighborhoods of the 1960s and 1970s, the social crisis revolved around the degradation of the communal space and the critical conditions of buildings that determined the overall obsolescence of the settlement system. In this case, regeneration, reclamation and recycling become strategic actions as well as techniques which have a strong ethical basis, linked to sustainable economy and the safeguarding of resources and territorial values (undeveloped land protection), which must measure their economic rationality and convenience in relation to a potential morphological redefinition of the related contexts: recycling is best when there is an added value due to the form and quality of the preexisting asset. Drawing "lifecycle maps" can be an interesting cognitive exercise to generate new geographies: it can be a form of reading the threshold values of urban metabolism, improving our expertise of the territory in relation to its ecosystems, in order to identify the resources and issues that the urban planning project needs to address, decode, measure, represent, and organize. According to the history of urban planning and its failures in the last 70 years, therefore, the central theme of the contemporary planning is "knowledge" - how to read the city and interpret its collective meaning as an expression of a shared project. This topic requires the attention of urban planners and regional analysts alike, to drastically reform the ways of reading and representing the geographies of change in relation to the phases of the life cycles of the materials on the territory. Urban planners must define the terms of an innovative reading of the city, involving institutions and governments in the process of transformation. A different knowledge of the city will enable us to identify the transformations of its metabolism, the growth and decline of the life cycles of its components, the transformability of some parts, to create projects for the reinvention of space or - even better - ways through which such space (an obsolete structure, for example) can be rethought in relation to the life of its residents and communities, or the people who could move there.
Urban planning is thus transformed within a renewed field of action by identifying materials, techniques and project strategies that have a new dimension compared to traditional practices, as they relate to new sets of values and goals. Resilience, metabolism, safety, risk reduction, energy saving, waste reduction, integration between waste cycle and land use are the goals and objectives of the public project, which have been relegated for a long time to the background of urban planning: they are the terms that can radically redefine the weights and polarities of the contemporary cities issues. Urban planning has to move in an adaptive way,

View of the Naples Business Centre from the SS162 viaduct.

Lido Santa Lucia, Torre Annunziata (Naples)

Boscoreale station (Naples), on the dismissed National Railway (FS) Torre Annunziata-Cancello.

On the previous page, view of the Torre Annunziata port.

On the opposite page, ex "O.P.G. Maddalena", Aversa (Naples).

based on the awareness that the future of the city derives not only from the rules of the plans but above all from their ability to address individual and collective behaviours: this requires a mapping of the land purposes based on the constant assessment of the transformability of its elements and on the analysis of its ecological, economic and urban systems, their reproducibility and obsolescence. Addressing the issues of resilience, climate change and natural and man-made risk, working on the water, waste and energy cycles in relation to the shape of the territory, recycling spaces and waste structures, does not simply mean resorting to 'new techniques of urban planning'. It also means innovating the public discourse on urban planning on the basis of a selective and collective reading of the territory, its developmental limits and its potential resources, understood as 'project materials'. The urban and territorial planning does not lose contact with his 'fundamentals': space, its organization and its design; architecture, the landscape, and the elements of the habitable space. It's quite the contrary. The challenge that can be tackled is the contextual and targeted use of new materials and new techniques in a project that can produce and modify the space for the wellbeing of those who live in it, but above all the activation of collective and community energies to achieve these goals. Only by establishing innovative forms of dialogue in public decision-making processes can urban planning regain control of its social and cultural foundations, based on the assumption that the construction of habitable spaces shapes the city, taking public and collective space - the "land design" - as the basis for the planning of the districts and areas of a city. Following these directions, urban planners have the opportunity to open up to new forms of learning, practices and techniques that make it possible to consciously approach the cities and their dwelling communities so that the future can, once again, take the form of a project.

SEZIONE
LIVI

Differentiate waste / The urban fabric

Fabrizia Ippolito

Let's think about applying the mechanism behind separate collection to landscapes: when properly sorted and differentiated, their waste material may become useful to start a recycling project. On the basis of urban literature[1] in relation to the Italian situation, such 'scraps' may include water, soil, agricultural areas, natural areas, the urban fabric, industrial areas, quarries and landfills, and infrastructure, each one with their specific declinations. Among these elements, however, the urban fabric plays a pivotal role in matter of quantity.
In a fragmented situation like the Italian one[2], the available data on urbanization and construction launch alarms on soil consumption, more than doubled since the mid of the 20th century. Therefore, investigating the potential of the amount of discarded materials, their quality and the reasons behind disposal can serve to go beyond the generality of those alarms in order to carry out a more complex intervention on urban landscapes. Some urban elements are not currently in use because of specific choices. They were deprived of their function, damaged by an accident, disadvantaged and excluded, or are made unlawful and illegal. They are the other side of the process of urbanization and landscape construction, that can be influenced by the recycling project.
An analysis of the discarded urban fabric, in the light of the recent urban debate and in the current Italian situation, can provide a better understanding of cities under this aspect and suggest applications for the recycling project.
Such fabrics can be classified and listed as follows:

Residual - marginal city / reconsideration

Some urban fabrics are residuals of choices that go in other directions. In this case, the 5,000 small Italian municipalities at risk of abandonment, of which 2/3 are located in the inland mountainous terrain excluded by urban expansion - which invests 60% in plains and is 30% faster on the coast. In addition, there are also parts of extended cities that are

1. For a taxonomy of reference about the definition of drosscape see A. Berger, "Drosscape. Wasting Land in Urban America", Princeton Architectural Press, New York, 2007.

2. For an approach to Italian landscapes from the point of view of fragmentation and as a reference for the data provided in the following paragraphs see F. Ippolito, "Paesaggi frantumati. Atlante d'Italia in numeri", Skira, to be published.

emptied due to a contrary retraction process. They are the places touched by 6,000 kilometers of disused railways, or crossed by the slow lines of the 16,000 km-long Italian railways - of which 1.250 kilometers are TAV (High-Speed Train lines). Or crossed by the state, regional and provincial sections of the Italian 200,000-km road network - of which 6500 km are highways.
They are the industrial areas and factory districts emptied out due to the outsourcing or relocation processes of the industrial economy. It is a marginal city, which requires new attention- marginal with respect to the directions normally taken by the city in recent years. These urban fabrics can thus be reconsidered as the place of a possible alternative to such directions. Italy is a two-faced country: part of it was built on the plains and coasts, is connected to the main cities, with a concentration of activities in large cities. Its counterpart is a country whose hills and inland are abandoned, connected by slow vehicles, with the distances between the Adriatic and Tyrrhenian sea filled by individual means of transport, dotted with widespread sheds, brownfield sites and empty houses. Taking action on this other Italy with a view to recycling can mean re-discussing the development models confronting the local peculiarities, and recognizing contamination between the city and the countryside while rethinking cities with a view to the landscape, identifying the inert elements and possibly integrating dissipation in this analysis. This approach may merge theories ranging from the degrowth to the Southern thought movement, including landscape urbanism and urban dissipation[3] as well as projects of revisitation of the abandoned places (S. Goethel Campbell, "Cloudspotting Detroit", 2010). Thus, its application to the recycling project in the Italian context can open up new perspectives on landscapes. In Italy, as the narratives on marginal landscapes and their celebrations progress, research on abandoned places and slow landscapes increases and new experiments in local

3. On degrowth see S. Latouche, "Breve trattato sulla decrescita serena", Bollati Boringhieri, Torino 2008 (original edition 2007); on meridian thought see F. Cassano, "Pensiero meridiano", Laterza, Roma-Bari, 1996. On landscapes and the city see C, Waldheim, "The Landscape Urbanism Reader", Princeton Architectural Press, New York, 2006; M. Ricci, "Nuovi paradigmi", ListLab, 2013, C. Gasparrini, "In the City/On the City", ListLab, Trento, 2015. On waste / dissipation see K. Lynch, "Wasting Away. An Exploration of Waste: What It Is, How It Happens, Why We Fear It, How To Do It Well", M. Southworth editor, Sierra Club Books, San Francisco, 1990.

communities are carried out (Feld 72, "Million Donkey Hotel", Prata Sannita, 2005; many Living Lab, including Cilento's LabScape, "Innesti", Sapri 2014)[4]. These open projects, together with active relations with the citizens, characterize recycling operations which, starting from the marginal areas, offer an alternative vision of the landscape and innovative transformation processes. With its high concentration of marginalized areas, the South may soon be reconsidered.

Unused - abandoned city / reuse

Many urban fabrics, or some of their parts, are unused due to the absence or disqualification of their function. They are the residential fabric in the dispersed cities or in small towns, where much of the 7,000,000 empty houses in Italy are found; the productive fabric in the industrial areas, in which about 700,000 warehouses are abandoned; the mixed fabric of the compact city where offices are closing down, 1 out of 10 in Milan, cinemas, 50 in Rome, and other assets not yet listed in the Census in many Italian cities. In addition, the military areas - with nearly 2,000 assets in decommission, of which nearly 500 have already been decommissioned - and the abandoned industrial areas, 9,000 hectares in total; mining sites, with nearly 3,000 abandoned mines. Fabrics in which the indeterminacy of their function can leave room to new uses.
To handle such a wide phenomenon, linked to the economic crisis and social fragmentation but also to the technical obsolescence of the structures, different reuse measures are being taken. In many occasions, they trigger processes of adaptation and reinvention that take advantage of the different degrees of uncertainty. Different realities emerge, such as parks in abandoned industrial sites, museums in the mining sites, hotels in the villages, shopping malls in cinemas and theaters. Industrial or commercial buildings are converted into homes, public spaces - sometimes private spaces too - are designated for social and cultural activities, or forms of cohabitation and co-management which change the ways of living such spaces. Reuse as a reinvention device acts on the existing fabric, capturing its latent potential and adopting adaptive modes of intervention.
In the current debate on architecture there are several relevant examples, such as the exhibition about architecture, cities and the landscapes of recycling, the new atlases of urban uses, or the history of reuse in the city over time[5]. They outline a mode of reusing the landscape that includes the amphitheatres of Arles, Nimes, Lucca and Split, transformed in cities through the invasion of everyday life, whose subsoils are inhabited by the architecture without architects of the Chinese, the ruins of Bosra and other Roman cities occupied by houses, the Italian historic centers re-populated by foreigners , the abandoned infrastructure of New York or Paris re-adapted into walking paths, the landfills in New York and Barcelona turned into landscapes by progressive projects, the repainted facades of Tirana, the quarries of Curitiba turned into public spaces,

On the opposite page and here above, the juvenile correctional facility Filangieri "Scugnizzo Liberato", Naples.

4. For more information on the stories and celebrations of marginal landscapes, see F. Arminio, "Vento forte tra Lacedonia e Candela. Esercizi di paesologia", Bari-Roma, Laterza, Roma-Bari, 2008; and the celebration of country life in Aliano, always organized by Franco Arminio. For research on abandoned or slow landscapes see V. Teti, "Il senso dei luoghi", Donzelli, Roma, 2004; and E. Lacerini, "Territori lenti", PhD thesis, Iuav University of Venice, 2004.

5. On the exhibition about recycle see P. Ciorra, S. Marini (edited by), "Re-Cycle. Strategie per l'architettura, la città e il pianeta", Electa, Milano, 2011. Among the atlases of urban uses :Motoelastico, "Borrowed City", Damdi, Seoul, 2013; David Trottin, Jean Christophe Masson, Usages, French Touch, Paris 2011. For examples of memory within the city see A. Ferlenga, "Città e memoria come strumenti del progetto", Christian Marinotti, Milano, 2015.

Santa Lucia beach, Torre Annunziata (Naples)

Twila Resort, Castel Volturno (Caserta)

Villaggio Coppola, Castel Volturno (Caserta)

the public housing project designed for the cities of Lima and Iquique (that is yet to be completed).
The Neapolitan area can have a prominent place within this scenario, with the mapping experiments (MappiNa, http://www.mappi-na.it), specific projects (Dianarchitecture, RS Architettura, "Restart", Casal di Principe 2015) and managing modes (L'Asilo, Napoli), but especially with a practice of daily adaptation[6].

Accidented - broken city / repair

Some urban fabrics are accidented due to unexpected problems. This includes: territories destroyed by earthquakes, such as the Belice or L'Aquila area, or by landslides, like Sarno, that are being reconstructed; 80% of old buildings and houses that date back to over 40 years ago and of which 20% are in poor condition; public works that are interrupted due to progressive lack of funds or function - currently 700 projects are registered as unfinished - but also a multitude of private buildings suspended for personal or family reasons, or for legal reasons. Fabrics that have suffered a damage, lost integrity, whose damage can be turned into a project.
While the destroyed areas wait for renovations or reconstruction interventions, old buildings await adjustment and interrupted work require completion, the accident, which interrupted the normal course of events, can also be an occasion for their reinvention. More than the restoration of the initial state or project, in the case of recycling an intervention of repair[7] - which may use post-production techniques and employ the principles of "up-cycle"- implies a manipulation of the existing material and the introduction of new functions[8]. In the cultural context of a renewal project, terminology identifies an approach in terms of a repair intervention on countries affected by natural disasters that may exceed the museification models, restoration or new foundation to work on the reinvention of urban life. The structural or energetic adaptation of degraded buildings may become a way of innovation for the function of such spaces, and the completion of the interrupted work can transform it into a project that is open to new locations and configurations. And if the project scenario provides a repertoire of repair interventions, ranging from the creation of new communities after a cataclysm (T. Ito, "Home for All", Rikuzentakata, 2011) to the investigation on housing experiences after modernization (Lacaton & Vassal "Tour Bois le Prêtre", Paris 2011), including the interpretation of the 'skeletons' as living spaces (Coloco, "Squelettes à habiter", Paris, 1999) and their transfiguration project (C. Gambardella, "House bunker", Itri , 2010-2012), the territory can offer, besides a number of opportunities for action, a legacy of experience and the aptitude to repair rooted in its ways of living. This is especially true for the Neapolitan territory: if in Naples things start to work when they are broken, or they were born broken in order to have the possibility to be transformed[9], then the Neapolitan philosophy of the 'broken', more than its aesthetic, can inspire the implementation of the project.

6. For examples of the landscapes of reuse and the Neapolitan attitude towards adaptation see F. Ippolito, "Tattiche", Il nuovo melangolo, Genova, 2012

7. On the reparation project see M. Navarra, "Repairing Cities", Lettera22, Siracusa, 2008

8. On post-production see N. Borriaud, "Postproduction. Come l'arte riprogramma il mondo", Postmediabooks, Milano, 2004 (original edition 2001); on "upcycle", that is a recycling technique implying a positive transformation in the condition of the object, see M. Braungart W. McDonough, "Cradle to Cradle: Remaking the Way we Make Things", North Point Press, New York 2002, e "The Upcycle: Beyond Sustainability. Designing for Abundance", North Point Press, New York, 2013

9. A. Sohn Rethel, "Napoli. La filosofia del rotto", Alessandra Caròla Editrice, Napoli, 1991

Disadvantaged - excluded city / urban revolt

Many urban fabrics are disadvantaged because of poverty. Primarily the housing estates, where 20% of accommodations are subject to rent arrears (40% in the south), and illegal occupation reaches 6% (9% in Central- Southern Italy). Their maintenance costs 26% of collection, and they are on sale at the rate of 56,000 every 10 years, or leave a controversial architectural legacy, such as the Biscione area in Genoa, the Corviale in Milan, the Vele in Naples, the Zen in Palermo. There are also improvised living solutions such as illegally occupied slum areas, like Saraceno in Castelvolturno, and other inhabited premises that are not houses (54,000, more than doubled in 10 years); in addition, the refugee and the nomad camps, which host about 40,000 Roma, and the temporary housing facilities, which accommodate more than 30,000 migrants, as well as the farmland with houses rented to foreigners illegally. In addition, there are emergency settlements such as the slums of Messina, dating back to the earthquake in 1908; but also the widespread underprivileged areas in many urban centers. It is the city of the poor, excluded from the city of the rich[10], which can start the revolt. If in the city of the poor - in the context of a compact city, and in a view of the city based on pairs - the counterpart of the outskirts is the center, in the dispersed city, with an approach oriented towards transformation[11], the widespread peripheral landscapes are opposed to the urban ones; and if in the metropolis rebellions occur through conflicts between the urban parts, in the landscape city a revolt can be caused by conditions of indeterminacy. In the recycling project, revolt can be expressed through the reinvention of fruition spaces: the redevelopment and the demolition of public housing, their manipulation; beyond the irregularity of informal housing, the revelation of their latent planning; beyond the urgency of the illegal arrivals, the imagination of the migratory landscapes; beyond the emergency caused by calamities, the planning of the temporary settlements. In the Italian landscapes, especially those of the Southern Mediterranean, these revolts are already evident. In parallel, a dissenting project takes shape in the wake of the debate on cities and architecture. It ranges from the theories of the '70s on the right to have a city based on alternative / survival architecture, to the current proposals for rebellious architectures and cities[12]. The necessity for an architecture of poverty is being raised again, thanks to our increased awarness of the occupation phenomena (Think Thank and I. Baan, "Torre David", Venice Biennale 2012), by researching living conditions on the border (T. Cruz, Casa Familiar, "Living Rooms at the Border ", Tijuana, 2008), and by creating reports from the militant front lines of the project (A. Aravena," Reporting from the front ", Venice Biennale 2016) and projects for the precarious humanity (S. Anderson," Insecurities ", MoMA , 2016). All of the above-mentioned elements are symptomatic of an ongoing revolt. If a revolt is the subversion of the state of things, rooted in practice and free from a predictable outcome[13], the architecture of the Mediterranean, which

Ex O.P.G. "Je so Pazz", Naples

10. B. Secchi, "La città dei ricchi e la città dei poveri", Laterza, Roma Bari, 2015

11. B. Secchi, in "Un progetto per l'urbanistica", Einaudi, Torino, 1989, differentiates the topics of the debate on the city and the interpretive approach.

12. On the two positions on cities and architecture, see Italian new editions of texts from the Seventies such as H. Lefebvre, "Il diritto alla città", ombre corte, Verona, 2014 (original editon 1968); C. Ward, "Architettura del dissenso. Forme e pratiche alternative dello spazio urbano", in G. Borella (edited by), Elèuthera, Milano, 2016 (the original dates back to the Seventies); Y. Friedman, "L'architettura di sopravvivenza. Un'architettura di povertà", Quodlibet, Macerata, 2009 (original edition 1978); and some more recent material such as Al Jazeera documentary, "Rebel Architecture", website: http://www.aljazeera.com, reviewed by "Domusweb", 28 August 2014, website: http://www.domusweb.it; D. Harvey, "Città ribelli. I movimenti urbani dalla Comune di Parigi a Occupy Wall Street", Il Saggiatore, Milano, 2013 (original edition 2012)

13. P. Amato, "La rivolta", Cronopio, Napoli, 2010

manipulates what it finds, expresses its subversive nature on a daily basis.

Illegal – city rehabilitation

Many urban fabrics are illegal due to the way they are built or used. They are the illegal buildings concentrated in Central / Southern Italy, in dispersed cities, especially on the coasts, where construction increased by almost 20,000 new buildings in 10 years despite the Galasso environmental law; most of them are residential, for a total of about 1.7 million houses, and more than 10% of those are newly built; premises subject to demolition ordinances, 47,000 in 10 years, of which 5,000 have already been executed, or subject to retrospective planning permission with the largest record of practices in Campania. They are the productive fabric of the black economy, which produces a GDP of 35% of the national total and a rate of irregular workers of over 20% in the South. But they are also the 11,000 properties confiscated from the mafia, 80% of which are in 4 regions of the South, intended for social re-use and assigned with an average 10-year waiting time. This is the illegal city, and it requires rehabilitation. It's in the shadow because it is invisible on maps and in the censuses, but also because it was removed from the urban debate. The widespread illegal city influences the quality of landscapes, particularly in the South. If fought with a legalistic approach, based on the environmental culture, it is subject to restrictions and demolitions; if approached with a project based on recycling, it may become an opportunity for projects and processes of transformation and a recipient for project implementations[14]. Mostly made up of houses, indifferent to public space and to the landscape, built to not last, it can inspire projects ranging from the clarification of the public space, to increased density of constructions and the re-naturalization of the residues, echoing the adjustment and repair hypothesis, densification or dissolving of the urban sprawl[15]. Oscillating between rules and exceptions and between necessity and speculation, and based on individual interest, it may encourage the use of incentive procedures, negotiation, sharing, and its ability to negotiate its own redevelopment. Made to meet different the needs and opportunities, through waste, exemptions and exceptions, it can express a wealth of skills to be invested in other situations: from recipes to bypass the rules (S. Cirugeda, "Recetas Urbanas", www.recetasurbans.net) , to tactical planning guides ("tactical urbanism", www.tacticalurbanismguide.com), to collections of spontaneous interventions ("spontaneous interventions", www.spontaneousinterventions.org), these new manuals envisage urban resistance and innovation. In the wake of the topicality of this debate with a new focus in the south of Italy, the South - and Naples, the city of the exception - can become a workshop of rehabilitation[16].

Above and on the opposite page, Rovigliano beach, Castellammare di Stabia (Naples).

14. For an environmental / legalitarian viewpoint on illegal activities see A. Cederna, "I vandali in casa", Laterza, Roma-Bari, 1956, and, more recently, P. Berdini, "Breve storia dell'abuso edilizio in Italia. Dal ventennio fascista al prossimo futuro", Donzelli, Roma, 2010. For a planning perspective see F. Zanfi, "Città latenti. Un progetto per l'Italia abusiva", Mondadori Milano, 2008

15. For hypotheses on the dispersed city / urban sprawl see the american debate on the topic, with approaches derived from New Urbanism, such as E. Dunham-Jones, J. Williamson, "Retrofitting Suburbia. Urban Design Solutions for Redesigning Suburb", John Wiley and Sons, New York, 2009 o G. Tachieva, "Sprawl Repair Manual", Island Press, Washington, 2010. Also, see Landscape Urbanism-oriented texts such as the work about Detroit by G. Daskalakis, C. Waldheim & J. Young, "Stalking Detroit", Actar Barcelona, 2001 a C. Davidson and M. Ponce de Leon (edited by), "The Architectural Imagination", American Pavillion, XV Architecture Biennale of Venezia, 2016, or oriented towards densification,such as M. A. Ray, R. Sherman, M. Zardini, "The Dens(e-c)ity", Quaderni di Lotus 22, 1999, o towards re-imagination, such as B. Bergdoll (edited by), "Foreclosed: Rehousing the American Dream", MOMA, New York 2012.

16. See TAMCLab, Laboratorio di ricerca sui territori dell'abusivismo nel Mezzogiorno contemporaneo, https://tamclab.wordpress.com

The life cycles of soil and waste

Massimo Fagnano

Soil Knowledge

In order to preserve and protect a resource, we must first know it. Unfortunately this is not to be taken for granted in the case of soil. In the global context and according to the European legislation[17], the definition of soil is the top layer of the Earth's crust (1-2 meters), formed by mineral particles, organic matter, water, air and living organisms, and subject to rapid degradation and extremely slow formation processes.

Unfortunately, in Italy, the Consolidated Environmental Act (Legislative Decree no. 152/06) contains a rather ambiguous definition of the soil resource (Art. 54, Part III, Italian Law on soil protection and against desertification): the purpose of this section is to define soil as: the land, the soil, the subsoil, the settlements and the infrastructure.

This definition does not take any account of the soil functions within the ecosystem. Instead, it considers it a mere amorphous surface available to any 'constructive' interest.

Soil functions

In facts, soils carry out a series of irreplaceable functions for the balance of the entire ecosystem, representing the environmental matrix in which lithosphere, hydrosphere, biosphere and atmosphere interact, ensuring the equilibrium of the entire ecosystem.

There are different conceptualizations and schemes to evaluate the functions and the services performed by soils in the ecosystem, but here is the best schematization provided so far[18]:

1. Purification of the water and reduction of the contamination of soils (fertilizers, pesticides or other contaminants are absorbed by soil aggregates through clay particles and organic matter, and then degraded by microorganisms).
2. Direct effects in mitigating and / or regulate floods (The infiltration of rainwater in the soil reduces surface runoff, the risk of flooding, erosion and soil loss).
3. Sources of biochemical and pharmaceutical products. The different microorganisms in the soil can be a source of new

17. European Commission (EC), COM 2006/231, "Thematic Strategy for Soil Protection", in "Communication from the Commission to the Council, the European Parliament, the European Economic and Social Committee and the Committee of the Regions", Brussels, 22.9.2006, source: http://ec.europa.eu/environment/archives/soil/pdf/SEC_2006_620.pdf

18. Finvers M.A., "Application of e2DPSIR for analysis of soil protection issues and an assessment of British Columbia's soil protection legislation", M.Sc. Thesis. Cranfield University, UK, 2008.

molecules that can be used in the pharmaceutical industry. They break down and recycle dead plants and organic waste, releasing nutrients available to plants.
4. Seizure of Carbon (Carbon - from more labile forms to more stable forms of organic matter - is stored and recycled in the soil, which is the main carbon deposit of the land surface) and control of other greenhouse gases (CH_4, N_2O).
5. The cycle of nutrients to support plant growth (primary production), which ensures the production of food and fiber, necessary to other animal organisms including man and the human society.
6. Support to human infrastructure.
The term itself, "fertility", is commonly used to define the quality of soil. It refers to its ability to generate life - not only plants (and this is quite intuitive), but also animals that eat the plants and humans, who draw nourishment and economic support from soil products.

Soil Formation

Soil formation occurs over millennia of interactions between climate conditions, bedrock and the biological activity carried out not only by microorganisms, but also by the roots of higher plants, which turn the original matrix (eg. rock, volcanic deposits, etc.) into a new substrate capable of promoting the growth of more plants. Therefore, this resource can be considered nonrenewable, at least in times of human society.

Degradation forms

According to the European Community, the main threats to the soil and thus to the quality and quantity of its functions in the ecosystem are:
- reduction of organic matter (favoured by intensive farming techniques, such as the aggressive tillage. It influences most other causes of deterioration);
- local or widespread contamination (derived from industrial activities and disposal / illegal burning of waste. It can

19. European Commission (EC), COM 2006/231, "Thematic Strategy for Soil Protection", in "Communication from the Commission to the Council, the European Parliament, the European Economic and Social Committee and the Committee of the Regions", Brussels, 22.9.2006, source: http://ec.europa.eu/environment/archives/soil/pdf/SEC_2006_620.pdf.

Drosscape [+] soil consumption

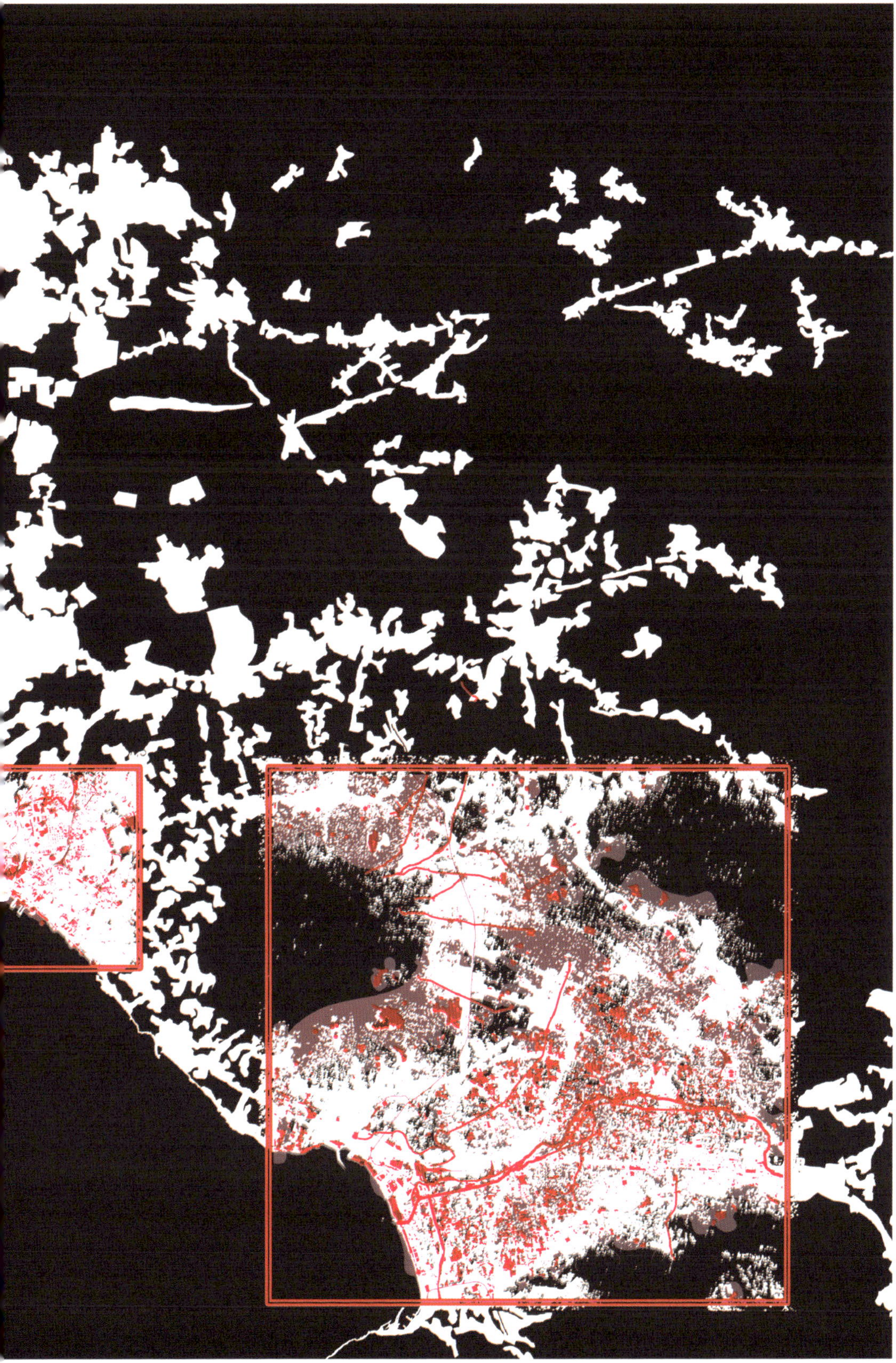

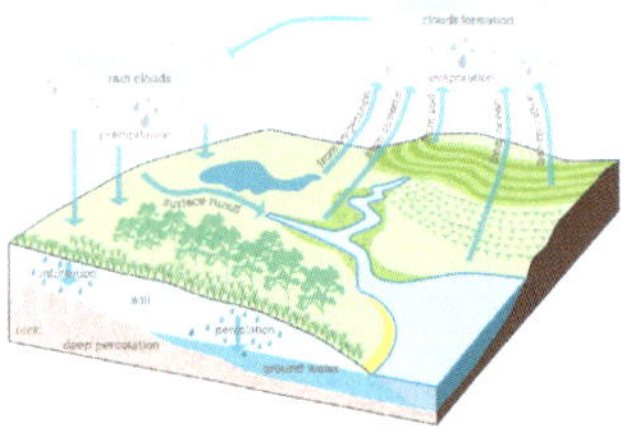

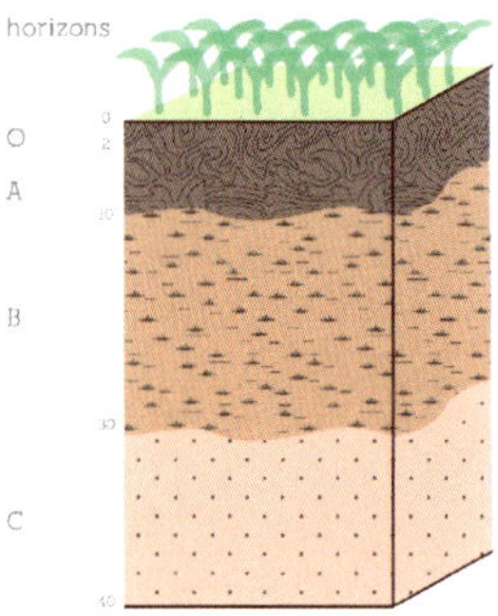

Soil in relation to the other components of the ecosystem.

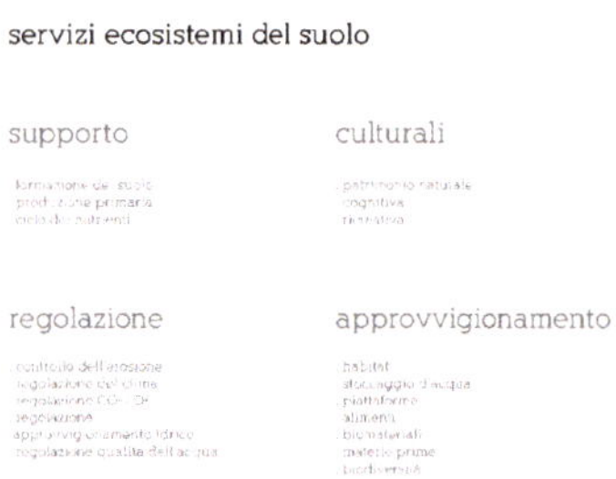

Soil's ecosystemic system. Source: F. Terribile, 2013.

threaten the human health through the dispersion of polluted particles and the production of contaminated food);
- compaction (due to the trampling effect on the soil in high humidity conditions, by mechanical means and agricultural operators. It can reduce the circulation of air and water);
- salinization (due to irrigation with brackish waters, or the intrusion of marine strata within the coastal areas. It may destroy the structure, with consequent reduction of porosity and of the air and water circulation);
- erosion (downstream movement of the most fertile surface particles, favoured by the lack of vegetation cover and by the loss of organic matter. It can lead to desertification);
- landslides (due to the absence of adequate plant coverage and retention works in the case of excess water);
- biodiversity decline (due to all the above-mentioned causes. It reduces the ability of soil to support plant production and react to climatic changes);
- desertification (due to all the above-mentioned causes. It expresses the soil's inability to support plant production);
- impermeability (it represents the most dramatic event of permanent elimination of the soil resource and all of its functions in the ecosystem. Until the '60s, it was determined by the housing needs of a society whose population was growing. In recent decades, it continues at an accelerated pace regardless of the housing needs[20].

Soil preservation

The goal of' 'modern agronomy' is to protect the soil resource, and - where possible - to treat its various forms of degradation. The main instruments are[21]:

- **Organic fertilizers** (compost). It favours: microbial biodiversity, porosity and a correct balance between the gas and liquid phase, availability of nutrients for the vegetation, carbon storage and mitigation of climate change, stability of structural aggregates and resistance to erosion, protection of the groundwater from nitrate pollution.
- **Conservative machining** (with less aggressive tools) reduce: the degradation of the organic matter and the compaction phenomena as well as the emissions and the pollution caused by the fuel used by agricultural machinery.
- **Vegetation cover** : a soil is in balance when it is covered by vegetation for 12 months a year; to ensure continuous ground cover, with different species (eg. grasses and legumes) encourages biodiversity, reduces the soil loss due to erosion, reduces compaction and protects the groundwater from pollution
- **Phytoremediation**: in the case of contaminated soils, vegetation can be used to preserve and enhance the positive functions of soil in the ecosystem and, at the same time, to reduce health and environmental risks (to reduce aerial dispersion and the migration of the

20. Terribile F., "Consumo di suolo e servizi ecosistemici", in Atti del Convegno "Il consumo di suolo: lo stato, le cause e gli impianti, Rome, 5th February 2013; source: http://www.isprambiente.gov.it/files/eventi/2013/convegno-consumo-del-suolo-2013/Terribile.pdf

21. See Antonio Di Gennaro's contribution to the LIFE-Ecoremed project, available at: http://www.ecoremed.it/images/stories/23giugno2015/Risorsa.pdf

polluting particles to the groundwater, and for the remediation of contaminated soils (progressive removal of the bioavailable fraction of heavy metals; in the case of organic pollutants, the roots, along with the soil microbes, can biodegrade the pollutant molecules).

Evolution of soil consumption in the Province of Naples between 1861 and 2000. Source: A. Di Gennaro, 2015.

Grassing of a vineyard used as a cattle pasture. Picture by Roggero.

Trees and grass help reduce the risks linked to contaminated soil.

Jeopardizing and regeneration of the water cycle

Rocco Lafratta

Introduction

It's not easy to write about water and not be pessimistic. This premise is needed to remind how vital the importance of the water resource is, and how it is at risk of real compromise, perhaps even irreversible, due to the overexploitation of the aquifers, massive collections and the absence of effective programs for the protection and safeguarding of the natural water tanks - of excellent quality - present in the mountains of the Apennines, predominantly calcareous, and in the Alps.
In addition, the constant input of contaminants has caused a widespread pollution of the surface and groundwater at various depths, depending on the depth of the soil and the geological component, where the geological layers function more or less as a filtering element.
Today we are used to opening a tap without asking ourselves where the water comes from, or wondering about the efforts and costs to ensure its drinking standard - it is readily available in our homes. We find it important only when we are denied it. In addition to the tap water consumption in Italy, 12.5 billion liters of bottled[22] water are consumed every year without any comparison with the tap water - the so-called "water of the Mayor" -in terms of quality, control and cost per cubic meter. Bottled water is now available anywhere. The water collection necessary to produce bottled water can also cause the piezometric levels of the more superficial aquifers to fall, causing the wells to dry out.
Great civilizations were all born where there was plenty of water along the waterways, in lakes, or in springs. Water is vital for drinking, washing, irrigation, for the production sectors, to produce energy, as a driving force, for water transport, and so on.
In the past, even recently, he had a profound knowledge of the water sources and the natural phenomena associated with them. They passed from generation to generation, through collective memory. There has always been a deep respect and sense of protection for the relevant sources of supply and the water-related environment.

22. Source: Legambiente

Rivers and spaces for natural flooding were preserved. If settling down on the banks was considered dangerous, stilts (still used today in the interior areas along the great rivers of southern America) or floating islands were built, which adapted to the variation of water levels. Men relied on an ancient experiential wisdom and profound knowledge of nature, of the "earth" and of its "bizarre behaviours".
The Egyptians, five thousand years ago, were anxiously waiting for the flood of the Nile because the water transported and spread fertile silt, ensuring the necessary elements to obtain a rich food production.
Nowadays at every flood there is a great deal of material damage and, very often, a loss of human lives too.

The water cycle

To understand why it has become indispensable to protect the sources of water supply and rebalance the use of water resources, it is essential that know how the water cycle works. This is to confirm the importance of water and of the process that connects the whole cycle, in order to understand the strong and well-recognized risk that surface and underground waters run. In fact, they are subject to considerable over-exploitation and concentrated pollution. Both those aspects are widespread to a much greater extent than it is generally believed. Water collects and transports everything.
In nature, the water cycle is very simple and is characterized by a modest air gas reservoir and a large liquid reservoir of marine waters and (very little) terrestrial waters, with a great self-regulating power.
The sun's energy moves the cycle through evaporation and the winds that carry moisture to the land surface, which then receives rainfall due to the condensation of moisture.
Through rainfall the water returns to the ground and thanks to the surface waterways and underground basins, it guarantees life, by returning to the organisms and feeding springs, streams and rivers. When the water reaches the ground, it collects loose or eroded chemical elements from the soil and

Confluence of the Alveo Comune Nocerino in a canal of the Sarno river, Scafati (Naples).

distributes them to other living organisms or pours them into the receptive bodies from which they will then return to the soil following their specific cycles. On the earth surface some of the water is partially retained, and part of it keeps flowing to evaporate again into the atmosphere. Thus, between the atmosphere and the earth's surface, a continuous exchange is established that determines the circulation of water in a continuous cycle: the rain falls and feeds the surface and ground waters that generate and / or feed the rivers and lakes, flow to the sea, then begin to evaporate again, leading to cloud formation and subsequent rainfall. It is clear that there are multiple sources of pollution: unmonitored waste water discharged directly in rivers and lakes, pollutants from the agricultural and zootechnical sectors, evaporation that also carries light polluting molecules, marine aerosols, particulate matter dispersed in the air, etc. All liquid or gaseous pollutants end up in the waters directly or through precipitation. This is why it is absolutely necessary to protect the sources of water supply. And we must prevent pollution, rather than purify the water afterwards.

Water consumption and quality

The quality of the water - regulated by its natural cycle, which has always been of vital importance - is today greatly compromised[23]. As mentioned above there are numerous and well-known sources of pollution affecting its cycle that, by collecting and concentrating existing elements, cause an imbalance of concentrations and compromise the quality of the superficial, underground, fresh and salt water. The pollution of the water resource started with the industrial development, mainly due to the transition from natural agriculture to intensive farming and the intensive use of chemical derivatives, in addition to the gradual increase of the population, especially in large cities. The increased demand for water supply was met without a proper adaptation of the water supply network, and its drainage and sewage. Both solid and liquid pollutant products have progressively polluted the soil with concentrations that were far above the natural self-purifying capacities. This development, from the post-war era to nowadays, has raised the need to supply more drinking water as well as water that is not drinkable, but still of good quality. This increased demand has led to an imbalance in the altered withdrawal / consumption cycle and in the natural recharging of the aquifers, due to massive catchments from springs and the use of deep wells. In these few decades we erased thousands of years of history, experience and memory. We have had the presumption of being able to govern the natural elements, we even tried subjugating them, and we were punctually proved wrong, with significant economic damage and often even with the loss of human lives. And in the name of non-deferrable necessities, we have triggered a process that has slowly led to the water being taken off the rivers, with the massive catchments of water from the springs; this has led to an overexploitation of the aquifers, with massive

23. In-depth information about the illustrated topic can be found in the following volumes: "CIRF Manuale" and "CIRF Decidere l'ambiente"

collections for industrial, agricultural and civil purposes, and the consequent depression of the ground water levels, resulting in irreversible subsidence phenomena. In addition, the lowering of the piezometric levels has allowed the saltwater to enter the seashores, triggering irreversible desertification processes. The industrial sector required a huge amount of water for production cycles, initially collected from surface waterways, then from surface aquifers, from 15 to 30 meters in depth. After the intervention, the water was discharged into the surface waterways, but without due attention to the purification process, which would have returned the collected water in the same quality conditions as at the time of withdrawal. With the progressive pollution of the surface water, surface wells were abandoned and collection levels increased in the so-called second and third aquifers - 50 meters deep, up to 200-300 meters. Intensive crops required more water for irrigation. This was also taken from wells, generally superficial. Major productions have been supported by an increased use of synthetic chemicals, which have considerably contributed to water pollution. The massive use of fertilizers, herbicides, agrochemicals and pesticides has generated a concentration of byproducts that bacteria and plants cannot assimilate with their natural absorption processes. After rainfall and leaching processes, excess products accumulate in the soils, and are then transferred into rivers, lakes, and groundwater until they reach the sea. Finally, the demographic growth of cities has also led to a considerable increment in the consumption of spring and surface water, increasingly collected to meet the raising demand for drinking water. The increased water extraction from springs has led to a reduction in the water levels in rivers, which are often completely dry during the summer months. Nevertheless, the depuration scheme provides that the purified waters end up in rivers and streams. Generally, the sewage system is mixed, which means that sewage and rainwater are collected together. This causes the purifiers, designed to receive only sewage, to go haywire after any heavy rain. Thus, the purifier does not work as it should. In many cases, purifiers are totally absent. So, over the summer period, dry rivers receive water that has not been properly purified, or not purified at all. This results in a continuous cycle of pollution of both surface and groundwater.

Water recovery, treatment and reuse

As mentioned above, the exploitation of the water resource has created several imbalances, in some cases in a reckless or even deliberate manner, which have considerably affected and damaged the water quality and the soil. Nowadays, there are also new imbalances caused by the drastic reduction in water consumption, which causes a new set of problems. This too is becoming a cycle, this time a completely artificial one. And the real bet is to be able to rebalance these new upheavals, often causing huge damage and losses. It is necessary to connect what has been done in the past with the related current imbalances and today's need to solve the new problems that are emerging after the inversion of the consumption trends. First of all, we should avoid carrying out further interventions on new soil, and work with what has already been built, or with what has already been extracted from the natural environment. There is a need for a much wider vision with a long-term perspective oriented towards the catchment basins, with shared participation and planning, and the help of local communities together with all interested parties. The project must absolutely be multi-purpose. Only in this way can we hope to regain what is now compromised. The crisis-related decrease in water consumption in the industrial sector, the abandonment of agricultural crops and the discontinued use of water from wells for civic purposes have led to a significant reduction in the collection and usage of water from the aquifers. This has led to a rise in piezometric levels. This, on the one hand, is a positive outcome as it reduces the effect of concentrated pollutants and the aquifers are richer in water. On the other hand, especially in flood areas, it can produce negative effect as everything that has been done in recent years in the civil engineering and infrastructure sectors is based on a range between 3 to 5 meters minimum depth from the land surface. Now, with the reduction of the withdrawals, the aquifers are just are a few dozen inches from the land surface, if not above. In many areas, these phenomena are exacerbated by the concomitant adverse effects of subduction. This is leading to a significant reduction of the rainwater accumulation in already saturated soils and to an increase

in floods, with considerable damage to crops, built areas and infrastructures. For these reasons, there are constantly flooded structures and areas where sump pumps have to be operated continuously, to ensure the availability of such infrastructures by avoiding permanent water damage. However, these areas are still subject to floods in the event of heavy rains. Today, a drastic change of direction is needed with a global view of the surface and underground water-related issues. Good quality waters are concentrated in specific areas, in the hinterland and in the mountains. Their transportation requires long-distance movements - many kilometers within the water supply networks. Along these networks they undergo huge losses, which in some cases reach almost 50%. We need to start a drastic waste reduction campaign on the water supply networks.
At the same time, information campaigns aimed at reducing consumption are also much needed. It is necessary to stimulate the realization of appropriate treatments for agglomerations of at least 2,000 inhabitants (Legislative Decree 152/2006 as further amended), which include decentralized treatments such as constructed wetlands. These should be structured with a primary treatment such as Imhoff tanks, followed by phytodepuration tanks with horizontal and / or vertical flow system, depending on the wastewater to be purified. Purified waters can be reused for all non-drinking purposes: irrigation, washing, flush lavatories, etc. We need to plan the use of the water resource and its collection from groundwater and springs, making sure we do not damage the aquifers by respecting the natural recharging times of the natural reservoirs. It is of vital importance to restore water in the rivers, especially during the dry summer season, guaranteeing a flow that can maintain the minimum level of watercourses to ensure the presence of aquatic fauna as well as aquatic and hydrophilic vegetation. The higher the levels of environmental flow, with its presence of aquatic and hydrophilic vegetation, the higher degree of self-purification can be achieved. It is necessary to create more areas for the natural treatment of waters such as wetlands, both large and small, located downstream of the purifiers, with multiple objectives such as: the regulation and mitigation of overflows, further wastewater purification services, replenishment of the aquifers, increasing the specific vegetation for depuration purposes, and creating areas for the natural repopulation of specific fauna, such as amphibians, reptiles, Odonata, birdlife, etc.
This is also aimed at increasing biodiversity and all the ecosystem-related processes connected to it.
Today, in order to avoid further soil consumption, we need more areas of natural flooding in association with the recovery of flood plains and with multifunctional features. In particular, we need to mitigate the risk of flooding downstream areas and to restore the necessary space for rivers. It is also indispensable in new settlements and - where possible - in built-up areas, to intervene on the collecting systems by separating the rainwater disposal networks from the wastewater, thus reducing the rainfall discharge in the sewerage system.

Regi Lagni canals, Castel Volturno (Caserta).

A series of small, widespread interventions on the built environment, which would reduce and roll rainwater, both in sewers and in waterways, such as: garden terraces that retain water, partially by absorbing it, partially by returning it gradually; rainwater storage tanks, to be reused to irrigate; elimination of the extensive water-proofing in car parks and squares in favour of draining and absorbing materials, and finally the construction of drains along urban and suburban roads. All these single actions would reduce the burden of rainwater into natural disposal networks such as rivers, canals and ditches, and in the municipal sewerage network, improving the performance of purifiers.

Archaeological area of Longola, Poggiomarino (Naples).

Taxonomy of waste.
A brown, grey and green atlas

Anna Terracciano

Drosscapes in the coastal plain of the Campania region

The unstoppable growth of the urban sprawl[24] at the disadvantage of natural landscapes leads to a continuous process of production of drosscapes[25] as a manifestation of the urban development and its cyclical crises. This dynamic leaves us with a legacy of socially, ecologically and economically disadvantaged landscapes, associated with a continuous spatial redefinition that represents an immense patrimony, with varying potential and opportunities for adaptation, reuse and recycling. But each specific context requires a rethinking of this field of research in both its morphological and relational structure and will inevitably be faced with the spectrum of the concept of drosscape, in its forms and its meanings.
In the Campana plain, drosscapes do not only outline a low-density, widespread geography, but also intercept the continuity of the infrastructure networks, the reticular dimension of the ecological connections and the risks associated with climate change, the porous structure of a territory[26] where the settlement patterns overlap with unstoppable soil consumption processes within a multiplicity of "figures". Increased pollution, multiple contamination factors, ecological degradation and the environmental emergency generated by the alterations in the waste cycle, and more generally by the pervasiveness of waste produced by out-of-control industrial chains, constitute not only a background condition but the main elements that characterize this phenomenon. Moreover, the lack of a political vision, a project of economic and space development as well as the absence of an effective strategy for abandoned areas have prevented the emergence of virtuous recycling processes. These factors have led to a systematic exploitation of natural resources that has profoundly affected the urban design, the culture and fate of local communities. Soil consumption has created urban sprawl and decentralized dwelling areas, diluting them in an increasingly wide territory and causing unpredictable outcomes.

24. F. Indovina (edited by), "La città diffusa" (The Scattered City), Iuav - Daest, Venezia, 1990

25. A. Berger, "Drosscape. Wasting Land in Urban America", New York: Princeton Architectural Press, New York, 2006

26. C. Gasparrini, "Città da riconoscere e reti eco-paesaggistiche", in PPC Piano Progetto Città, n. 25/2011

Over-dimensioned infrastructure has designed a hierarchical, single-mode, and unstructured system, generating over time a constellation of barriers, residual spaces and urban machines that are functional to the city metabolism and that the city itself tends to eject, along with their relationships and flows. In addition, the continuous dynamics of dismissal, abandonments and reuses, constantly mixing materials and relationships, have produced a labile urban structure in which many ecosystems that are currently compromised still coexist. In addition, progressively and in very few years, a huge amount of superficial and profound "waste" has been produced, affecting the waterways and the urban and suburban soils, as well as the residual and marginal areas of residential settlements[27] .
In these areas, however, there is also a variety of actors (institutions, specific operators and associations) whose skills and demands often overlap in a limited amount of places and artifacts, mostly related to the waste cycle crisis or to environmental emergencies. These places, in a suspended state due to sometimes conflicting decisions, often become the subject of informal appropriation practices, even through improper uses on the verge of illegality. This variety of informal uses and demands, which are difficult to spatialize, can easily originate a reverse city that the physical city is unable to tell.
The inevitably collective, political and systemic nature of the issues at stake (and in particular the environmental issue), together with their degree of interconnection and interdependence with the production, decline and disposal of "waste" impose not only a necessary redefinition of the research areas but also and above all a change in how to look, interpret and describe this new territorial dimension as a "deposit of multiple cycles of life"[28]. This is not only becoming the key to any possible analysis, but "is also a first step to build a theory that considers the city and the territory as renewable resources."[29]

27. C. Gasparrini, "Waste, Drosscape and Project in the Reverse City", Pavia R., Secchi R., Gasparrini C. (edited by), "Il territorio degli scarti e dei rifiuti", n. 8, Aracne, Roma, 2014

28. S. Dryzek John, "Rational ecology: environment and political economy", Oxford, Oxfordshire UK New York, New York USA: B. Blackwell, 1987

29. P. Viganò, "Cicli di vita, energia, riciclo in Re-cycle Italy. Nuovi cicli di vita per architetture e infrastrutture della città e paesaggio", n. 3 Aracne, Roma, 2013

Research areas and main issues

The Campana plain extends over an area of 1,350 square kilometers, divided into three parts by the volcanic complexes of the Campi Flegrei and Somma-Vesuvius: in the north we find the lower part of the Volturno river, innervated by a vast system of streams and drainage channels crossing the ancient wetlands. In the centre lies the ancient Sebeto valley, in the plain to the east of Naples and the artificial hydrographic system of the Regi Lagni, dating back to the Borboni rule. Finally, the plain of the river Sarno extends in the south, with its dense network of irrigation channels and -in a barycentric position - the artificial stream 'Canale del Conte'. The aquifers of the plain, together with the three above-mentioned hydrographic basins, the minor waterways, and the areas due to water rehabilitation make up a large "water machine" that further combines with landfill sites and the urban sewerage network. A system that is the result of a progressive modification of the territory, implemented through hydraulic control, and strictly related to the management of the resources and their exploitation. For centuries, this system has fostered the growth of diverse economic and social structures, shaping the design of the territory, taking care of it, and creating a sustainable balance. The water remediation of the plain has thus contributed to its productive and political evolution by defining the most extensive and complex infrastructure of the region, which is crucial for the agricultural use of soils and urban structures. A potential ecological armor that, unfortunately, has become a main risk factor due to climate change related events and its action as a vehicle for the pollution caused by the poor disposal of urban and industrial wastewater. This is extremely dangerous as the hydraulic risk is combined with pollution, the jeopardizing and mismanagement of the territory and the waste areas, buildings and infrastructures, more generally called "drosscapes", in a context of increasingly uncontrolled urban metabolism. Progressive and traumatic alteration of the virtuous cycle of waters has deposited multiple materials on the surface and in the stratigraphy of the soils. Although they have exhausted their life cycle, they can still constitute potential regeneration elements. The historical continuity of ecological networks is now in contrast with their fragmentation, while the right distance between the anthropic systems and the water contrasts with a condition of overlapping and mixed elements. Although the fundamental value and identity of the water resource is well recognized, this awareness contrasts with widespread discontinuity, obliteration and uncontrolled exploitation. The meaning and aim of this text is not only to outline some fundamental issues that are essential to the contextualization of the drosscape phenomenon in Campania, but also to understand that the history of the plain has been strongly impacted by the metamorphosis of its waters, since they have not only shaped its form and its landscapes but have also played a central role on a social and economic-productive level. This awareness,

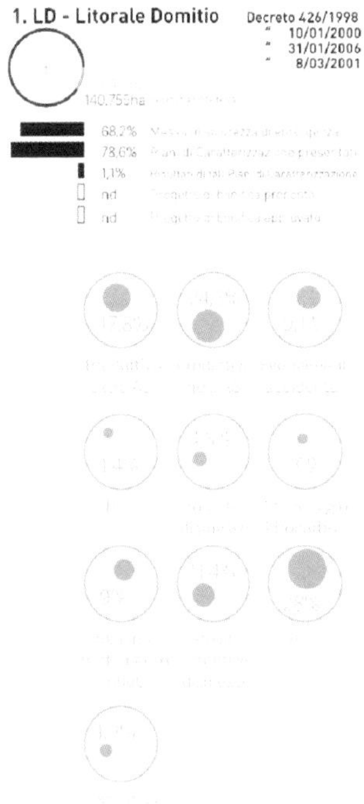

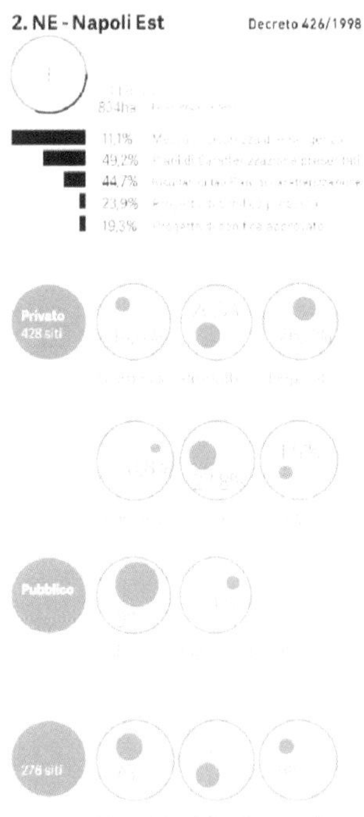

65

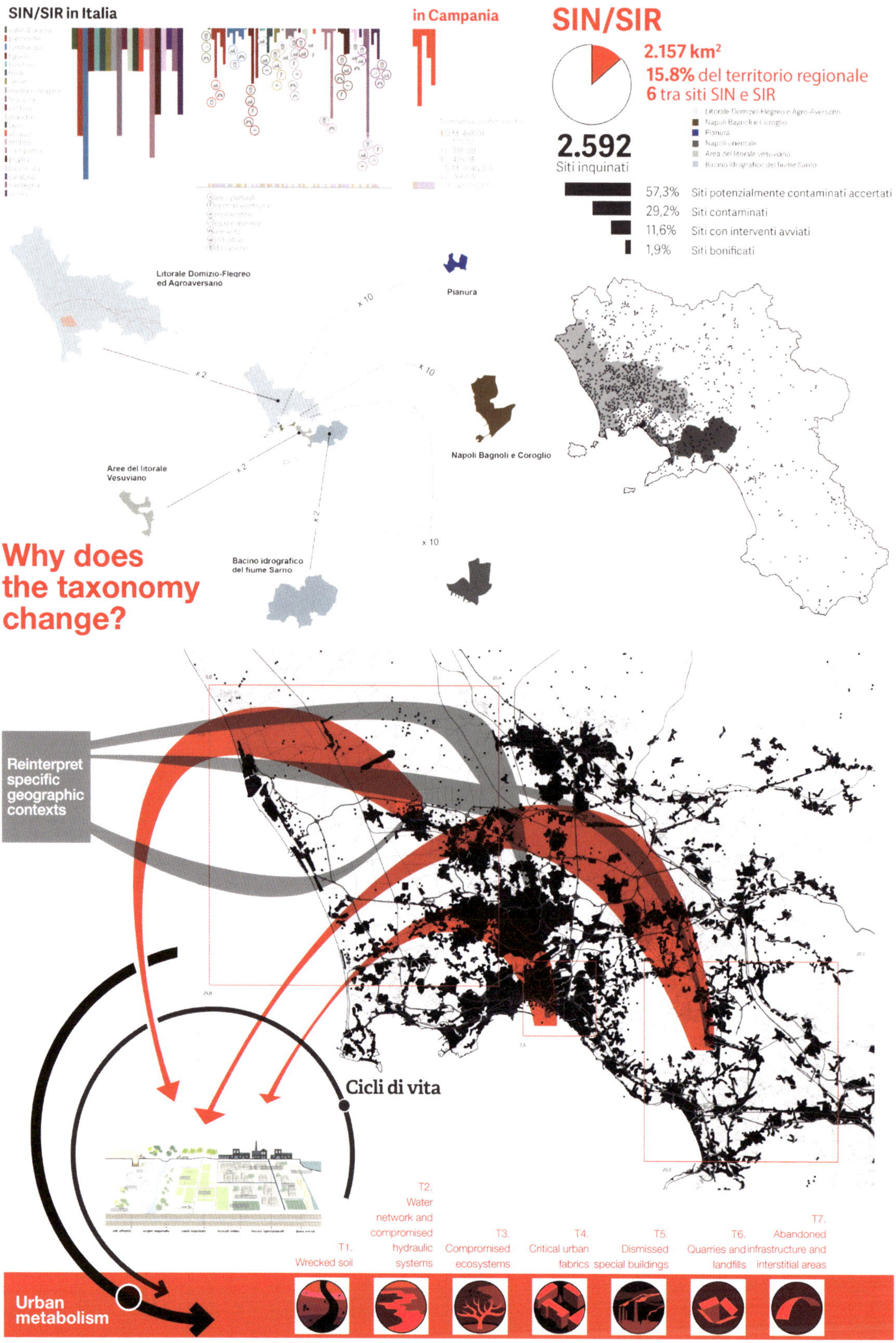

SIN/SIR in Italia
in Campania
SIN/SIR
2.157 km²
15.8% del territorio regionale
6 tra siti SIN e SIR
2.592
Siti inquinati
57,3% Siti potenzialmente contaminati accertati
29,2% Siti contaminati
11,6% Siti con interventi avviati
1,9% Siti bonificati
Litorale Domizio-Flegreo ed Agroaversano
Pianura
x 10
x 2
Napoli Bagnoli e Coroglio
Aree del litorale Vesuviano
Bacino idrografico del fiume Sarno
Why does the taxonomy change?
Reinterpret specific geographic contexts
Cicli di vita
Urban metabolism
T1. Wrecked soil
T2. Water network and compromised hydraulic systems
T3. Compromised ecosystems
T4. Critical urban fabrics
T5. Dismissed special buildings
T6. Quarries and landfills
T7. Abandoned infrastructure and interstitial areas

and the need for a cognitive framework that deals appropriately with the importance of the issues exposed - also capable of overcoming the sectoral approach that considered the phenomenon of pollution independent of the territory and social dynamics - has led researchers to identify the above mentioned hydrographical basins (identified as Sites of Community Importance)[30]: the Domitio-Flegreo Coast, Naples East and the Plain of the Sarno River. These three areas are extremely different in size, density, history and economy, and for that reason they are good subjects to analyze the processes of conceptualization and generalization in the context of a taxonomy of "drosscapes" in the Campana Plane.

Towards new taxonomies and geomapping of drosscapes

Given the vastness and diversity of the research areas, the Campana plain was overlapped with a grid (with a 2.5x2.5 km module) as a conventional measurement system to make the different parts of the territory comparable. This has allowed us to circumscribe three large squares that constitute the "focus" of the areas corresponding to the three river basins. Besides redefining the research field, a redefinition of the concept of "drosscape" was needed, since in the context of Campania it is not limited to the traditional critical porosity of abandoned areas and buildings, brownfields and residual spaces. In fact, it has been necessary to expand and clarify a taxonomy capable of involving other urban and landscape materials affected by the exhaustion of the economical, productive and eco-systemic life cycles, emphasizing not only their stratigraphic aspect (topsoil, soil, subsoil) but also their relational and multiscale dimension (uses, flows, social demands, and so on.)
The ecological, urban and economic effects of drosscapes extend beyond the compromised sites, as they intercept a multiplicity of spaces and cause a different degree of compromise to other elements such as: agricultural soils affected by the pollution of nearby landfills, the surface and ground water network, the dune eco-systems altered by aggressive soil consumption, the critical fabrics resulting from the exhaustion of the cycle of the tourism industry buildings, many of which are now abandoned or reoccupied for other purposes[31]. This necessary change in perspective returns a multidimensional mapping where the descriptive neutrality typical of traditional geographic maps is compensated and enriched by the interaction with a multitude of interpretative components that elude the zenithal gaze. Increased inequalities, a multiplicity of local realities in which new identities, lifestyles and economies[32] are emerging, and other dynamics, are not only opposed to the growth and the metropolitan dimension of the city, but also translate into different and contrasting production practices, use and appropriation of spaces and places, especially in the more marginal ones. The difficulty of recognizing "drosscapes" in a unique and immediate way through the (real or virtual) site inspection and photographic investigation, as was the case with Berger's search[33], and especially the impossibility

3. PS - Piana del Sarno Decreto 266/1995

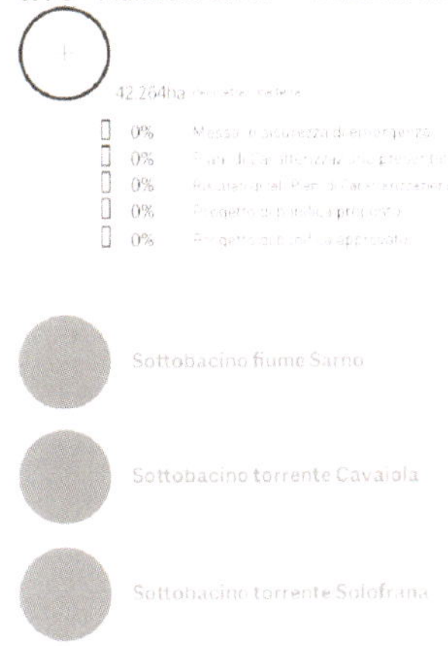

When comparing data from the SIN/SIR in the three contexts, delimitations are very different. The ones regarding East Naples are based on reliable data, proving a recognizable industrial past. For the ones concerning more than one municipality, such as the Domitio-Flegreo coastline and the Sarno Plain, the legislator, in the impossibility of a shared reclamation program, calls for preliminary investigations and sub-delimitations, which reduce them however to the creation of a Census of Polluted or Potentially Polluted Sites. Graphs in these pages are based on data from: Federambiente (2010), Legambiente (2012) and Campania Region (2003, 2005, 2010, 2011).

30. Established in 1998 (l. 426); according to article 252, Legislative decree 152/2006 these sites were identified by the Ministry for the Environment, Land and See (MATTM), "the Areas of National Interest are extended areas where widespread, deep and dangerous pollution can compromise the development of areas of strategic importance." (ARPAC 2008).

31. C. Gasparrini, "Waste, Drosscape and Project in the Reverse City", R. Pavia, R. Secchi, C. Gasparrini (edited by), "Il territorio degli scarti e dei rifiuti", n. 8, Aracne, Roma, 2014

32. G. Amendola, "Le forme urbane della paura", in "Urbanistica", 104. Roma: INU Edizioni, 1995; ID. (edited by), "Scenari della città nel futuro prossimo venturo", Roma-Bari: Laterza, 2000; ID., "La città postmoderna", Bari, Laterza, 2003

33. See Alex Maclean's photographic contribution in A. Berger, "Drosscape. Wasting Land in Urban America", New York: Princeton Architectural Press, New York, 2006

Drosscapes in the Campania Plain

T1. Wrecked soil

T2. Water network and compromised hydraulic systems

Hydraulic system

Coastal pollution

Città Domitio-Flegrea

Abandoned land, Giugliano

Artificial canals, Regi Lagni

Città Orientale

Landfill, "Ambito 13"

Emerging aquifer, "Ambito 13"

Città Sarnese

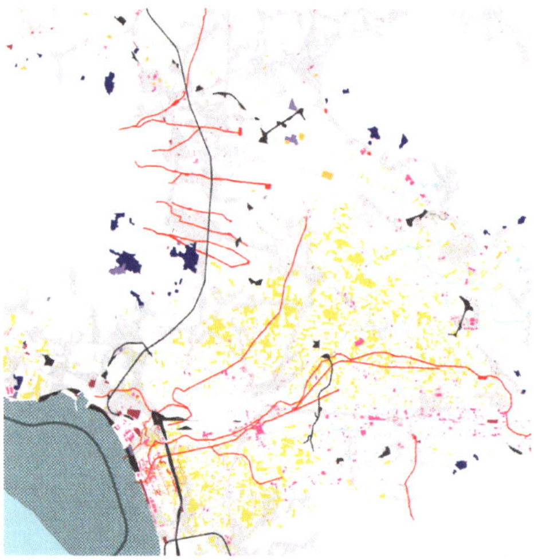

Agricultural land, Scafati

Industrial discharges, Sarno

T3. Compromised ecosystems

Dunal systems

T4. Critical urban fabrics

Tourist real estate cycle

T5. Special buildings

Abandoned

Functioning

T6. Quarries and landfills

Quarries

Landfills

Temporary storage sites

T.6 Abandoned infrastructure and interstitial areas

Disused infrastructure

Interstitial areas

Coastal dunes,
Castel Volturno

Villaggio Coppola,
Castel Volturno

Artificial canals
of Regi Lagni

Taverna del Re landfill,
Giugliano

Unused free areas

Greenhouses,
"Ambito 13"

Gypsy camp,
Gianturco

Gasometer,
Gianturco

Infrastructural beam,
"Ambito 13"

Greenhouses and agricultural land

Greenhouse and
agricultural land,
Poggiomarino

Artificial canals
of Regi Lagni

Novartis,
Torre Annunziata

Vesuvio quarry,
Terzigno

SS268 road,
Angri

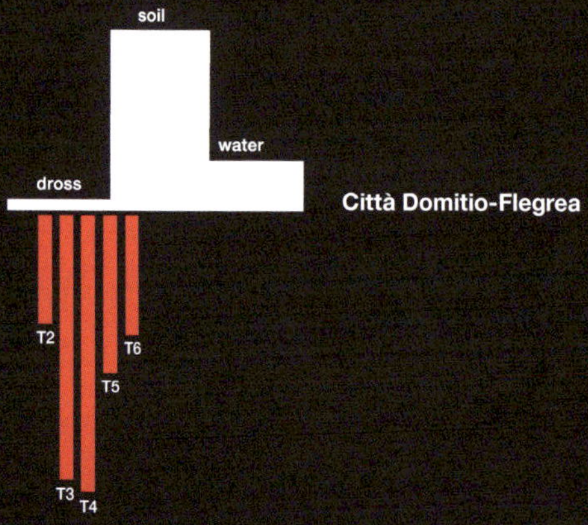

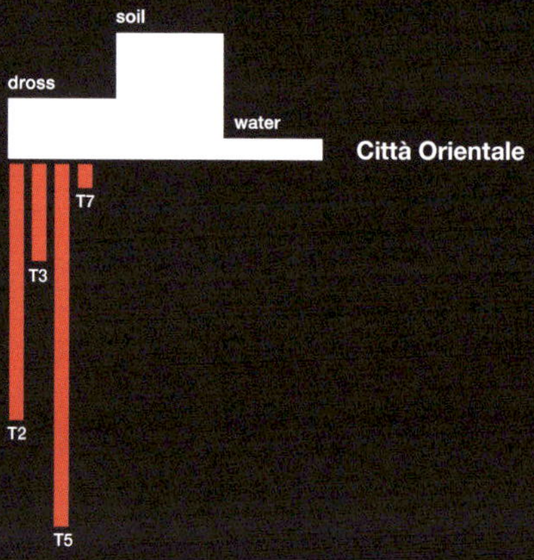

Dross [+] Infra/Water/Rural/ scape

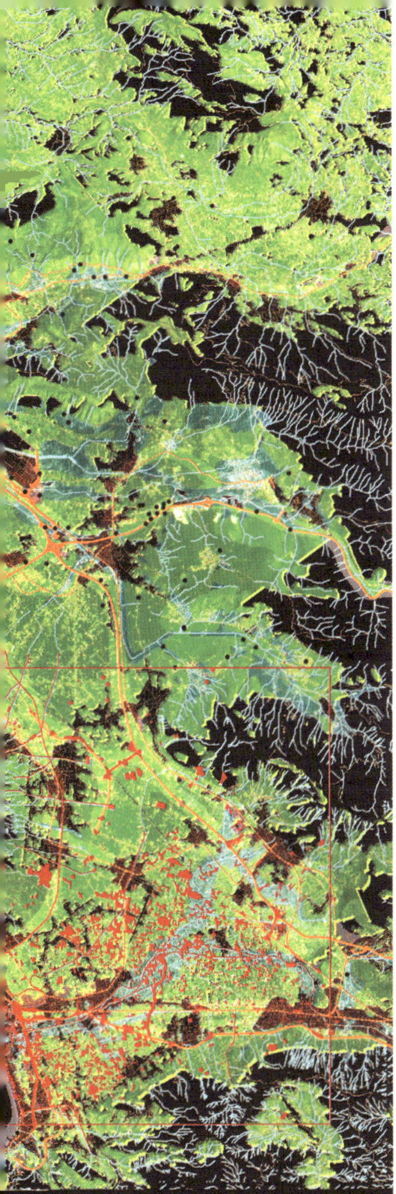

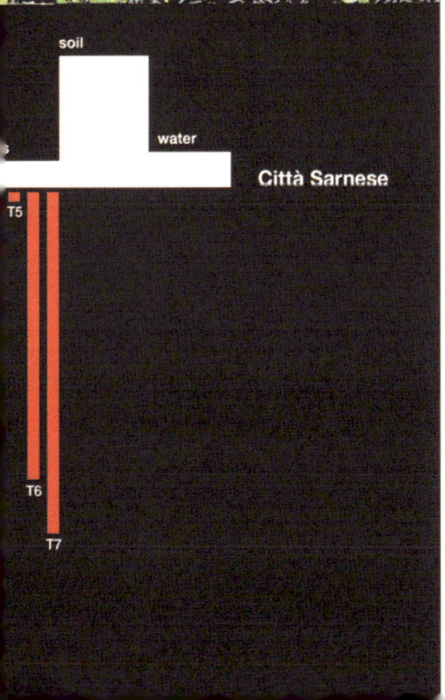
soil
water
Città Sarnese
T5
T6
T7

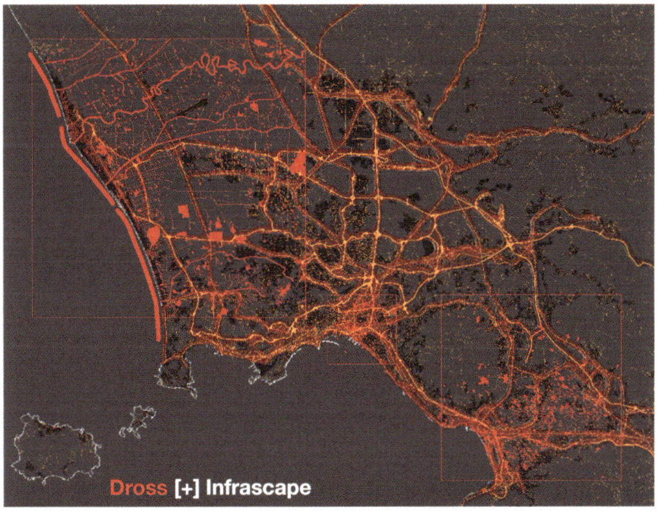
Dross [+] Infrascape

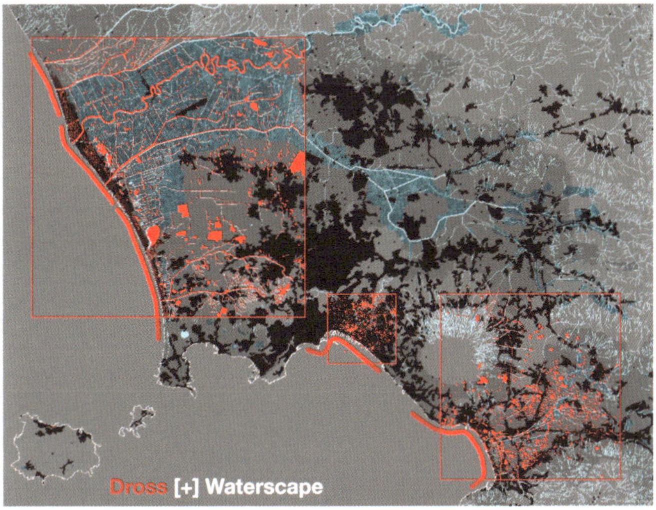
Dross [+] Waterscape

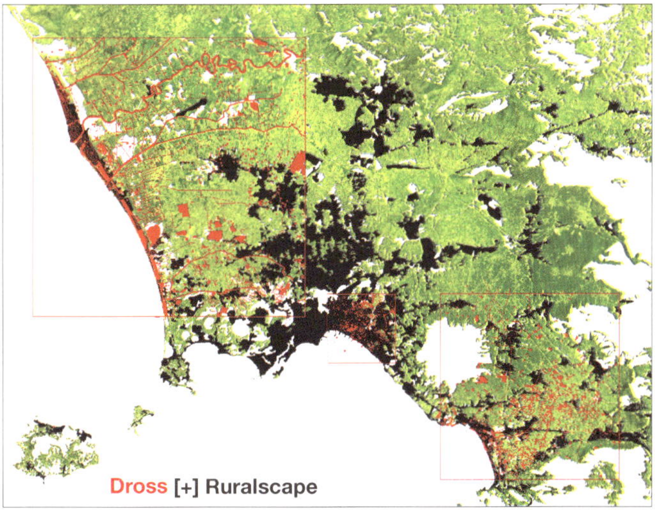
Dross [+] Ruralscape

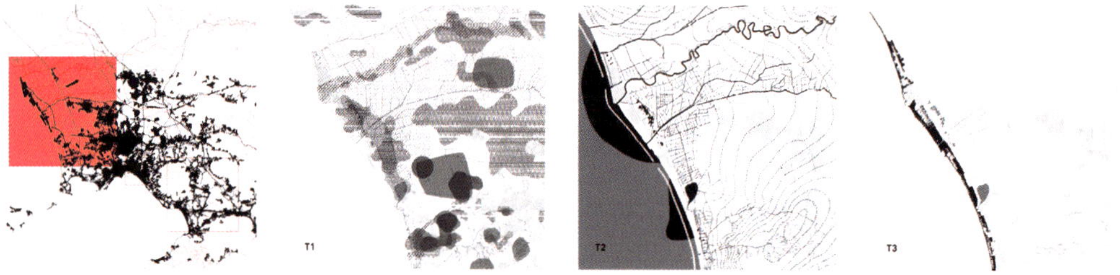

Città Domitio-Flegrea

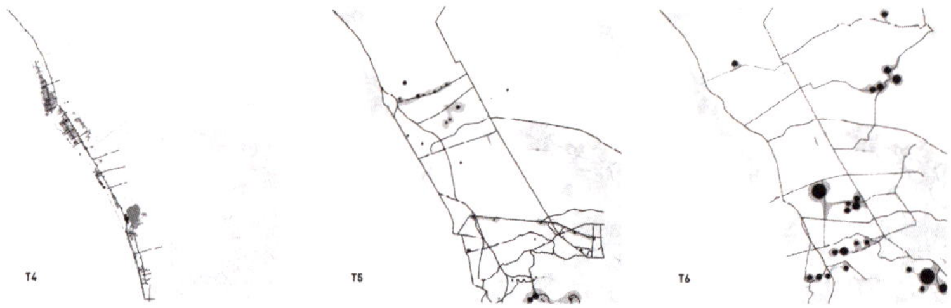

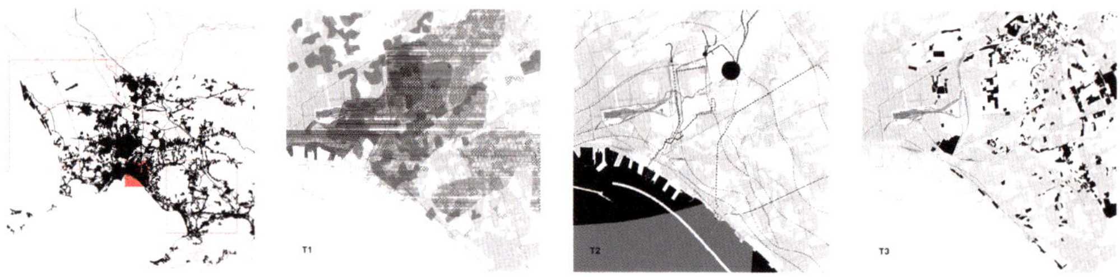

Città Orientale

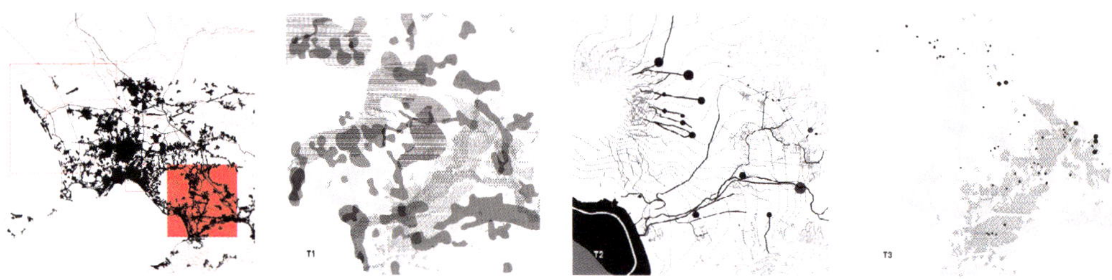

Città Sarnese

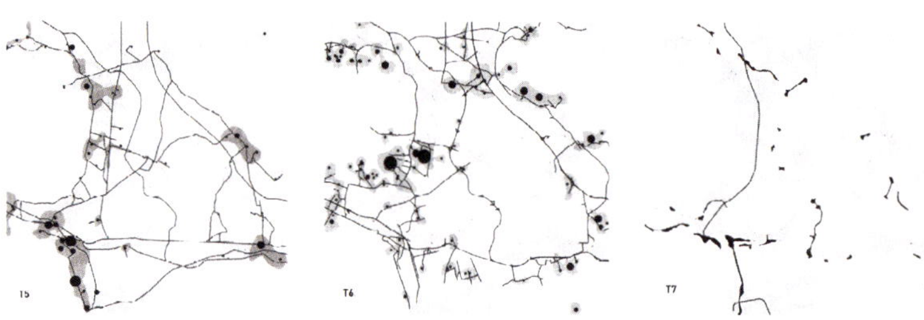

of representing the state of the life cycles and the different degrees of resilience in a drosscape (with reference to the intensity of the phenomena in progress and their evolution) in traditional zenithal or diagonal representation modes has led to a reflection on how to build a dynamic system capable of returning the complex and multiscale nature of drosscapes. It is based on the virtuous combination of a G.I.S (Geographical Information Systems) database and Multi-Criteria Analysis of both collected and produced data, using a specific set of indicators for each drosscape theme (T), articulated into a tree system generated from the combination of Resources, Uses, and Values. The goal is to obtain an instrument capable of viewing, interrogating, analyzing, interpolating and monitoring data, while at the same time building simultaneous narratives capable of identifying the most critical areas and conflicts, as well as the levels of susceptibility to recycling and the priorities for intervention. This interpretive complexity must be able to identify, besides the size and diffusion of dross-capes, also new "figures" able to point out the dynamics of their flows and relationships. Without claims of exhaustiveness, research has thus rebuilt a "geography" of places[34] able to bring back in the map a phenomenology that deforms space, going beyond the traditional forms of representation. A new structure takes place, made of bonds, areas of influence and relationships networks, which is the inverse figure of this territory and not simply its negative. "Figures" of burnt territories that like black holes[35], together with their system of rules and relationships, reveal another space-time dimension. Thus, the seemingly fragmented, confused, and potential- lacking 'latent design' emerging in these territories can actually explore fertile spaces for the project[36]. In conclusion, the meaning of such maps is to synthesize the most relevant dynamics of the 'life cycles' of drosscapes and their exhaustion, and the new trajectories for the activation of 'new cycles', but also to identify priorities and outline the potential relationships of "landscape networks".

On the opposite page, mapping of the punctual and relational dimensions of drosscapes under the three research frameworks.

34. F. Farinelli, "L'invenzione della terra", Sellerio Editore, Palermo, 2007

35. A. Terracciano, "Geografie dello scarto vs geografie del riciclo. Disegni di una traiettoria possibile", Pavia R., Secchi R., Gasparrini C. (edited by), "Il territorio degli scarti e dei rifiuti", n. 8, Aracne, Roma, 2014

36. R. Brown, "Mapping the Unmappable, Knowing the Unknowable", in "Public Interfaces", C. Ulrik Anderson and G. Cox eds., Aarhus: Aarhus University, 2011

Part 2

Geomapping, datascapes, drosscapes (drossdatascape)

"In the last decades, pervasive rhetorics on landscape and the cartographic space-time simplifications have not been able to describe the contemporary City's urban material in its concurrent miniaturization/bigness and overlapping/distancing dynamics; nor to describe its crossing flows' dynamics, making it impossible to decipher the latter's relationship with the places and their uses from variable combinations of self-referential social subjects. Those rhetorics and simplifications have therefore resisted fiercely and intentionally even to the re-emerging of those monsters connected to the insane dynamics of urban metabolism, those waste and refusal areas often invisible to the distant mechanical look of aerial photography or standard topographic coding, both above and below the "earth's skin". Today, however, those monsters-drosscapes need to find a new place in our descriptive interpretations and this requires a radical effort in rethinking the traditional tools in order to reverse the priority order, established with modernity, of the scales in relation to the values; or, indeed, to try to reconcile in a virtuous manner these two options of urban narration so distant and conflicting until today".

Carlo Gasparrini

DrossDataScape in the Campania Plain

f_t4ldr_05

Notes for a cognitive approach to drosscape mapping

Marina Rigillo

The word "drosscape" is a suggestive term, coined by Alan Berger[1] to describe the landscapes of waste, including not only abandoned places, but a wider taxonomy of spaces that have been used by urban metabolism until they had exhausted their beauty and functionality. These "discarded spaces" are not only a by product of the loss of the productive value of the site, but also (and perhaps above all) represent the vestiges of contemporary barbarity, resulting in a spontaneous and non-rational sentiment in a community that wants to keep these places distant, "Not in my back yard," and instinctively considers them dangerous and "obscene."[2]
In facts, their image we see is obscene and so is the risk that is often associated to them, but above all, the decision that originated such places is obscene - whether it is the result of informal practices, illegal motives, or institutional choices - since it is anyhow enslaved to blind logics of consumption, and ignores all space production regulations.
The specificity of the drosscape can actually be found in its genesis: such places were created to meet the development needs of the territory with a capitalistic approach. Those spaces, already identified by Manuel Castells in 1989 as belonging to the post-industrial society, were "Not a space of places, but a space of flows."[3]
What distinguishes drosscapes from other types of abandoned areas is the absence of a relationship between the use of space and the surrounding area - a net, yet painful cut with the history and structure of the territory, and with that "genius loci" which also allows us to distinguish the charm of ancient factories and industrial ruins, and which leads us to perceive the drosscape as 'the wrong side of the ruins.'[4]
Conceived as spaces to host streams (of commodities, waste, raw materials, people), drosscapes escape from every strict principle of taxonomic description. Created to meet sector-specific needs, they are by definition mono-functional places whose identity is revealed only in relation to the productive and / or social process that generated them. Subsequently,

1. A. Berger, "Drosscape. Wasting Land in Urban America", New York: Princeton Architectural Press, New York, 2006

2. K. Lynch, "Wasting Away", Sierra Club Books, San Francisco, 1990

3. M. Castells, "The Informational City: Information Technology, Economic Restructuring, and the Urban-Regional Process", Basil Blackwell, Oxford, 1989; already quoted in M. Neuman, "Space and Place, Haste and Waste", in the Berckley Planning Journal 7 (1), 1990)

4. C. Woodward, "Learning from Detroit or "the wrong side of the ruins" in A. Jorgenson, R. Keenan, "Urban Wildscapes", Routledge New York, 2012

the suffix "scape" refers to a landscape idea that overcomes any aesthetic preconceptions in order to be rooted in the representation of everyday life, sometimes very distant or even totally different from any idea of beauty[5]. The "scape" suffix also allows a further possibility of interpreting these places by attributing them a semantic recognition that makes waste an emblematic function of our age: such spaces become landscapes, as long as they can be decoded through a set of recurring signs which also testify the consumption process that generated them, as well as the meaningful values attributed by society to their resources and history. From this perspective, drosscapes can be traced back to those "spaces of representation"[6] that identify the myths and symbols of contemporary civilization, not because they perform prestigious functions, but due to the fact that they are a significant part of the production of goods and services that nourishes and distinguishes the culture of mass consumption, paradoxical "landmarks" of a society that expresses a vigorous propensity to destruction. Desolate landmarks, but still a tangible sign of our time, expression of new needs, and of a different - and often destructive - experience of the territory[7].
If we accept the idea of drosscapes being part of the contemporary landscape, as spaces created to respond to production processes typical of a capitalism-generated "de-regulation", we must also consider the systemic character of this process and the particular geography of the (not just spatial) relationships established by drosscapes, both globally and locally. These are, in fact, the essential identifying features which, in the first place, allow us to isolate drosscapes as an object of scientific interest that should be reinterpreted according to descriptive codes aimed at a recovery operation. In this sense, the definition of a methodology for the description of a drosscape - and its transposition into a drossmapping process (a mapping system that takes into account the specific qualities of the object) - defines the theoretical contribution of the research, placing the mapping issue within

5. EU (2000) "Convenzione Europea del Paesaggio", source: http://www.convenzioneeuropeapaesaggio.beniculturali.it/uploads/2010_10_12_11_22_02.pdf

6. H. Lefebvre, "La production de l'espace", Editions Anthropos, Paris, 1974

7. R. Mistrach, K. Orff, "Petrolchemical America", Aperture Foundation NY, 2012

Dune system, Domitio-Flegreo coastline

Dismissed quarry, Vesuvio national park (Naples)

a new epistemology of complexity that increasingly appears to be the distinctive feature of contemporary land design: "Our failure to perceive and cope creatively with these vast and interrelated systemic issues is to some extent a cognitive problem"[8]. An approach that marks a change of strategy, 'the signal of the emergence of a will [...] of re-appropriation and re-meaning of the land as a collective value to be shared.'[9] In the logical framework outlined by the research unit, the description methods, the typology and the quality of the selected elements, as well as the logical structure that holds the cognitive process together, define the prerequisites to handle the complexity of the information related to drosscapes in terms of hypertext: their special features, their impact, the current and the original uses, the tangible and intangible values deriving from their presence on the territory. Such elements are essential to initiate that process of knowledge without which one cannot imagine a future for these spaces. Thus, knowledge is the indispensable support of that "planning constructivism" that allows one to see drosscapes as places for new opportunities.[10]

The lifecycle of the man-made environment is another essential point of reference for the mapping project. It's rewritten in a resilient perspective, directly built on the concept of adaptive cycle as a regeneration process based on the combination of four main phases (growth / consumption, status quo preservation, collapse / release, reorganization) during which the processes of adaptation and transformation of the socio-ecological systems alternate[11].

Researchers use the concept of "adaptive cycle" to bring it back to the specificity of the design, in a context where the term "re-cycle" can be defined as the ability / opportunity to direct the collective capacity of a territory towards new functional behaviours, coherent with the dynamics of organization and resilience of the territory itself. The research also elaborates on the notion of "hypercycle", developed in chemistry and biology as a principle for the circular connection of positive, self-catalytic feed-backs, which - along with the process of transformation of the territory - become the agents that configure new adaptive cycles aimed at increasing the performance of both the natural and the built environment: "the hypercycle is the tool for integrating long-restricted self-replicative entities into a new stable order, which is able to evolve coherently."[12]

In this sense, the definition (and testing) of new methods to organize knowledge represents the specific scientific contribution of the PRIN experience to the study of drosscapes. Knowledge is in fact understood as 'a structure that connects' a variety of possible design approaches, holding together the complexity of processes and visions of the territory.

The idea behind the project is that of a cognitive map that, through the organization and the setting up of a hierarchy of information, generates associations and interpretative syntheses according to non-deterministic logic while, on the

8. R. Mistrach, K. Orff, "Petrolchemical America", Aperture Foundation NY, 2012

9. R. Mistrach, K. Orff, "Petrolchemical America", Aperture Foundation NY, 2012

10. E. Vittoria, "Il costruttivismo progettante", in R. La Creta, C. Truppi, "L'architetto tra tecnologia e progetto", Franco Angeli Milano, 1994

11. Gunderson L.H., Holling C.S., *"Panarchy: Understanding Transformations in Human and Natural Systems"*, Island Press, Washington, D.C., 2002

12. M. Eigen, P. Schuster, "The hypercycle. A principle of natural self-organisation", Springer, Berlin, 1979

other hand, it fosters independent research and projects. Description is therefore seen as an integral part of a process of cultural advancement aimed at identifying new modes to produce knowledge, while accepting the intrinsic problems associated to the object and the relationships it establishes. This kind of multidisciplinary knowledge is oriented towards both the physical aspects of the site and towards the material and immaterial flows that generated it; it's an inter-sector knowledge, designed to meet the needs of multiple parties who are involved in the decision-making process aimed at the physical and productive recovery of drosscapes; it's also a multi-scale knowledge that can structure functional cognitive grids at the desired decision / transformation level. In facts, there in an increasingly urgent need to produce knowledge for the project with an innovative approach, to handle both the information and the decision making process by developing cognitive logics designed to establish a substantial continuity between the moment of description, that of knowledge, and that of decision and control.
The definition of methodological and operational systems aimed at developing a knowledge of the place is, in fact, a kind of *vis directiva* (force directing the action) to the transformation of the site[13]; a knowledge that expresses local rules, functional to the nature and history of the place, and also acts as the guiding principle behind the project, so that the proposed solutions are commensurate with the real, and become objective recovery and regeneration possibilities arising from the cultural, social and economic conditions of the local community. The construction of new cognitive apparatuses for the project is therefore aimed at the definition of functional descriptive codes to identify systems, which refer to specific technological and space repertoires. They also make reference to constraints inherent to the objectives of the intervention, and to combinatorial rules made objective by the interpretive structure chosen for that particular context. In addition, the public and open nature of the knowledge process becomes an element of further control over the project, as it is derived from a background of sharing scenarios, all of which relate to a scientifically supported information system, promoting an approach based on the comparison between the different "visions" of the change and between the various actors involved in the decision making process. In this sense, the methods and tools to manage such complexity of knowledge are chosen according to a "safe-to-fail" design approach, namely to verify the pre-existence of the assumptions for the implementation of the choices in the structural conditions, setting out the very idea of intervention: '[...] the successful design of each of these depends on having adequate knowledge of the relevant geographic conditions, as well as respecting the constraints and taking advantage of the suggested by those conditions'.[14] Such an approach also ensures that decisions are made objectively in a system of transparency, meeting the need for a new political and

Regi Lagni canal, Castel Volturno (Caserta)

Fornillo basin, Terzigno (Naples)

13. G. Ciribini, "La normativa dell'impatto ambientale", Alinea Firenze, 1990

14. W.R. Miller, "Introducing Geodesign: the concept, ESRI, source: https://www.esri.com/library/whitepapers/pdfs/introducing-geodesign.pdf, 2012

"Taverna del Re" landfill, Giugliano (Naples)

professional ethos for the territory, giving the community - and all social actors involved - the possibility of understanding "why" and "how" a given decision was made[15].
Producing such knowledge in matter of drosscapes becomes therefore a key issue in PRIN research, and one of the main themes of the research itself, whose aim is to develop methodological and operational tools to deal with the information/decision relationship, using the cognitive approach to defining the key elements of the process of describing drosscapes.
It is important to emphasize once again that it is not a deterministic process whose goal is to obtain a univocal solution. The goal is to identify a knowledge-building method that works in hypertext, combining different information levels and absolutely autonomous interpretative synthesis. Through the hypertext model, the drosscape mapping process is organized as an operation to produce and capitalize knowledge, in order to organize and govern information, and evaluate the resulting design ideas through alternative scenarios, to facilitate the monitoring of interventions, policies and processes of regeneration and to produce a continuous updating of the knowledge itself.
The innovative nature of hypertext as a theoretical model for knowledge is encapsulated in the logic of geographic information systems (GISs), which represent an important opportunity to map and store a huge amount of data through a predetermined cognitive architecture. In fact, the GIS inventory project has a recognized theoretical, as well as practical, value that makes it a knowledge-producing tool for managing both the information and the decision-making processes, developing a substantial continuity between the phase of description, that of knowledge and that of planning and monitoring. Through the use of GIS, a knowledge model is built on the specificity of the object of knowledge (namely, a drosscape): a dedicated, up-to-date and multi-purpose information system that allows the verification (or falsification) of the decision-making process. It is possible, in a hypertext logic, to combine the GIS inventory architecture with the use of some MCDA multi-criteria evaluation software (ILWISS, in the case of PRIN research) that can support decision-making through the construction of meta-design scenarios useful to compare possible alternatives within a grid of rules defined downstream of the knowledge process.
In this sense, the combined use of G.I.S. and MCDA software can update and amplify the essential teaching of the mapping overlay promoted by Ian McHarg[16] who, with tools that were appropriate at the time, anticipated the conceptual terms of the hypertextual description of the territory and the importance of referring to multi-levelled, as well as multidisciplinary and multi-scalar, knowledge systems.
Thus, is important to emphasize that the use of GIS in the perspective hereby described is not limited to the production of digital maps or a more efficient data management. It also presupposes the construction of a logical model to

15. G. Ciribini, "La normativa dell'impatto ambientale", Alinea Firenze; Campagna M. (2014), "Geodesign from theory to practice: from metaplanning to 2nd generation of planning support systems", TEMA 2014 Special Issue, 1990

16. I. McHarg, "Design with Nature", Doubleday & Co, 1969

support the architecture of the inventory, in order to place the obtained knowledge in a cognitive network that allows the management of information with a proactive approach, based on decision-making. In this sense, the proposed mapping and evaluation system responds to the need for contemporary research to produce new interfaces for the project by working through software (GIS-ILWISS) in ways that are "infra-disciplinary" rather than interdisciplinary[17]. In addition, there is an essential point of reference for PRIN research: the scientific advances made in the field of "geodesign" by Carl Steinitz's work[18], which identifies a method for the support of environmental design based on a conceptual and operational cognitive framework: "you can't start with data because you don't even know what data to collect unless you have a question and a framework for saying, Why do we need the data? So, you basically start in reverse. What do the decision makers need to know to be able to say, what should we do? What impacts, therefore, do we have to study? What designs do we have to create as options? Why should we make a design if things are working well? To know they're working well, I need to know how they work. To know how they work, I need data, and now I know what data I need».[19]
In the light of the described theoretical-methodological framework, the construction of a hypertextual architecture that supports drossmapping should firstly focus on determining the criteria for selecting and organizing data. It should be taken into consideration that one of the most serious risk for researchers is to process the complexity of the description phase in terms of redundancy of information or, conversely, of excessive simplification and de-structuring. In fact, such data are:

- heterogeneous, as they result from different disciplines but also from different description systems (objective and interpretative);
- disomogeneous, because of a variety of scalar conditions and multiple sources (statistics, iconography, surveys, characterization campaigns, etc.);
- fluid, because they are representative of dynamic processes, which vary in time and space.

We must therefore reflect on the question of the quality of the information, the reliability and reproducibility of the data, their "cost" - in terms of acquisition, and in terms of management and updating - and the possibility (and usefulness) of establishing performance requirements to select the information. The idea behind is to rely on a reasonably limited number of functional indicators to answer (key) questions identified by researchers as the basis for the creation of GIS; indicators that must be matched with accessible data sets that can be used to implement a multi-scalar drosscape knowledge. In addition, the research establishes the pre-requisite of "measurability" as a key factor in the selection of information which, although not always directly referable to existing databases, must always be based on specific criteria to verify / refute the achieved

Old nursing home "Villa Eleonora", Castel San Giorgio (Naples)

Ex O.P.G. "Maddalena", Aversa (Caserta)

17. Schiaffonati F., Mussinelli E., Gambaro M., (2011) "Tecnologia dell'architettura per la progettazione ambientale" in *Techne Journal of Technology for Architecture and Environment* n.1

18. C. Steinitz, "A Framework for Landscape Planning Practice and Education", in Process Architecture n.127, 1995; C. Steinitz, "A Framework for Geodesign", Esri Press, Redlands CA, 2012

19. Wheeler C., "A conversation with Carl Steinitz". Source: http://www.esri.com/news/arcwatch/0412/a-conversation-with-carl-steinitz.html, 2014.

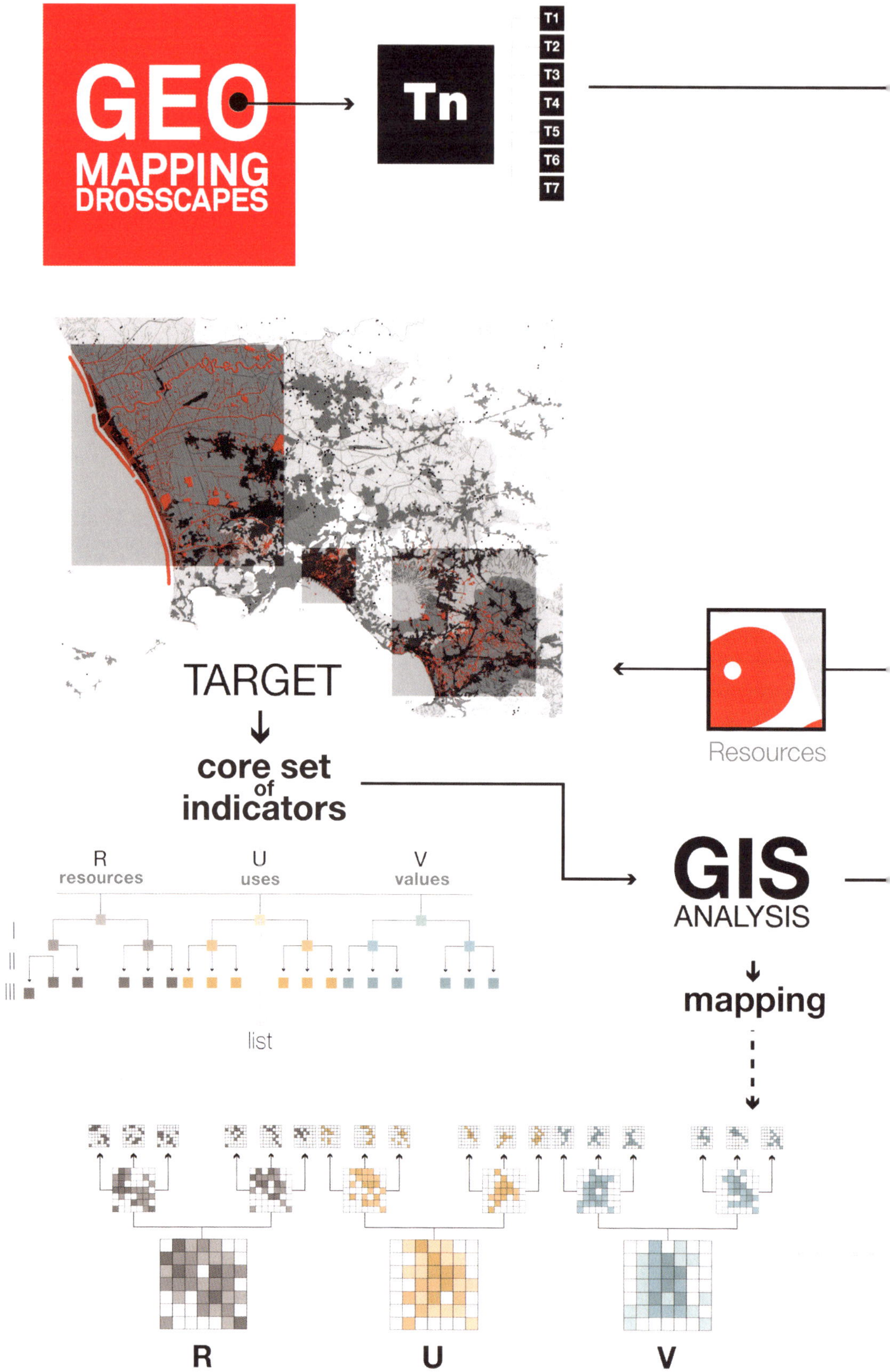
GEO
MAPPING
DROSSCAPES
Tn
T1
T2
T3
T4
T5
T6
T7
TARGET
core set of indicators
Resources
GIS
ANALYSIS
R
resources
U
uses
V
values
I
II
III
list
mapping
R
U
V

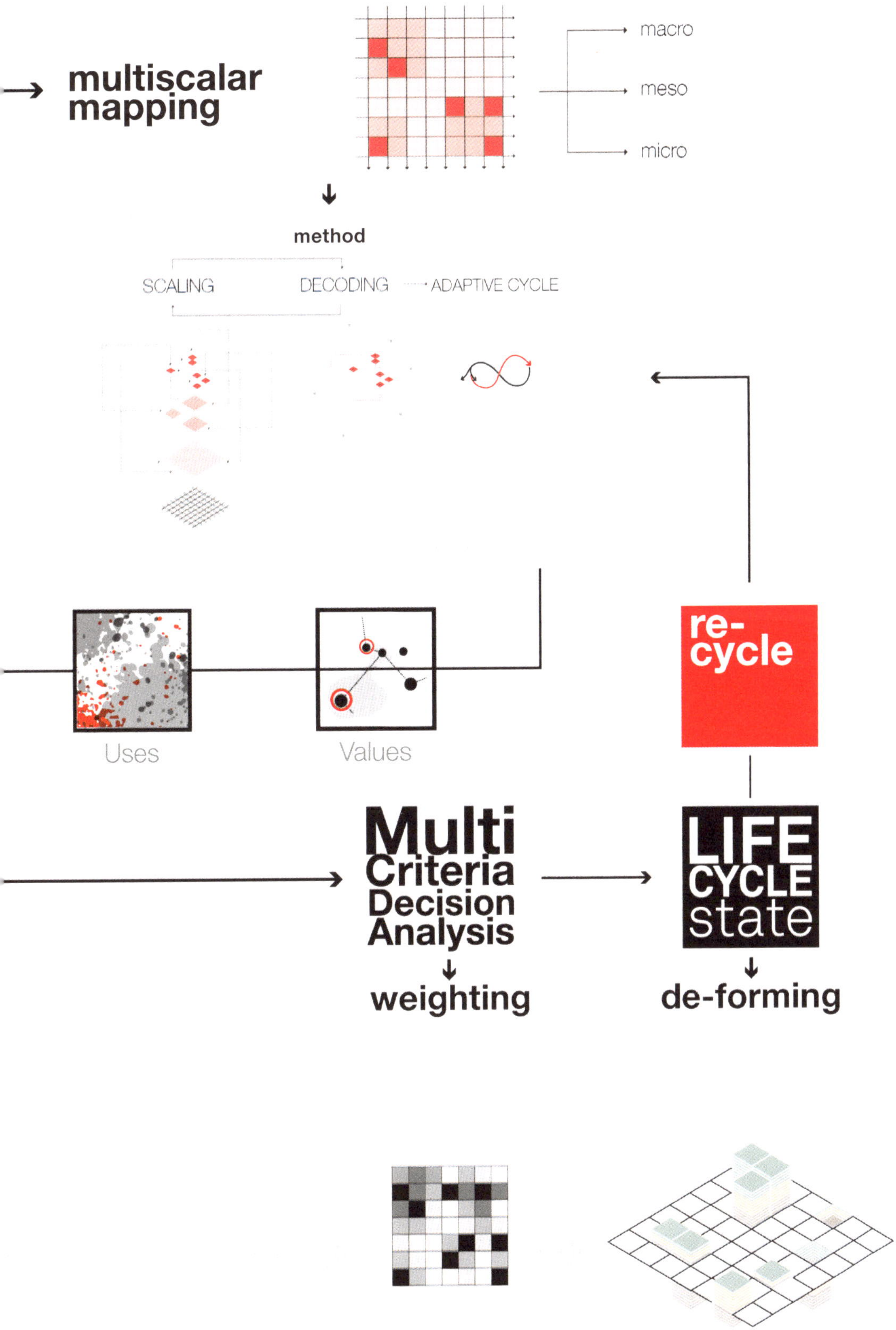
multiscalar mapping
macro
meso
micro
method
SCALING
DECODING
ADAPTIVE CYCLE
Uses
Values
re-cycle
Multi Criteria Decision Analysis
weighting
LIFE CYCLE state
de-forming

goals and progress. From this point of view, research is oriented towards the definition of a drossmapping architecture designed to introduce two different "rationalities" of knowledge: the metric, one-dimensional, objective rationality (the rationality of measurement, numbers, statistical samples) and the semantic, interpretative, predictive rationality, ingrained in the design approach; a rationality that Giuseppe Ciribini[20] defined as the hermeneutical, multidimensional component of the recovery process, aimed at composing its functional, perceptual, relational and cultural aspects. On the basis of this choice, the second step was to define the thematic areas within which to place the concept of "hypercicle" applied to drosscapes, identifying the transformations produced by / on the drosscape in terms of land use, resource consumption and modification of tangible and intangible values of waste areas as engines that generate a new, potential life cycle. Drosscape mapping is therefore divided into three main systems (Uses, Resources, Values), which are organized according to a tree structure that is subdivided into components, sub-components and indicators. The indicators tree defines the necessary amount of information to initiate this reflection in terms of design, articulating the hypertext model according to the further stages of data spatialization, their integrated evaluation, and the return (in terms of feedback) to the design hypothesis based on the determination (objective and transparent) of the susceptibility to transformation of the relevant area. Unlike traditional approaches, elements of originality and innovation of this method reside in the cyclical nature of the cognitive process and in the possibility of separating the moment of description (choice of indicators and spatialization of the latter) from that of the evaluation, explicating the decision process through a preliminary statement of the criteria through which the result of the description itself is interpreted. In this sense, PRIN research is very much based in how the mapping of the indicators is accomplished. It must meet the need to represent the context of the drosscape according to a multi-scale ("micro-scale" and "meso-scale") and multidisciplinary perspective. Prior to this, the selection of a core-set of data is considered an essential step for the site knowledge and the determination of the type of drosscape. Such data derive from the specialization of the three description systems into component and sub-component that work as descriptive codes of the system itself, a kind of "invariants" of the cognitive process that makes it possible to determine a drosscape specificity and identify the features that distinguish that particular site from all others, read in the light of the principle according to which: 'the organisation of a system is the organization of the difference. It establishes complementary relationships between the different parts, as well as between the parts and the whole[21].' It is interesting to point out that the distinction of the cognitive process in four steps (systems, components, sub-components, indicators) is useful in determining categories of descriptors that make it possible to compare the description for different types of drosscapes and territories. At the same time, this approach

Old warehouses of the Royal State Railways, "Ambito 43", East Naples.

20. G. Ciribini, "La normativa dell'impatto ambientale", Alinea Firenze, 1990

21. E. Morin, "Il Metodo, Ordine, disordine, organizzazione", Feltrinelli Milano, 1983

ensures the presentation of the specific (functional, semantic, and perceptive) features of the type of drosscape in question. The use of a double reading code allows us to recognize and distinguish the "endogenous" and "exogenous" characteristics of the analysed drosscape, while trying to describe the nature of the (material and immaterial) flows that originated it and the "uniqueness" of the site, key elements to form the substrate of knowledge that is essential for its transformation into an adaptive and resilient logic. At this stage, it is important to specify the type of information used and its quality in terms of reliability and accessibility. In fact, we worked with two categories of data:

- data from official sources, attributable to a metric, one-dimensional, objective notion of knowledge
- and data that were produced during the Re-Cycle Laboratory of the Neapolitan Research Unit, based on a kind of semantic description that is of interpretative nature and sensitive to difference, therefore very much intended to represent the specific conditions and flows that characterize the drosscape and its territory. This distinction is an important and risky passage for research. In fact, the subdivision of information into a macro group leads to discontinuity in the generation of the corresponding thematic maps, distinguishing between vector maps that connect the punctual data and its representation in a one-to-one relation, and raster-type maps that are not connected to tabular data. Such spatialized information does not allow direct overlap, and it was therefore necessary to integrate the data management into a single G.I.S., through software selection and the standardization of vector files.[22] The ability to overlay maps derived from different data systems, together with the ability to distinguish the descriptive process according to two types of descriptors (invariants and "gender"), defines the original contribution of the research, which aims at establishing a new, scientifically verifiable use of GIS as a cognitive process, making it a useful method to define meta-design scenarios to support decision-making, rather than just a tool for information storage and management. The project for the tree of the criteria is an essential element of the research, as it holds the description process and the evaluation process together and is structured on the basis of the system of requirements already identified by Malczewky[23] to define a cognitive framework that is:

- **complete**, thus capable of covering all the different fields of the description
- **operative**, that is, meaningful for the decision-making process
- **reducible** to sub-criteria so as to facilitate a targeted attribution of the weights and values for the evaluation
- **not redundant**, so that a description criterion is not "weighted" more than once
- **essential**, thus resulting from a minimum number of description / evaluation factors.

Boscoreale station (Naples), on the dismissed National Railway (FS) Torre Annunziata-Cancello.

22. See Giuseppe Servillo's contribution

23. J. Malczewky, "GIS and Multicriteria Decision Analisys", Wiley and Sons, New York, 1999; J. Malczewky, "GIS-based multicriteria decision analisys: a survey of the literature", Journal of Geographical Information Science 20 (7), 2006

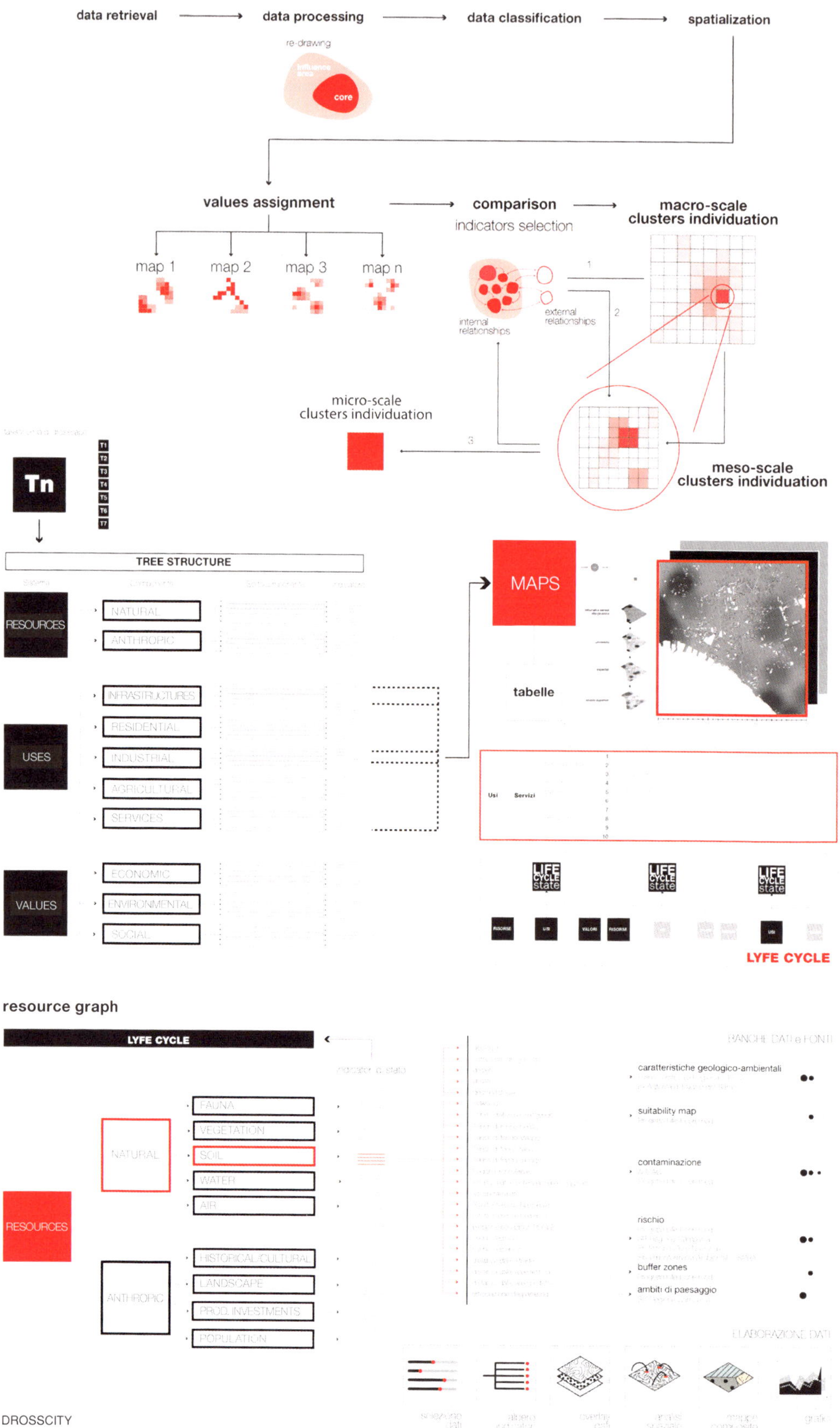
data retrieval
data processing
re-drawing
core
data classification
spatialization
values assignment
map 1
map 2
map 3
map n
comparison
indicators selection
internal relationships
external relationships
macro-scale clusters individuation
micro-scale clusters individuation
meso-scale clusters individuation
Tn
TREE STRUCTURE
RESOURCES
NATURAL
ANTHROPIC
USES
INFRASTRUCTURES
RESIDENTIAL
INDUSTRIAL
AGRICULTURAL
SERVICES
VALUES
ECONOMIC
ENVIRONMENTAL
SOCIAL
MAPS
tabelle
Usi
Servizi
LIFE CYCLE state
LYFE CYCLE
resource graph
LYFE CYCLE
NATURAL
FAUNA
VEGETATION
SOIL
WATER
AIR
RESOURCES
ANTHROPIC
HISTORICAL/CULTURAL
LANDSCAPE
PROD. INVESTMENTS
POPULATION
BANCHE DATI e FONTI
caratteristiche geologico-ambientali
suitability map
contaminazione
rischio
buffer zones
ambiti di paesaggio
ELABORAZIONE DATI

In this perspective, the "multi-criterion" evaluation forms an integral part of the knowledge system; the values attributed by the decision maker (as well as by the designer or the actors involved in the process) are easily attributable to the original data and the cognitive reference framework, according to an immediate and transparent process that can be represented and communicated objectively. Finally, the ability to discern between the cognitive data (represented and mapped in the description process) and the value assigned to it (subject of further thematic mapping) allows us to envision alternative and comparable scenarios for the decision-making process[24]. The operational experimentation of the knowledge system was directed to two case studies, involving a different type of drosscape and a different reference territorial scale. This step meets a twofold objective: on the one hand, the need to verify / refute the processed model, also by triggering an interactive feedback process for the setting up of the tree of criteria and the selection of indicators; on the other hand, the desire to explore the applicability of the model to multiple conditions of waste and, above all, to the different territorial scales to which this difference refers. The research group identified two particular areas that are significant under the viewpoint of both the consumption process that modified the area of the plain and the multiplicity of drosscape typologies: the "domitian city" and the "east city". The reasons for this choice also lie in the desire to identify the different opportunities to transform the territory arising in the two areas and the different ability to generate positive and adaptive "hypercicles" from this specific context conditions. Finally, when referring to the three systems of identification of the cognitive model, the research has aimed at returning the results of such knowledge in a "style" of representation that would allow an immediate comparison between the different levels of consumption / opportunity resulting from a combined reading of the uses, resources and values, while attributing an essential value to perceptual synthesis in terms of communication and sharing of the cognitive model. In the light of the above considerations, it is also necessary to point out the limits of the work that has been done so far. First of all, we refer to the construction of the tree of criteria that was tested and validated for some and not for all the drosscape categories identified in the taxonomy. The partial application of the model, of course, determines a residual level of indetermination regarding the choice of the components (but especially of the sub-components) through which the model is articulated. A further, and perhaps greater, grade of uncertainty also applies to the choice of indicators that, although functioning within the drosscape category, were not currently identified and classified for all description systems. Consequently, it is necessary to verify the accessibility and reliability of the data in question and of the related sources. The lack of experimentation in the field of the model makes it impossible to reach an objective evaluation of the

24. M. Rigillo, E. Cervelli, "Mapping Urban Vulnerability: the case studio of Gran Santo Domingo, Dominican Republic", Advanced Engineering Forum 11, 2014; Cervelli E., S. Micheletti, M. Rigillo, "Proactive risk management and integrated knowledge for the governance of urban systems: urban vulnerability in the case of Gran Santo Domingo", in Procedia – Social and Behavioral Science 223, 2016

LDF The coastal city
(T1+T2+T3+T4+T5)

LDF The city behind the dunes
(T1+T2+T4+T6)

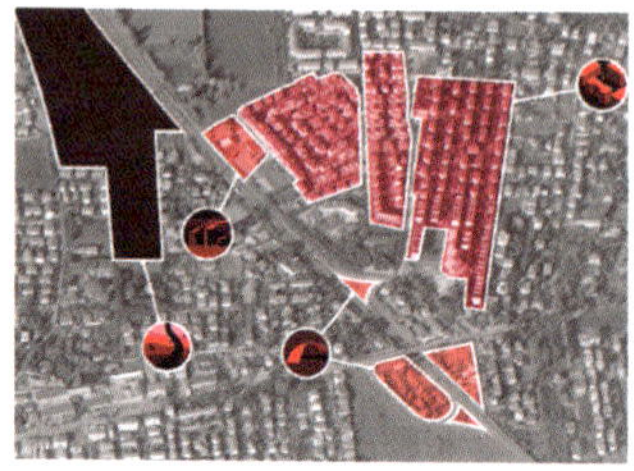

LDF The polycentric city
(T4+T5)

LDF The bypass road city
(T1+T4+T5+T6+T7)

LDF The blurred city
(T1+T2+T6)

Drosscape landscapes

effectiveness of the product / process achieved, due to the lack of an effective simulation of the communicative potential of the system in comparison with a sample of stakeholders and decision makers, and with the communities involved. The feedback and observations resulting from this comparison are to be considered essential for the definitive model, but above all this passage is indispensable to deliver the outcome of the work according to forms and modes that are functional to the transfer mechanism of the theoretical structure of the research to the specificity of the territory. These limits are to be regarded as physiological in comparison with the characteristics of the PRIN research, which are inherent in the theoretical quality of basic research and can be judged as future lines of in-depth research.

On the opposite page, picture of the fuel pellets from the "Taverna del Re" landfill, Giugliano (Naples). Picture by Riccardo Siano; Source: http://napoli.repubblica.it/cronaca/2015/12/23/news/ecoballe_via_libera_alla_gara-130060419/

Below, "drossmapping" and analysis of the lifecycle end for the drosscapes in the Domitio-Flegreo coastline, including Patria lake and the "Taverna del Re" landfill.

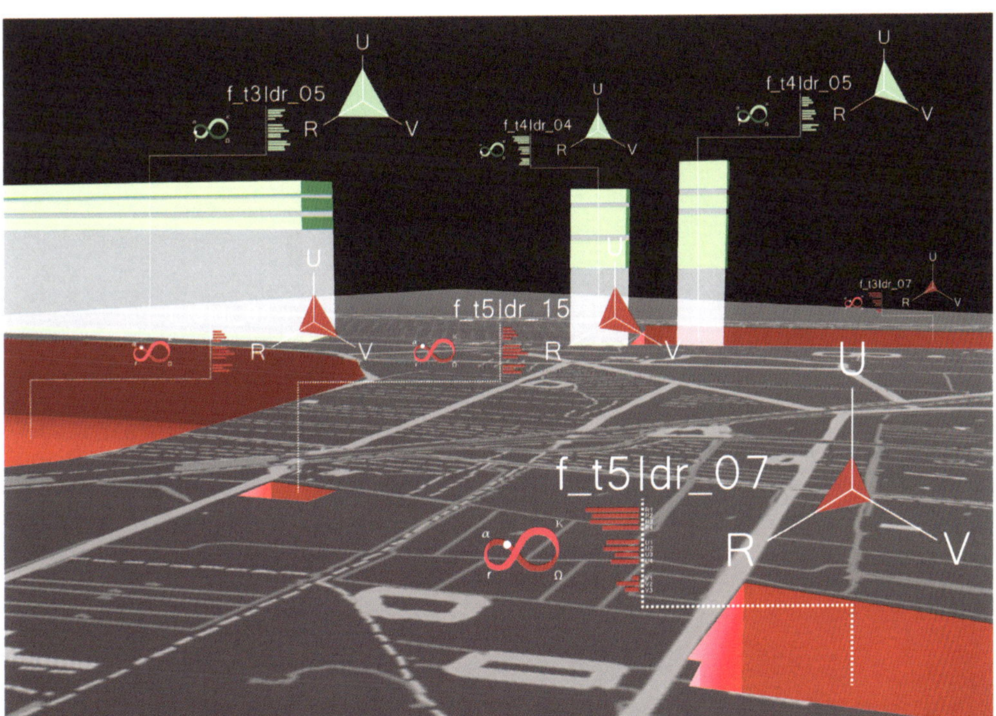

Drossmapping in East Naples: the relational potential of waste areas

Daniele Cannatella

To consider reintroducing waste areas into the metabolism of an urban system, one has to take into account a geography that consists of a system of complex urban parts[25], each one with its own features and peculiarities. While, on the one hand, it is true that one of the main causes for their state of abandon is their mono-functional nature, on the other hand their condition of heterotopic[26] spaces improves their transformative potentials and their relational strength. In addition, the conditions of mental freedom and the potential for action that they are able to generate thanks to their abandoned areas status[27] represent a major component of their charm: it is in these kind of spaces that creative processes are triggered. The difficult part, however, is to identify the underlying rules and dynamics; it's not easy to understand the interactions they create with the different "layers"[28] that make up the territory, and with the multitude of elements that make up each level.

Therefore, the implementation of a lifecycle assessment tool for disused areas had to take these aspects into account, as they determine the complexity of the phenomenon as well as its multi-scalar and multidisciplinary dimension. Like any model, this also provides a simplified - but not simplistic - view of the functioning of a territory and its parts, identifying the stratigraphic and relational aspects as important tools to re-read and interpret the phenomena that, eventually, lead to the deterioration of some parts of the city. Such knowledge will help establishing the prerequisites that are necessary to reactivate them. This is the main strength of waste areas: they are in the reorganizational phase of a system[29], when new forms of innovation and structuring can lead to the creation of new adaptive cycles.

In addition, a tool of this magnitude also has to deal with the intrinsic limitation of the available data and thematic maps which, in fact, is a major obstacle to the creation of a model that is scalable and transferable, and at the same time exhaustive. In this sense, the selection of data must be determined by well-defined criteria in order to avoid interrupting the logical

25. C. Gasparrini, "Waste, Drosscape and Project in the Reverse City", in R. Pavia, R. Secchi, C. Gasparrini (edited by), "Il Territorio degli Scarti e dei Rifiuti", Aracne Editrice, Roma, 2014

26. M. Foucault, "Different spaces", in J. Faubion, "Aesthetics, method, and epistemology: Essential works of Foucault", Volume 2, pp. 175-185, London: Penguin, 1998

27. K. Lynch, "Wasting Away", Sierra Club Books, San Francisco, 1990

28. S. Nijhuis, M. Pouderoijen, "Mapping Urbanized Deltas", in H. Meyer, S. Nijhuis, "Urbanized Deltas in Transition", Techne Press, Amsterdam, 2014

29. C. Holling, L. Gunderson, "Panarchy: Understanding Transformations", in "Human and Natural Systems", D.C.: Island Press, Washington, 2001

continuity that connects the phases of description and analysis with the design phase and, finally, the monitoring.
In the light of the importance of the relation drosscapes establish with the territory, the eastern part of Naples is located in an extremely interesting area. The analysed quadrant is in the Sebeto Valley. A land historically linked to production, it is located southeast of the Campana plain, close to the consolidated city and the coast, bounded to the east by the Somma-Vesuvius system and to the west by the hills of the city. The evolution of the territory of Eastern Naples has been marked by a large number of changes over time. After the first reclamations, that took place during the mid-fifteenth century to regulate the surface hydrographic system, the eastern part of the old town entered a phase of major agricultural prosperity, followed by a gradual conversion to industrial use from the late 18th century. Such a progressive shift of functions has affected several parts of the territory through slow and non-linear transitions, characterized by the appearance of the first Borboni factories along the coast line and their coexistence with the cottage system, only to mark a brief phase of change that resulted in the final conversion of the area into a major industrial manufacturing centre, after the construction of the first refinery in 1937. In the '70, the progressive decay of the industrial sector begun, and continued over the next twenty years. As a first consequence many factories were transformed into warehouses, but they eventually closed down. At the same time, the expansion dynamics of residential areas (begun in the '50s north-east of the industrial area with the Ina Casa districts of Barra, Ponticelli and via Stadera) resulted in an almost total anchorage of the residential development to the industrial area.
Over the centuries, the process of anthropization in this area has been characterized by a slow departure from the land-water combination that had historically characterized it and that contributed to the existence of a single production settlement system[30]. This nearly resulted in the total disappearance of the surface hydrographic system. Poor water management -

30. Barca S., "Napoli orientale: la città del rischio", in "I frutti di Demetra", Bollettino di Storia e Ambiente, n.7, pp. 33-39, 2005.

Drosscape nel quadrante di Napoli Orientale.

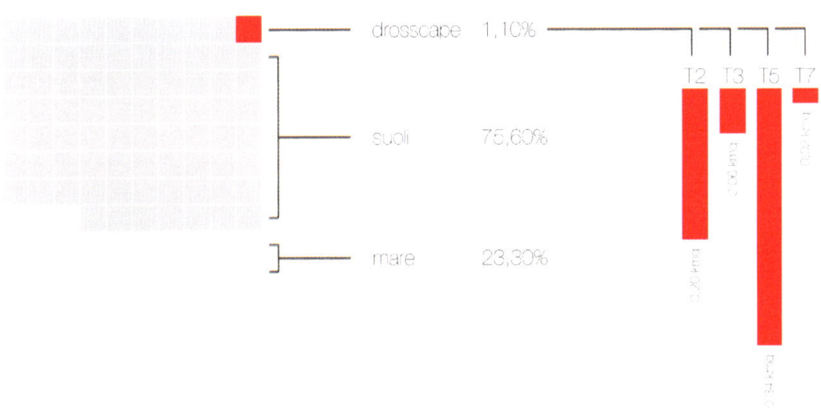

of both shallow and deep waters - is the result of the arbitrary juxtaposition of road infrastructure to the reticulated system of existing roads and water networks and the excessive collection of groundwater during the industrial period. Nowadays, the territory carries a high level of hydrogeological risk, associated with water contamination and outcrop phenomena. In fact, pumping has caused the pollution of the groundwater; the subsequent cessation of the activity has encouraged the recharging of the water table, whose depth, to date, is on average 2-3 m from the land surface. These factors contributed to exacerbate the flood phenomena caused by the critical conditions of the surface hydrographic reticulated system, in a context of high environmental degradation.

Abandoned junction of the SS162, East Naples

Secondary roads, "Ambito 13", East Naples

On the previous page, Old warehouses of the Royal State Railways, "Ambito 43", East Naples.

Drosscapes in the East Naples quadrant

East Naples is currently characterized by the presence of a myriad of abandoned fragments, featuring different spatial characteristics in terms of extension, grain and shape. These spatial characteristics can be identified in every drosscape category within the area of the Sebeto river plain. They are inseparably linked to the use of the soil and the different forms of production, and were generated by the non-articulated and non-linear succession of the various forms of production. Within the analysed area - a quadrant that extends just over 56 square kilometers - one can outline a reverse city geography, extremely complex and diversified, but governed by clearly defined rules. Here, the elements that make up the drosscape cover a total surface area of 470.98 hectares, or 8.37% of the quadrant surface; a percentage that rises to 12.34% if the surface covered by the sea is excluded. Therefore, the mapping process becomes an extremely significant tool to understand the dynamics and peculiarities of east Naples. Different hierarchies and relationships take place, starting from the central "core" - constituted by dismantled industrial areas and the related artefacts - and affecting the cultivated area that tends to spread northeast- southwest (the legacy of the old production system in the countryside). The crown of abandoned residential areas and the complex of interstitial areas linked to the transport infrastructures are also involved. Such dynamics can be used to describe the territory, but also to provide suggestions for the activation of the recycling strategies in which they themselves play a primary role. Even from a morphological point of view, the shape and size of the fragments of each family are the direct result of the processes that led to their formation and / or abandonment. This is the case, for example, of linear spaces associated with the infrastructure system, or of the extensive elements that fall under the category of abandoned industrial areas. The importance of the mapping process becomes even more evident when it is necessary to go beyond a taxonomy that is exclusively based on land use: such a classification alone is not enough to return the full complexity of the phenomenon associated with the production of drosscapes. It is precisely the interaction with the different layers that constitute an urban system - from the "deepest" layers, characterized by slower cycles and rhythms, to the more "superficial" ones, characterized by faster evolutionary paths. The intangible relationships a drosscape establishes with the city can provide an exhaustive description of the life cycle of each element that makes up a drosscape in the eastern region. In the light of the relational aspect, it is therefore no coincidence that the mapping process is also influenced by such interactions. Thus, the boundary delimitation moves away from a purely cartographic conception in the direction of an output that is closer to a thematic map, in order to highlight the distribution of drosscapes as a phenomenon, rather than focus on their precise location and extension. This is even more evident in east Naples, where the spatial extension of certain phenomena such as surface and deep soil pollution necessarily coincides with the boundaries of industrial property, since the data used to process information were derived from the individual S.I.N. characterization plans in east Naples.

Under the SS162 viaduct, East Naples

Pipeline, "Ambito 13", East Naples

On the following page, pollution of surface and deep soils, East Naples.

Resources, Uses, Values and Dynamic Maps: drossmapping east Naples

The "drossmapping" phase has resulted in the discovery of about one thousand "fragments", concentrated on a fraction of the territory that is extremely small, especially when compared to the other two analysed areas. This density of abandoned or underused areas highlights the strength of the relational aspect inherent in the residues, both in terms of existing relationships and in terms of potential relationships. Subsequently, drosscapes have been classified into the seven proposed families (wreck soils, abandoned water networks and dismantled hydraulic devices, compromised ecosystems, critical settlements, discarded buildings, quarries and landfills, dismantled infrastructures and interstitial areas) and then evaluated through the three identified systems, namely "resources, uses and values", and the tree structures in which they are subdivided. The results obtained from the tests performed on selected "drosscape" families have highlighted some peculiar aspects of each system. From the point of view of the resources, the distinction in the natural and the anthropic component has made it possible to bring out extremely interesting information, and very diverse. Even if the "deeper" resources are, in fact, more compromised - due to the pollutants in the surface soil, in the deep soil, and in the groundwater - the vegetation makes the abandoned areas meaningful from an ecological point of view, to the extent that they can contribute to the conservation of biodiversity in the urban environment. Although a quantitative assessment of how they contribute is extremely complicated[31], a qualitative soil-based analysis is sufficient to provide a significant figure, even more when it becomes possible to identify invariants for each drosscape family. However, as already mentioned above, the reticulated surface water system has virtually disappeared.

31. M. C. Hall "Organising waste in the city", in M. J. Zapata, "The ecological and environmental significance of urban wastelands and drosscapes", page 21-40, Bristol: Policy Press, 2013

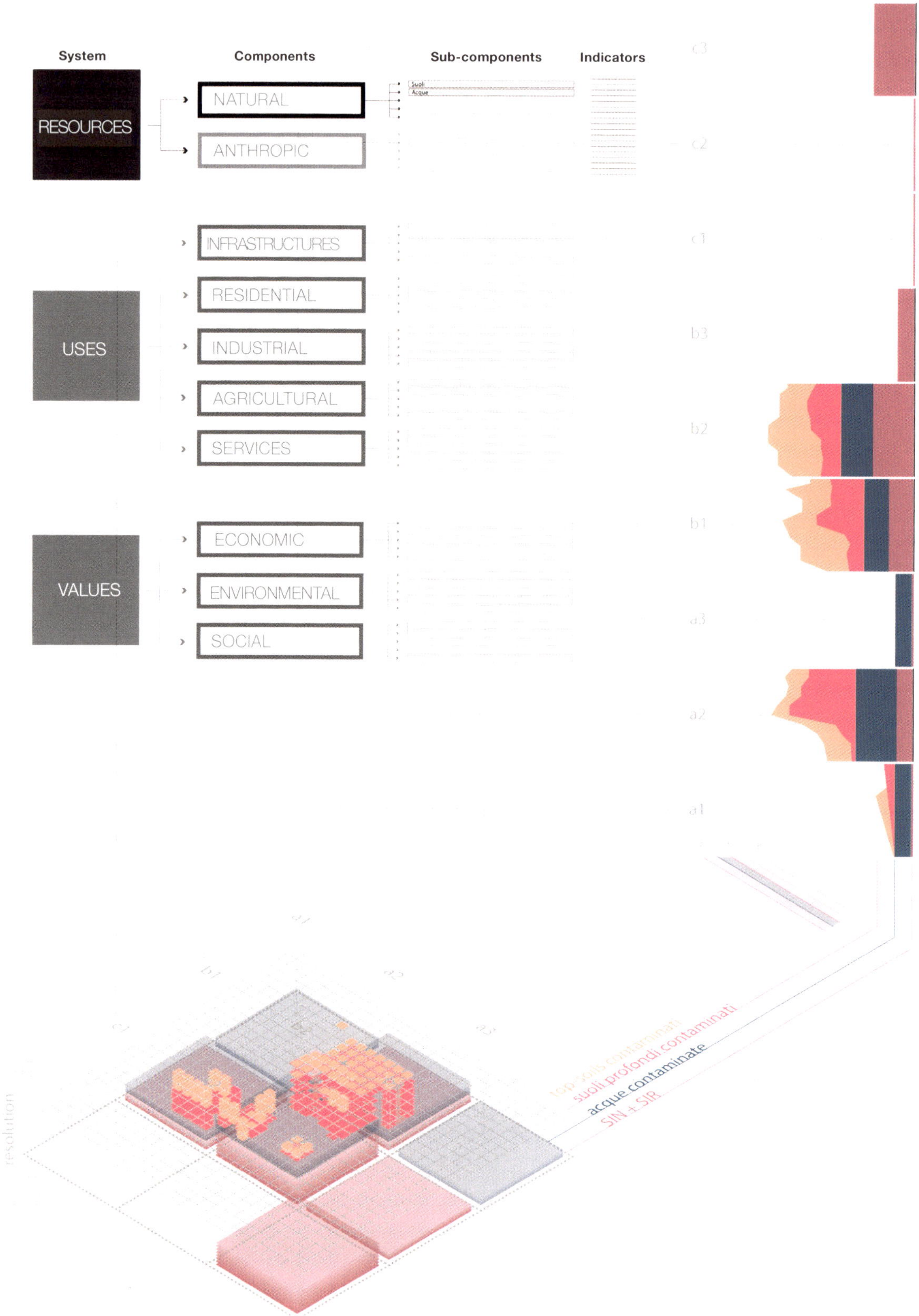
System
Components
Sub-components
Indicators
RESOURCES
NATURAL
ANTHROPIC
INFRASTRUCTURES
RESIDENTIAL
USES
INDUSTRIAL
AGRICULTURAL
SERVICES
ECONOMIC
VALUES
ENVIRONMENTAL
SOCIAL
top soils contaminati
suoli profondi contaminati
acque contaminate
SIN + SIR

0 km
2,5 km
5 km
7,5 km

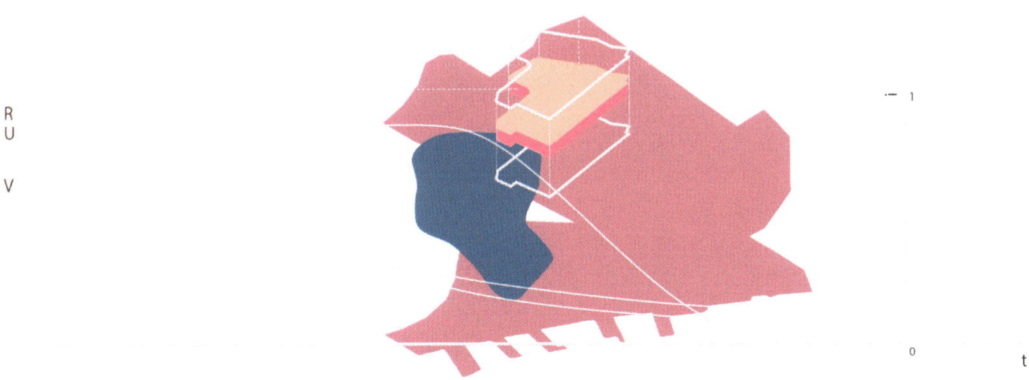
R
U
V
1
0
t

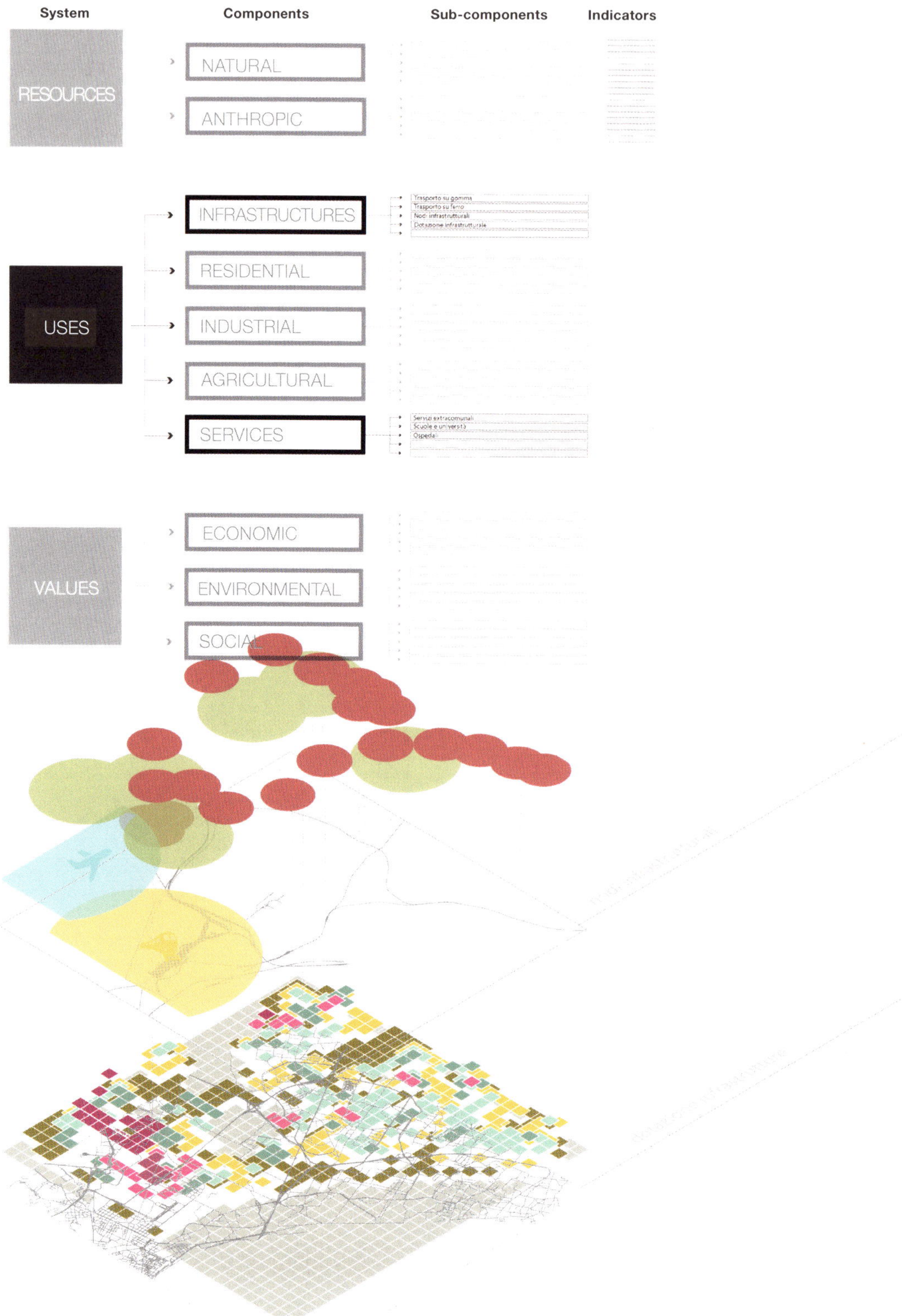
System
Components
Sub-components
Indicators
RESOURCES
NATURAL
ANTHROPIC
USES
INFRASTRUCTURES
Trasporto su gomma
Trasporto su ferro
Nodi infrastrutturali
Dotazione infrastrutturale
RESIDENTIAL
INDUSTRIAL
AGRICULTURAL
SERVICES
Servizi extracomunali
Scuole e università
Ospedali
VALUES
ECONOMIC
ENVIRONMENTAL
SOCIAL

0 km
2,5 km
5 km
7,5 km

scuole
universita
ospedali
tribunali

On the previous page, tree structure and "drossmapping" in the East Naples quadrant.

Within the valley area there are only a few traces left to witness its existence, mainly in the interior of the analysed area and completely disconnected from the Somma-Vesuvius hydrographic system. As far as the system of uses is concerned, drosscapes in east Naples are in a privileged condition in terms of infrastructure and accessibility. The close presence of residential areas - the old town in the west and the neighbourhoods in the north and in the east of the quadrant - in addition to the localization of numerous council and sub-municipal services, contributes to increasing the relational potential of the abandoned areas, hence their predisposition to recycling. Lastly, the system of values emphasizes a much more conspicuous top-down approach in comparison with the activities from below: in this quadrant more than 500 hectares are involved in public and private projects and initiatives, such as the P.U.A. in the Kuwait areas, which has an extension of about 95 hectares. However, relatively low presence of associations in this quadrant highlights how the social value attributed to drosscapes is not necessarily directly proportional to their ecological value. There are several possible explanations: one concerns scarce awareness in the population of the potential of abandoned areas for the provision of ecosystem services within heavily urbanized contexts; a further explanation concerns the way waste is perceived. In fact, the duality of the approach to these particular landscapes exemplifies the difficulty that exists in "accepting their appearance"[32], which often heavily counteracts their ability to attract attention in relation to their condition of urban "wilderness".

Conclusion

The case of east Naples is a striking example of how the evolution of cities (both dynamic and non-linear), when characterized by a "crazy" metabolism, is able to generate tangible signs on the territory: residues characterized by precise rules, both morphologically and relationally. On the other hand, the use of the term "dross" to describe these areas highlights a purely anthropocentric view of drosscapes. Dross is devoid of any utility for human purposes[33] and still linked to a capitalist legacy. This vision, however, ends up being restrictive, since it only returns a partial reading of a complex phenomenon, which can instead constitute a real resource for the city. Lynch himself says that when labelling something as dross, one has to wonder who actually considers it waste, and in fact most likely "objects that are no longer wanted by someone [may be] desirable for someone else."[34] Leaving too much space to waste areas as such would be a crass mistake: it can lead to overlooking the ecological potential of these areas[35], as waste areas within urban environments often offer a high degree of biodiversity, becoming "shelters for diversity [when] everywhere, elsewhere, it is driven away."[36] This is even more relevant in densely urbanized areas, where an increased level of biodiversity is a major prerequisite to improve the

32. W. Erz, "Wasteland: Land for conservation", in S. Strasser, "Waste and want. The other side of cosumption", page 29-34, Providence: Berg Publishers, 1992

33. K. Lynch, "Wasting Away", San Francisco: Sierra Club Books, 1990

34. S. Strasser, "Waste and want: the other side of consumption", in S. Strasser, "Waste and want: the other side of consumption", page 5-21, Providence: Berg Publishers, 1992

35. M.C. Hall, "Organising waste in the city", in M. J. Zapata, "The ecological and environmental significance of urban wastelands and drosscapes", pp. 21-40, Bristol Policy Press, 2013.

36. G. Clément, "Manifesto del Terzo Paesaggio", Quodlibet, Macerata, 2004.

quality of life[37], and at the same time becomes the basis for the development of mitigation and adaptation strategies to cope with the impact of climate changes. Such diversity can be found mainly in the social sphere. While it is true that the importance of a drosscape can only be fully appreciated in bottom-up processes[38], in the case of east Naples, however, this results in the presence of predominantly informal activities, sometimes at the limits of legality, almost to point out the marginality of these spaces. To make the recycling process virtuous and effective, it is necessary to implement a strategy that is able to keep up with the short-term needs (the upgrading of the social fabric as well as the physical fabric, with the consequent improvement of the quality of life in the area) with long-term goals, dictated by the need to ensure the safety of the area. The construction of a "hybrid machine"[39] capable of providing temporary and durable uses involves the reverse city, through all those fragments whose potential is extremely high in terms of reducing the ecological footprint of the city and delivering ecosystem services[40]. The critical condition of the area can give rise to conflict: the need to reverse the environmental degradation of the land and the need for adaptation and mitigation measures to contain the hydrogeological risk must be guided by a common framework. In this sense, drossmapping becomes a useful tool for stakeholders, able to provide a key to interpretation that not only explains the intrinsic features, but also clarifies the relationships drosscapes establish with the territory, while considering both the quantitative and the qualitative aspects coming from an extremely heterogeneous set of data, synthesized through the use of geographic information systems.

37. J.C. Savard, "Biodiversity concepts and urban ecosystems" in "Landscape and Urban Planning", vol. 48, 131-142, 2000.

38. A. Berger, "Drosscape", in C. Waldheim, "The Landscape Urbanism Reader", pp. 167-183, New York .Princeton Architectural Press, 2006.

39. C. Gasparrini, "Waste, Drosscape and Project in the Reverse City", in R. Pavia, R. Secchi, C. Gasparrini (edited by), "Il Territorio degli Scarti e dei Rifiuti", Aracne Editrice, Roma, 2014.

40. P. Del Tredici, "Spontaneous urban vegetation: Reflections of change in a globalized world", in "Nature and Culture", vol. 5, pp. 229-315, 2010.

Strategies and techniques of "drossmapping": applications in the Domitio-Flegreo City

Sabrina Sposito

Introductory notes: the approach and objectives of "drossmapping"

The proliferation of derelict areas in one of Italy's most fertile plains - the Piana Cam pana, on which this research endeavours to carry out an in-depth study - encourages a reflection on the critical outlook that the relationship between the production, use and perception of space has assumed in contemporary civilization. A distinctive sign of this relationship is the recursive and gloomy mechanism through which waste pours into cities and landscapes, accumulating with other fragments. In the context of this incessant reproduction and diversification, drosscapes acquire enough extension, density, and reticular structure to be configured as the typical fabric of a new territorial articulation.

Their wide and pervasive dimension is unequivocally one of the specific features of this phenomenon, which is increasingly intensified on the systemic scale, and manifests its effects far beyond the areas directly affected by conditions of dismissal, underutilization, abandonment or pollution. The waste is, in fact, a "device" that is closely related to the city and its urban metabolism: the dynamics of growth and transformation of the present society are more rapid and more vigorous than the ability of the natural and built environments to absorb and handle the change. Therefore, they encourage dynamics of 'linear metabolism'[41], where what is consumed and deteriorated is subject to continuous expulsion and rejection practices[42].

From the viewpoint of spatial disciplines, the concept of drosscape introduced by Berger[43] embraces a landscaping approach, and thus proposes a negative urban re-reading which, while drawing a comparison with the degraded territories of American post-modernism, searches a latent potential in waste through which to define strategies aimed at an ecological and productive growth.

Starting from this concept, the Naples Research Unit - operating within the National Interest Project (PRIN - Progetto di Rilevante Interesse Nazionale) "Re-cycle Italy"- identifies seven typological families (wreck soils, abandoned water

41. C. Kennedy, S. Pincetl, P. Bunje, "The study of urban metabolism and its applications to urban planning and design", in "Environmental pollution", 159(8-9), pp. 1965–73, 2011

42. K. Lynch, "Wasting away - An Exploration of Waste: What It Is, How It Happens, Why We Fear It, How To Do It Well", Sierra Club Books, San Francisco, 1990

43. A. Berger, "Drosscape: Wasting Land in Urban America", Princeton Architectural Press, New York, 2007

networks and dismantled hydraulic devices, compromised ecosystems; critical settlements, discarded buildings, quarries and landfills, dismantled infrastructures and interstitial areas) deemed to be suitable to create a drosscape taxonomy based on the specific features and the critical factors that have been identified as paradigms in three target areas of the Piana Campana. This sample defines, in fact, a set of neglected areas that need to be recovered through a systemic, adaptive and resilient approach, using their latent transformative potential[44].

On the basis of this categorization, and in support of a drossmapping operating tool, logical, methodological and interpretative structures have been defined, through which to dynamically simplify, associate, map and hierarchize information on the territory with the aim of describing and evaluating the types of drosscape in terms of life cycles and recycling opportunities[45]. In fact, since Holling's teachings[46] on the adaptive cycle of socio-ecological systems, this instrument relies on the hypothesis that drosscapes have their own life cycle, which may vary according to their intrinsic and extrinsic properties. In other words, it depends on the specific typology of waste and the material and immaterial relationships that the same waste directly or indirectly conveys within the context. Therefore, it is assumed that the life cycle is not necessarily completed and uniform in all derelict areas. Through the a multi-scalar and multi-disciplinary non-deterministic professional approach, it is actually possible to recognize, represent and update different degrees and intensity of exhaustion, to which possibilities and project priorities and potentials are progressively linked and managed according to the definition of "cluster" of transformability.

How can the exhaustion of the life cycle of a drosscape be measured?

According to the proposed method, for each typological category a cognitive tree structure is set up, based on three systems - Resources, Uses and Values - which are subdivided into components, subcomponents, and indicators. Through

44. C. Gasparrini, "Waste, Drosscape and Project in the Reverse City", in R. Pavia, R. Secchi, C. Gasparrini (edited by), "Il Territorio degli Scarti e dei Rifiuti", Aracne Editrice, Roma, 2014

45. For further research, see the methodological in-depth text that opens this thematic area of drossmapping, where the general nature of architecture as a the tool is explicitated through reference to the main theoretical principles.

46. L.H. Gunderson, C.S. Holling, "Panarchy: Understanding Transformations in Human and Natural Systems", D.C.: Island Press, Washington, 2012

a methodological and interpretative engagement in the construction of indicators, and the creation of significant maps and weight assignment, drossmapping implements an integrated process for the life cycle analysis, which is summarized in the intercourse between Resources, Uses and Values and allows a detailed query at each point in the structure and at any time. For example, a drosscape whose total life cycle was completed, and whose natural resources system is seriously compromised, might nevertheless still show potentialities in the social component of the value system, or in the infrastructure component of the system of uses, and so on. This structure is thus the very framework of the learning process and its operational interface, and it identifies a spectrum of thematic and spatial themes involving the drosscape on which to carry out strategic interventions and investments, in order to activate a new life cycle wherever the territory really demands it, if there are the conditions for doing so, and if it is useful to reactivate or increase its community's resilience. In particular, this contribution explicitly outlines some of the significant criteria for the selection and spatialization of indicators in relation to the landfill category, in the "domitian-phlegrea city"[47]. More specifically, we analysed a vast area called "Masseria del Pozzo-Schiavi" in the territory of Giugliano, in Campania.

New life cycles for landfills:

1. **Why choosing this type of drosscape and sampling area**
   The choice to test drossmapping on this specific type of drosscape comes from some evaluations that involve the methodological choices and the research context:
   - The landfill refers to a specific planimetric survey.
   - Sampling and monitoring campaigns provide a large amount of data on the state of environmental compartments, waste, and leaching (run-off streams/seeps).
   - It is a type of recurrent drosscape in the domitian-phlegrea city, and is one of the main factors that cause the area to be compromised. The domitian-phlegrea city is, in fact, part of the "Litorale Domizio Flegreo-Agro Aversano", a part of the coast considered a Site of Regional Interest (S.I.R)[48], which is characterized by the large presence of pollutants related to the waste cycle. It it interconnected to the so-called "Terra dei fuochi" (The Land of Fires), that is, the area where the illegal burning and dumping of toxic waste by organized crime has reached dramatic proportions.
   - The recovery of landfills, while still related to a limited number of specific elements, still has an impact on the health of the environmental networks and human settlements, thus "it contains important environmental, urban and productive consequences, involving a wide range of spaces that are part of it the regenerative process, and are related to it"[49].

47. In this volume, the expression "domitian-phlegrea city" means the portion of the hinterland north of Piana Campana - connected to the metropolitan city of Naples - in which the ways and forms of production and management of space and the interactions with socio-economic and ecological issues assume a high degree of complexity, and are problematic enough to configure a particular urban-territorial structure.

48. By Ministerial Decree issued on January 11, 2013 by the Ministry of the Environment, Land and Sea Protection (Official Gazette no. 60 of March 12,2013) the area is no longer included in the reclamation Sites of National Interest (SIN) identified in art. 252 of Legislative Decree no. 15 of 3 April 2006. The Campania Region is therefore responsible for the remediation of the area.

49. M. Rigillo, "Il progetto ambientale per il recupero delle discariche abbandonate o dismesse. L'esperienza del progetto SUFALNET e il caso studio della discarica di Cannetiello a Cava de' Tirreni (SA)", in M. Rigillo (edited by), "Oltre la siepe. Scenari di ricerca per il progetto ambientale", Napoli: Editoriale Scientifica, 2013

As for the sampling area selection, it was considered appropriate to select one of the five Broad Areas (Aree Vaste, AV Areas) that the Regional Plan of Reclamations in Campania[50] identifies within the perimeter of the SIR "Litorale Domizio Flegreo-Agro Aversano" coastline, that consists of relatively small areas containing a persistent number of polluted and / or potentially polluted sites, whose combination results in considerably high environmental damage. Albeit on a small scale, this selection contains a systemic intent, as it requires an investigation about reciprocal interactions that could potentially amplify the risk of pollution and the vulnerability of such ecological, social and urban systems.

2. Context of reference

Spreading from east to west, from the hinterland of the Campana plain to the Tyrrhenian coast, and from north to south between the low valley of the Volturno River and the mountainous reliefs of the Phlegrean Fields, the domitian-phlegrea city was born in the north-west of an extraordinarily fertile plain, known as "Campania felix" during the Roman domination. By flying over the depths of its irregular structure, it is evident how profoundly the interaction between the plain and the city has mutually transformed both of them[51]. The mapping process of waste areas within the analysed area - a quadrant extending for 900 square kilometers - returns the magnitude and intensity of the changes (Image1). The reclaimed irrigated lands with their minute plots of canals, carved transversally by the Regi Lagni, together with the Patria Lake, the soil texture, the old farmhouses and the dispersed settlements, intertwine with the gigantic infrastructure networks, massive commercial and industrial areas , and huge quarries and dumps, occupying a key position between the coastal strip (to the west) and the conurbations of Aversa and Giugliano (to the east). In this constellation of urban, rural and natural realities, the degree of compromise and fragmentation of natural and man-made resources has reached critical levels due the inclusion of settlements in precious parts of the woods and coastal dunes, as well as the dismissal and deterioration of the facilities and infrastructure resulting from water reclamation interventions, the presence of potentially polluted sites that are linked to abandoned industrial sites, abandoned landfills and quarries, illicit spills, waste treatment plants and special high impact production activities. The risk of contamination of environmental compartments is indeed a growing threat to agricultural activity. The first data provided by the "Terra dei Fuochi"[52] Workgroup show that in a sampled area of about 50 hectares, "there are 15,78 hectares of Class D (land with a ban on agriculture for food production purposes and silvopasture), 36.7% of it. The remaining 11.6 hectares, equal to 27%, are in Class B (land with restrictions on certain agricultural food products under certain conditions)"[53].

50. Adopted as final resolution by regional council board on 27 May 2013, n. 129 (BURC No. 30 of 5 June 2013), and approved by the Regional Council on 25 October 2013. Available at: http://burc.regione.campania.it

51. M. Fondi, "L'ambiente umanizzato", in CIRAM (Centro Interdipartimentale di Ricerca Ambiente) (edited by), "Rischi naturali ed impatto antropico nell'area metropolitana napoletana", Guida editori, Napoli, 1994

52. The Group combines the expertise of the Campania Regional Agency for Environmental Protection (ARPAC) with that of other institutions, in order to assess the suitability for agricultural use of land at risk of contamination.

53. M. Vito, C. Marro, "Terra dei fuochi: primi risultati dalle indagini", in "Arpa Campania Ambiente", Anno XI, n. 3, 2015, available at: http://www.arpacampania.it

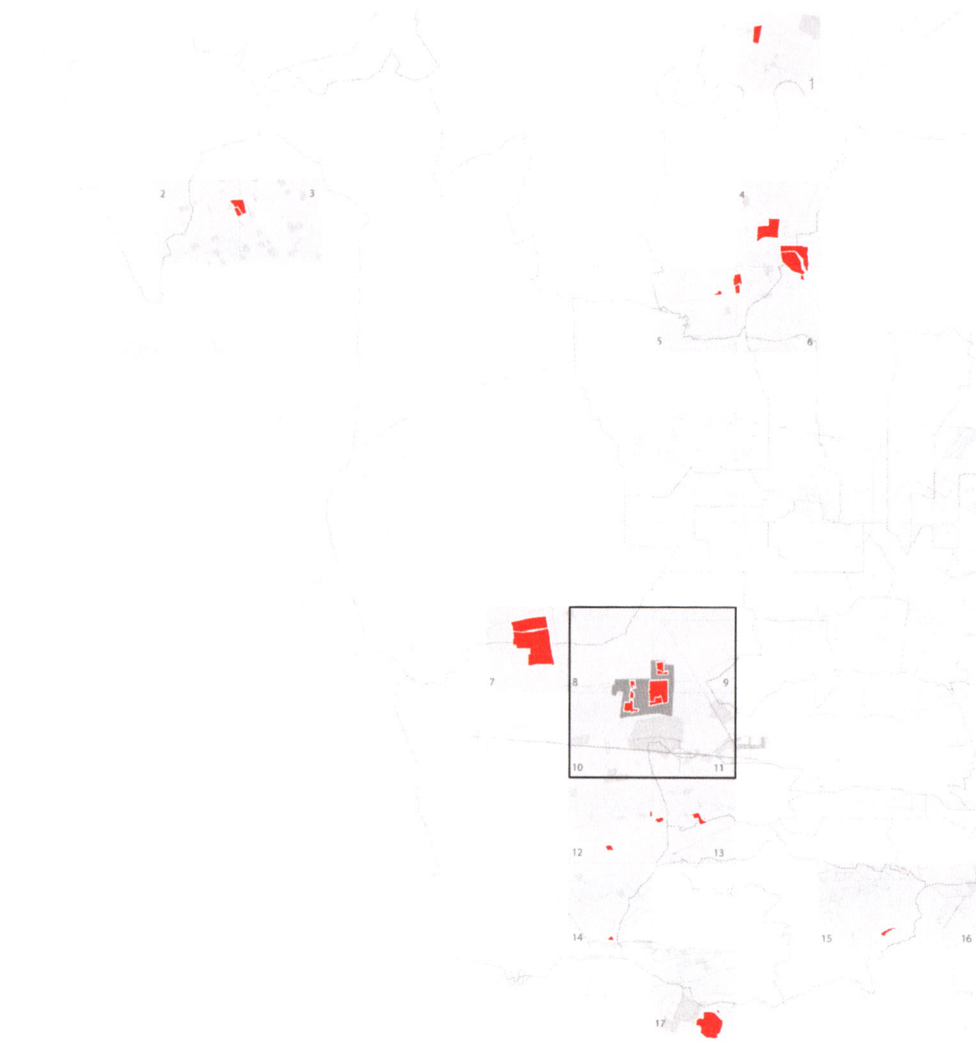
1
2
3
4
5
6
7
8
9
10
11
12
13
14
15
16
17

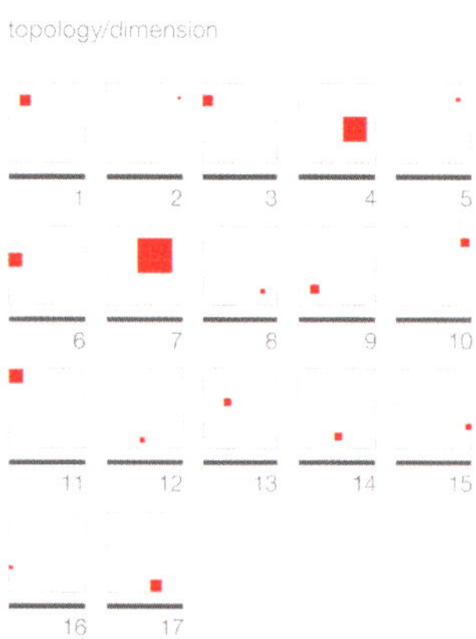
topology/dimension
1
2
3
4
5
6
7
8
9
10
11
12
13
14
15
16
17

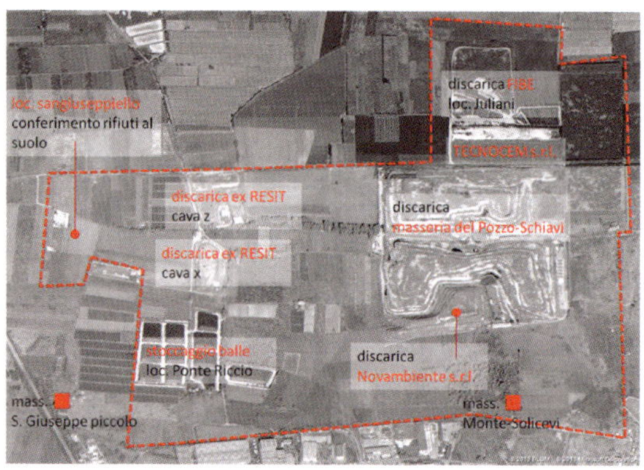
loc. sangiuseppiello
conferimento rifiuti al suolo
discarica FIBE
loc. Juliani
TECNOCEM s.r.l.
discarica ex RESIT
cava z
discarica
masseria del Pozzo-Schiavi
discarica ex RESIT
cava x
stoccaggio balle
loc. Ponte Riccio
discarica
Novambiente s.r.l.
mass.
S. Giuseppe piccolo
mass.
Monte-Solicevi

3. Disposal analysis

In this multitude of spaces of uncertainty, abandonment and contamination, the different types of sites that are legally dedicated to waste disposal show, in the context of the Domitian-Phlegrea city, a peculiar and exasperated character. In the context of an ideal division of the quadrant in 144 sub-quadrants on the side of 2.5 square kilometers, seventeen cells are affected by the presence of this drosscape or by part of it (Image 2). The surface extension is very diverse and reaches a maximum of 130 hectares with the "Taverna del Re Temporary Storage Site" between the municipalities of Giugliano in Campania and Villa Literno, which has been registered as a potentially contaminated site (Annex IV of the Regional Recovery Plan). Its dimension is slightly smaller in comparison with the area occupied by neighbouring coastal Lake Patria. Taverna del Re, along with other very extensive and controversial sites, is located in an extra-urban territory, still close to a crown of high-density settlements included in the major conurbations of the inland north of Naples (Image 3). This is where the Regional Recovery Plan surveys the "Masseria del Pozzo-Schiavi" Broad Area, which covers an area of about 210 hectares and is about two kilometres away from the town of Giugliano in Campania. The area includes a total of six sites, all authorized but unused, for an estimated total of slightly more than seven million cubic meters of waste (including special, toxic and harmful waste) (Image 4). The cognitive, treeview structure used for the analysis of the exhaustion of the landfill lifecycle covers the study of the stratigraphic-relational conditions (Images 5-6) measured thanks to the creation and crossing of appropriate indicators referring to Resources, Uses and Values systems. For example, in the case of the natural resources system it is essential to take into account the levels of contamination of the soil, water and air compartments, the environmental depletion and the environmental risk on the basis of some factors such as: the topography and geomorphology of the area, the rainfall, the depth of the ground, the site conditions, the characteristics of the waste material, and the interactions between the various sites in the Broad Area. When analysing the system of uses, the focus is on the completion of the economic cycle linked to the main relevant productive activities (agricultural, industrial or services), a decreased real estate demand, functional and social marginalization, and material depletion. Ultimately, in relation to the value system, life cycle depletion is analysed in a context of absent social dimension, and lack of protected areas as well as non-existing public / private planning for the site reclamation and recycling. A proper "dataset" is undoubtedly crucial. The database includes the organization of data from institutional sources (namely the EEA-European Environment Agency, the Campania

On the opposite page, view of all the different landfills in the Domitio-Flegreo quadrant. In evidence, the Area Vasta "Masseria del Pozzo-Schiavi" in the territory of Giugliano (Naples).

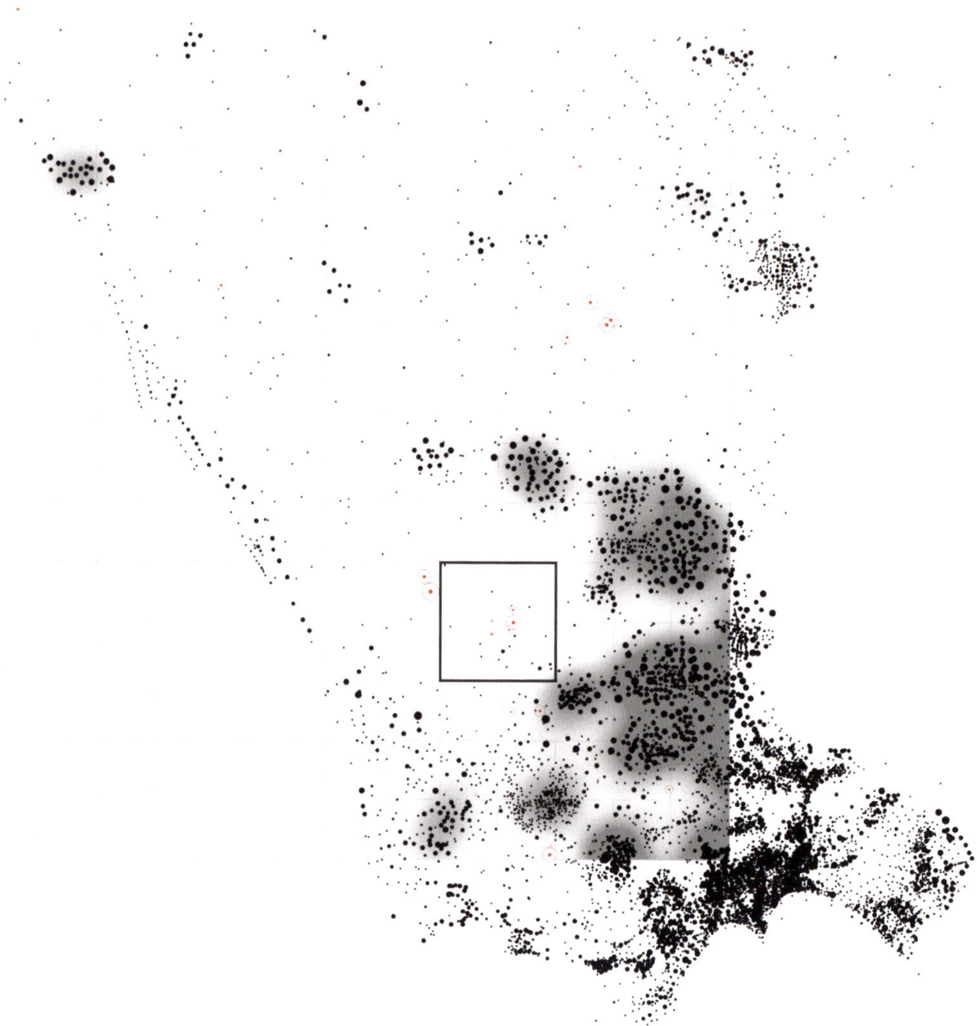

pattern (distribution)

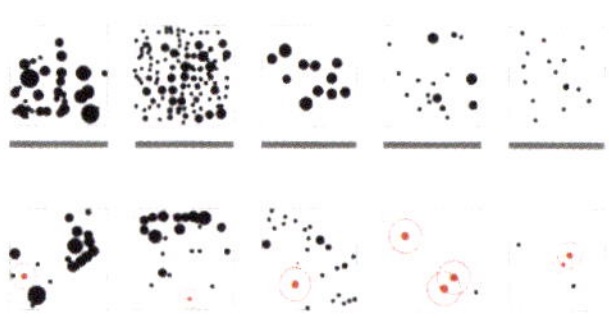

landfills (area in squared metres)

- 950 - 25.000
- 25.000 - 65.000
- 65.000 - 150.000
- 150.000 - 450.000
- 450.000 - 925.000

population (number of innabitants)

- 0 - 80
- 80 - 200
- 200 - 350
- 350 - 550
- 550 - 800
- 800 - 1100
- 1100 - 1600
- 1600 - 2300
- 2300 - 4000
- 4000 - 6800

Landfills in relation to the local population. In evidence, the Area Vasta "Masseria del Pozzo-Schiavi" in the territory of Giugliano (Naples).

On the opposite page, data mining from the systems' use with possible combinations. Processing based on data from: Urban Atlas (European Environmental Agency), Carta Utilizzazione Agricola dei Suoli Campania (Campania Region), ARPAC analysis, Reclamation Plan for the Campania Region and ISTAT statistics on population.

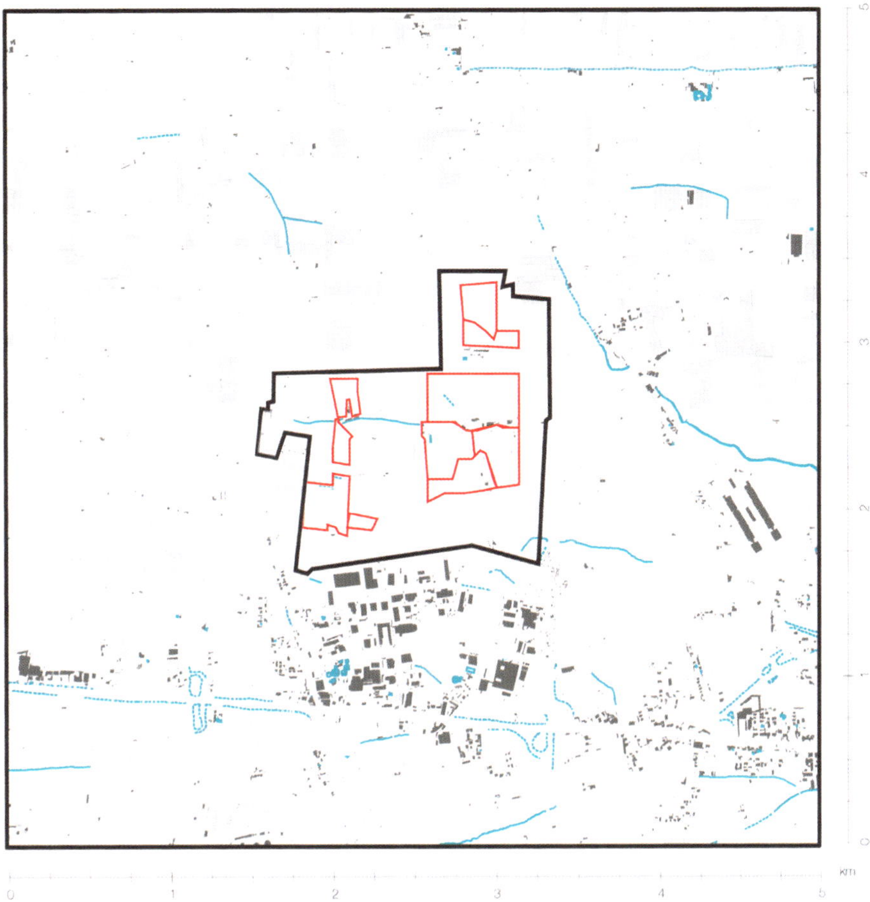
0
1
2
3
4
5
km

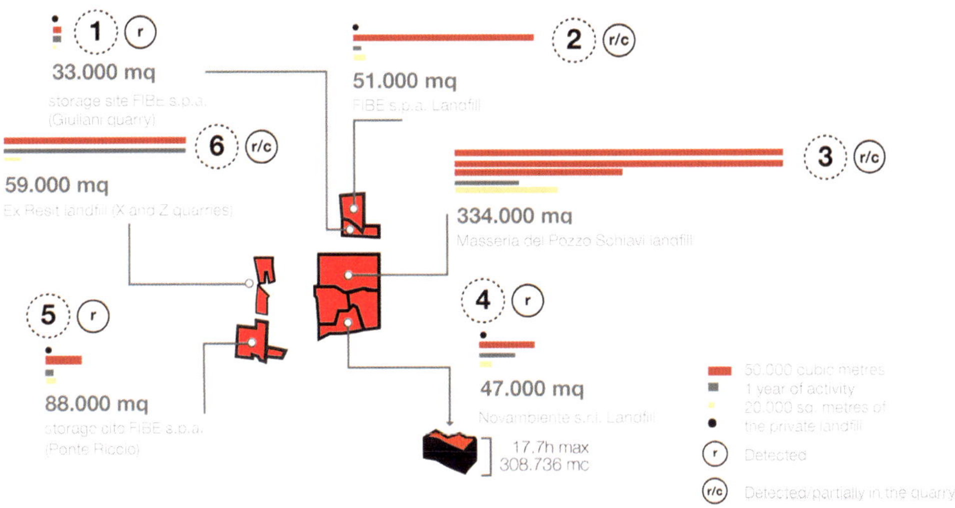
1
r
33.000 mq
storage site FIBE s.p.a.
(Giuliani quarry)
2
r/c
51.000 mq
FIBE s.p.a. Landfill
6
r/c
59.000 mq
Ex Resit landfill (X and Z quarries)
3
r/c
334.000 mq
Masseria del Pozzo Schiavi landfill
4
r
47.000 mq
Novambiente s.r.l. Landfill
17.7h max
308.736 mc
5
r
88.000 mq
storage site FIBE s.p.a.
(Ponte Riccio)
50.000 cubic metres
1 year of activity
20.000 sq. metres of
the private landfill
r
Detected
r/c
Detected/partially in the quarry

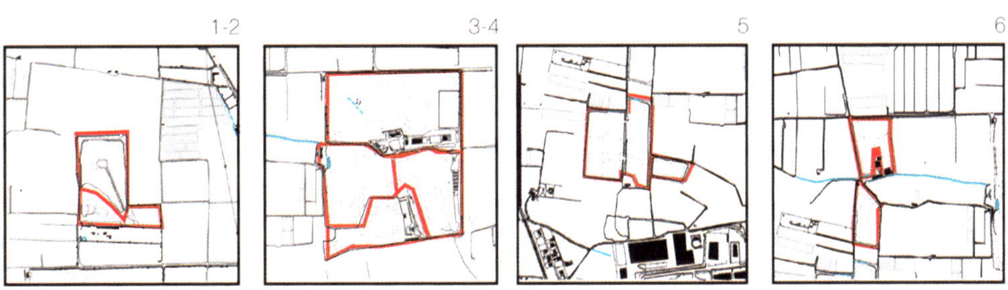
1-2
3-4
5
6

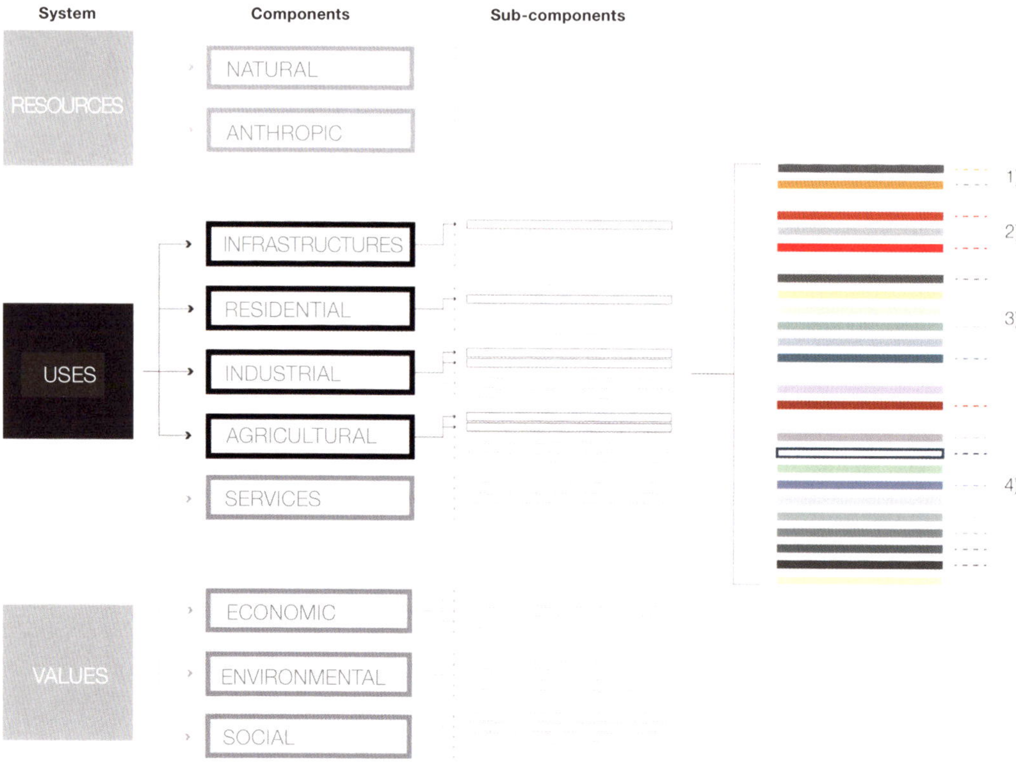
System
Components
Sub-components
RESOURCES
NATURAL
ANTHROPIC
USES
INFRASTRUCTURES
RESIDENTIAL
INDUSTRIAL
AGRICULTURAL
SERVICES
VALUES
ECONOMIC
ENVIRONMENTAL
SOCIAL
1)
2)
3)
4)

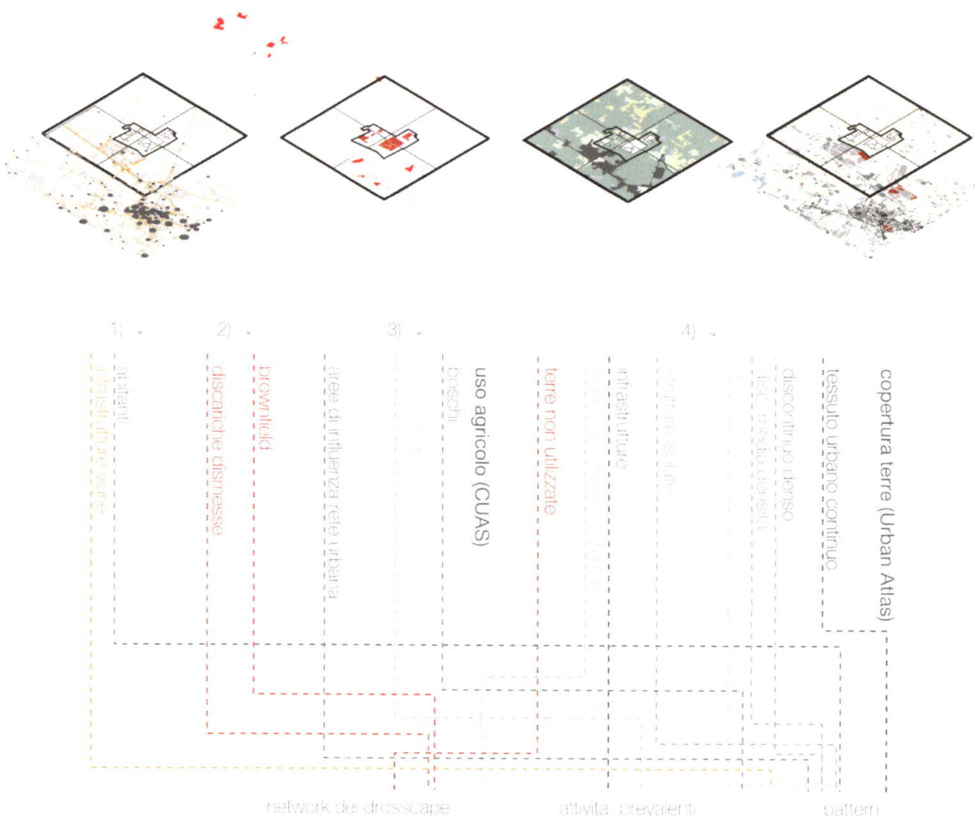
1)
2)
3)
4)
infrastrutture viarie
discariche dismesse
brownfield
aree di influenza rete urbana
boschi
uso agricolo (CUAS)
terre non utilizzate
infrastrutture
discontinuo denso
tessuto urbano continuo
copertura terre (Urban Atlas)
network dei drosscape
attività prevalenti
pattern

Region, the Central Campania Basin Authority, ARPAC-Regional Environmental Protection Agency Campania, the Life-Ecoremed Project, ISPRA - Higher Institute for Environmental Protection and Research, ENEA - National Agency for New Technologies, Energy and Sustainable Economic Development, and ISTAT, the National Institute for Statistics) as well as those specifically produced by the research group. The test on the sampling area shows that, although this drosscape constitutes a deep wound *ad imperitura memoria* and a strong reminder of the lack of connection between local economy and ecology, it still manifests an unexpressed transformative potential. The presence of infrastructure, the persistence of its historic-landscaping heritage, and the excellent quality of its agricultural production, together with the achievements of the land reclamation, the interest expressed by local communities for the decontamination and ecological recovery of landfills, the remediation interventions (planned or already in place) are some of the most important elements for the physical and cultural transformation necessary to start a recycling project. This means that the project is not focused on the mere technical interventions to ensure the safety of the landfill. It rather analyses all those elements and factors on which to act so that the landfill can effectively activate complex and varied processes and functions for the city. The aim of this contribution is to point out some basic steps of the descriptive phase, while showing some of the possible output maps. However, it is important to mention that, through the tree of criteria, it is possible to draw a comparison between the completion of a life cycle within each drosscape family and between different families. In addition, the priority assignment by the decision maker is subject to further thematic mapping of the evaluation phase. The latter is aimed at the construction of alternative and comparable scenarios for the definition of intervention priorities. Should the sampling area show potentialities, these should be compared to the degree of completion of other sites of the same type or of a different type, in order to identify the most significant and urgent intervention clusters.

Conclusions

In conclusion, among those analysed in the Piana Campana , the Domitian-Phlegrea city is the most extensive and complex and has required a thorough and extensive reconstruction of the reference framework and of the specific features of the landfill family in the sampling area. In fact, following the economic crisis and the waste issue, this area has been subjected to strong and persistent dynamics that were initially dictated by property speculation. One of the crucial issues to be tackled is how to undermine the common tendency to consider landfills isolated elements, thus failing to see the multiplicity of relationships and influences involved. Bringing our reflections to a more general level, the kind of "drossmapping" and the cognitive architecture that supports this hypothesis shows, in fact, that the investigation of the drosscape phenomenon is inevitable and constitutes an important investment opportunity in the area. Based on the three systems of Resources, Uses and Values, this approach assumes that the reference scale varies depending on the significance and representativeness of the data, that data are partly derived from official sources, and partly specifically produced, and that such data are gathered and represented through procedures that allow a combination of metrics and semantics. During the definition of the tools, in facts, it emerged that the currently available databases are insufficient to know and manage this emerging phenomenon. The creation of an adequate dataset presupposes alliances and interactions between different institutions and disciplines, with a wider involvement of local communities. This allows the use of G.I.S. as an important cognitive tool capable to interact with specific drosscapes and areas of influence in terms of urban metabolism, life cycles and recycling. It is clear that this approach operates within a topological and relational set of narratives as a means to define the transition of drosscapes to new three-dimensional geographies of adaptive, multi-functional and resilient landscapes.

The role of geomapping in decision-making

Giuseppe Servillo

The use of G.I.S. in the elaboration of the local profile outline, with reference to the endogenous and exogenous characteristics of the analyzed drosscape, played a significant role both in the creation of databases and in the layout of alphanumeric maps, as well as in the implementation of geomapping. The process, carried out in a G.I.S environment, has proved to be dynamic and able to continuously monitor the evolution of the activity, in accordance with the variability of the indicators. GIS software technology has also made it possible to link geographic information with data of various nature and typology, keeping such information constantly updated and marking great progress with respect to the classic "modus operandi".
The structure of the Information System, which forms the basis of the above-mentioned local profile outline, has been integrated into a single environment both for the management of alphanumeric data and for graphic and cartographic data. The data management has been closely linked to the logical model of the Information System and, consequently, to the choice of a desktop G.I.S. open source software called "Qgis", containing graphic documentation in numeric and vector format.
The process of creating the information archive (information layers) involved a number of operations: the collection of existing data, the verification of their reliability and consistency, the implementation of geomapping, the acquisition of new data, the setting up of thematic areas to organize the database, the activation of further analysis and surveys, and the predisposition of information layers. To outline the territory profile, we collected the documentation (in the bibliography) made available by governmental bodies and regional agencies such as the Ministry for the Environment and the Protection of the Territory and the Sea, the National Institute of Statistics, The Revenue Agency, the Campania Region agency, the Basin Authority of Central Campania, the Environmental Protection Agency of the Campania region and

the Naples Metropolitan Agency. In particular, thematic information databases have been consulted, as well as all those tools for the knowledge of the characteristics of the territory that represent issues or phenomena of local relevance, related to specific themes (hydrography, geology, geomorphology, environmental constraints, hydrogeological disturbance phenomena , soil features, soil value, property value, etc.). These data were consulted through OGC standards, Web Map Service (W.M.S.) or Web Feature Service (W.F.S.).
On several occasions, the Recycle Laboratory's work group, starting from the official geographic data elaborated by state agencies, has been able to re-elaborate further information on the territory, such as rainfall outflow estimates (runoff), real estate quotas for O.M.I. (Observatory of the Real Estate Market), agricultural values for the region, and demographic elements for the ISTAT census sections.
In the case of the annual maximum rainfall volume estimate, an empirical analysis has been carried out using data derived from the pluviometric graphs of the Agrometerological Center of the Campania Region and of the urban E.E.A. atlas created in the context of the Global Monitoring for Environment and Security project, a digital mapping tool that provides the data and maps of land use of the major cities in Europe.
The data provided by the Agrometerological Center of the Campania Region on the annual atmospheric precipitation were used to determine the approximate volume of run-off water in the surveyed areas. The above-mentioned data were crosschecked with the surface for a single type of soil use, followed by the different run-off coefficients indicated by the "R.I.E." project of the Autonomous Province of Bolzano. These coefficients are derived from specific literature, in particular: DIN, FLL, A.T.V., Frühling scale, UNI 11235. The different run-off coefficients have been attributed according to the type of surface being calculated. According to the methodology used, the following representation applies:

Estimate of annual maximum run-off = a x b x c
where
A = Analyzed surface
B = Maximum annual precipitation
C = Run-off coefficient by surface type

As already anticipated, further implementation activities have been included such as mapping individual OMI (Osservatorio Mercato Immobiliare=Real Estate Market Observatory, homogeneus territorial domain) areas by the Revenue Agency on the territory. These OMI real estate quotes refer to the first half of 2014. Real estate quotations were mapped for each area that was considered a homogeneous territorial domain (OMI area) and for each type of real estate, with market values ranging from minimum, maximum and average. Mapping the estimate procedure carried out by the Revenue Agency has been made possible by consulting quotations on a semi-annual basis for single municipalities, on a OMI area and usage scale, as well as by researching maps through GEOPOI (a cartographic framework) territorial navigation service, accessible via all browsers that allow full SVG graphic display (Scalable Vector Graphics). All activity was handled through a "personal geodatabase", geo-referenced in in the UTM-WGS84 coordinate system (33N). This became particularly useful when location data were communicated using Global Positioning System (GPS) techniques. In order to make the data provided by the various public administration entities geographically comparable, in some cases, geographical information has been migrated to the UTM-WGS84 reference system, thus securing a connection with all the databases gathered in this process. Ultimately, following consolidated methodologies of geospatial analysis, punctual and/or polygonal data derived from the official sources were interpolated. The results were elaborated in the light of a critical analysis carried out by the workgroup and subsequently structured according to customary encodings and specifications.

ACEN
monitoraggi
Open street map
progetti
ARPAC
website
Legambiente
site visits
google earth
Progetto Life-Ecoremed
PTR
SIT Regione Campania
MAPPE
Autorità di Bacino della Campania Centrale
Unione industriali Napoli
Protezione Civile Nazionale
CUAS
PTCP
Corine Land Cover
ISPRA
Confesercenti
Agenzia delle Entrate
dati
http://www.laboratoriocampano.org/
http://www.movimentoterradeifuochi.it/
Piano Regionale di Bonifica dei Siti inquinati della Regione Campania
Atlante geo-chimico
IACP Napoli
WWF Agro Aversano
PUA Ambito 13 Napoli orientale
CIG Consorzio imprese di Giugliano
Autorità portuale di Napoli
siti web
Camera di Commercio di Napoli
Ministero dello Sviluppo Economico
pubblicazioni scientifiche
Piano di caratterizzazione dell'area in loc. Masseria del Pozzo-Schiavi
http://www.naplest.it/
censimenti
European Environmental Agency
Corpo Forestale di Napoli
interviste
ISTAT
fotografie
Centro **I**nterdipartimentale di **R**icerca **A**mbiente
Lam Labortorio Ambientale Mobile Capitaneria di porto
social network
Rete Agrometeorologica Regionale– Assessorato Agricoltura

source cloud

Part 3

Recycling and features of the multi-scale project

"The drosscapes' territorial and landscape dimensions call for recycling multi-scale strategies, from the single fragment to the great urban and territorial networks, able to interpret - within scenarios of ecological regeneration and spatial reconfiguration of the contemporary City - the interactions between environmental, infrastructural and settlement challenges and the opportunities of transformation to build innovative landscapes, alternative economic models and sustainable energy cycles. Increasingly more projects, policies, resources, pervasive actions and unconventional practices - also for temporary use - converge around the drosscapes, in a structural phase of public funds scarcity. More importantly, this takes place in urban sprawls full of interstitial, abandoned or contaminated spaces that cannot be taken care of by the public sector alone. The multi-scale recycling project paradigms urge for the need to keep together - in space and time and through forms of multilevel governance - the local, fragmented and prevalent, dimension made of tactics with systemic adaptive and resilient centralized strategies. These paradigms also call for rules, guidelines and up-to-date forms of representation to innovate plans, programmes and projects in relation to compatible, temporary and long term, uses, diagrammatic simulations, parameters to evaluate the effectiveness of project proposals and modes of activation of direct and indirect resources."

Carlo Gasparrini

Plan de Végétalisation de la ville de Lyon

by Coloco & Gilles Clément

les dynamiques naturelles

Adopter une vision multiscalaire de la nature
définition stratégique et mise en place d'objectifs globaux
application technique et retours d'expériences locales
Visions de continuités
Du paysage au jardinage
paysages lyonnais
jardinages lyonnais

PARCS
VOIRIES

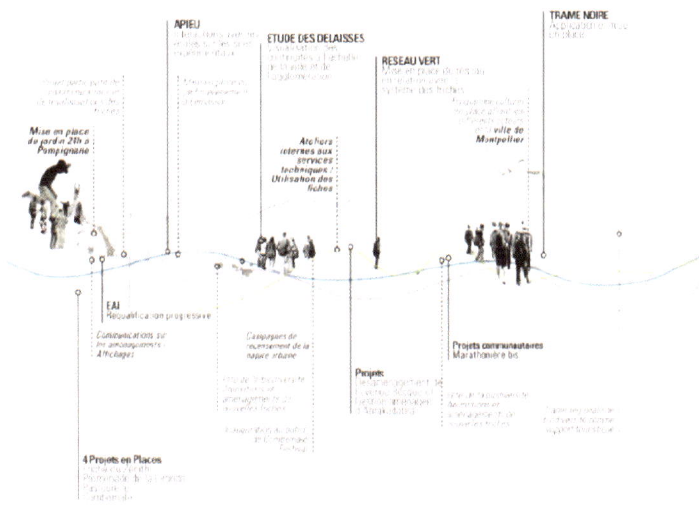
APIEU
ETUDE DES DELAISSES
RESEAU VERT
TRAME NOIRE
Mise en place du jardin 21h à Pompignane
Ateliers internes aux services techniques / Utilisation des fiches
ville de Montpellier
EAI
Requalification progressive
Communications sur les aménagements - Affichages
Campagnes de recensement de la nature urbaine
Projet
Projets communautaires
Marathonière bis
4 Projets en Places

Resilient urban landcapes. Cognitive and planning strategies to change contemporary urban design

Michelangelo Russo

Urbanism changes its rationality, materials, tools and cognitive strategies in a time of ever more radical change in conditions under the pressure of the global crisis. An economic crisis that, in the city, is not only visible in the slowdown of the transformation and growth processes, but rather in the difficulty of public administrations to generate "welfare" in the face of the inequalities produced by the concentration of flows and cumulative wealth in metropolitan regions (which are also the most energy-consuming) that contrast with poverty and the social and urban marginality of the world's cities' suburbs. This crisis takes on a profound environmental dimension due to the paradox of "unlimited growth", which creates inequalities in the availability and use of space, in the different forms and conditions of its livability. A space, that of the city, increasingly "unequal" in matters of security, due to external threats such as the risks caused by changes in the conventional rhythms of climate phenomena which determine rising tides, increasing the extent of weather events and the occurrence of periods of drought. These phenomena produce effects that cities struggle to cope with more and more. Entire populations are becoming fragile with respect to intense events that put some parts of the territory "structurally" at risk: the security of the territory represents, together with other factors, a scarce resource, not available to all in equal measure, causing conflicts between communities that have to compete for the use of non-reproducible resources, such as water.

These inequalities are the corollary of an environmental crisis that derives, especially in the economy, from a misunderstood notion of "growth" that has generated for several decades, and in particular since the post-war period, the drastic overcoming of limits and the alteration of the fragile balance of environmental systems and their ecological components. For decades, the idea of growth has been the beacon of the classical theories of the economy and it is still considered to be the progressive objective of unlimited development agendas: "unlimited growth" for an expansion of consumption without caution for the limits of the Earth. These environmental and ecological limits have often been violated, exceeded and removed to the advantage of an

economy that has caused, in cities, a generalized and widespread increase in the built-up and urbanized environment to the detriment of agricultural soils, open spaces and the conservation of environments and landscapes.
The concept of growth in these terms is a "dis-value"[1], and its consequences cannot be measured exclusively in economic, monetary and quantitative terms. In fact, for the territory, these are irreversible consequences, linked to the phenomena of building expansion and urban sprawl, where the expansive growth has been translated in some parts of the country, as in the South of Italy, into the degeneration of building speculation, and has become illegal with the destruction of the values of the landscape and the environment. Growth leads to the stratification of our urban territories where, over time, expansion has changed the relationships between the ecological components of long-lasting structures and landscapes.
Therefore, even for contemporary territories it is possible to think of a "prosperity without growth"[2] that does not mean "degrowth" (the "décroissance" of Latouche), nor does it allude to a perspective of retreat and contraction of a generic model of pauperism: it rather represents the possibility of considering the "collective dimension" of the social values of urban space as a principle of change, through a project aware of the limits of our ecosystem, of the times and the effects of their transformation. In economics, the concept of "decoupling" is the "traditional response to the dilemma of growth." In environmental terms, it means decoupling welfare growth from the production of "throughput" which represents the harmful effects on the ecosystem: it means, in other words, freeing economic growth from the pressure of ecological limits and of non-reproducible resources; "doing more with less; more economic activity with less environmental damage, more goods and services with less consumption and emissions", that is, improving efficiency in the use of resources.
"Throughput" is the "metabolic flow through which we live and produce", that is "the flow of energy and useful material that comes from environmental sources, goes through the economic subsystem (production/consumption) and returns to the environment in the form of waste[3]."

1. I. Illich, "Nello specchio del passato. Le radici storiche dei moderni concetti di pace, economia, sviluppo, linguaggio, salute, educazione", Borolli Editore, 2005
2. T. Jackson, "The Cinderella economy: an answer to unsustainable growth?", in "The Ecologist", Londra, 27 July 2012
3. H. Daly, "Che cos'è lo sviluppo sostenibile?", in "Lettera Internazionale", n. 92/2007, pp. 20-24

Turkana women waiting to access the wells during the drought of 2009, Rift Valley, Kenya. Picture by S. De Luigi.

Floods in the French Riviera, October 2015.

On the opposite page, Aral Sea, Kazakhstan. Picture by D. Ruef.

In urban planning, this principle can be translated into the perspective of decoupling growth (understood as development) from expansion. In fact, expansion refers to the growth of cities since the Second World War, to the construction of the public suburbs, but also to the pulverized spread of a low-density urbanization that has thickened along the infrastructure networks, along the traces of the historical territory in the rural peri-urban space of the Italian urban regions. But the spread of illegal building, not always dictated by actual need, has also invaded and modified parts of the Italian landscape.
"Decoupling" - in urban planning - means creating development and wellbeing on the basis of conditions other than expansion, where the transformation works on the existing stock as a "heritage" of resources to be regenerated. The impoverishment of environmental resources in Italy, in the "dispersed city" of Southern Italy, for example, has had a strong influence on the configuration of inhabited space, transformed by homologation and marginality into spaces without identity; streets without margins and pedestrian areas; factories without open spaces; fences without porosity; closed cities, difficult to access, fragmented and divided. Space loses continuity, which is not only ecological continuity, but also the pattern of relationships and forms of accessibility which, conversely, become obstacles, barriers and segments - infrastructure, construction, management - that represent signs of a deep discontinuity.
The solution is not a kind of urban planning "without growth", but a form of urban planning that knows how to acquire awareness of a new tension that exists between the social, economic and environmental components of the territory, which constitute a "new urban question[4]" where growth is to be understood as an increase in well-being in the relationship between the community and the territories to which it belongs, as a valorisation of the limits, in the awareness that "no subsystem of a finite system can grow indefinitely"[5] and that an ever-growing economic system cannot fail to compare itself with a finite ecological system. This implies that "we return to think about the collective dimensions" trying to "develop more democracy, reducing the inequalities of space"[6] thinking in new forms of the space of the city in relation to the landscape and the environment. Landscape and environment must be conceived as habitable space, sectors of new economies, devices of a new form of wellbeing and welfare.
In urban planning, decoupling expansion from development means rethinking the city's design as a set of "strategies of the possible", with a deep awareness of the existing, of the values recognized by the communities, concrete instances of people, societies and stakeholders.

4. B. Secchi, "La nuova questione urbana: ambiente, mobilità e disuguaglianze sociali", in "Crios", n.1/2011, pp. 83-92
5. T. Jackson, "Prosperità senza crescita. Economia per il pianeta reale", Edizioni Ambiente, Milano, 2011; ed. or. "Prosperity without Growth: Economics for a Finite Planet", Earthscan/Routledge, London 2009
6. B. Secchi, "La città dei ricchi e la città dei poveri", Laterza, Roma-Bari, 2013

Recycling means thinking about the future of the city starting from its material consistency, from its relations with multi-scale contexts, from its transformation potential, from its ability to maintain the continuity of life through innovative actions that raise the performance of space, in terms of development, quality, resilience. "Resilience", as the ability of cities to react to external shocks, is a term that today replaces and embodies a new, more pragmatic sense of the established concept of sustainability.
The importance and success of this concept probably lies in the fact that making cities safer and more resilient is the necessary precondition for new forms of development in cities, societies, economies, forms that are different from the past, and capable of utilizing innovative techniques and contents.
Urban planning is certainly one of the imaginable answers to the question of resilience, in terms of adaptation, as well as mitigation: it is not possible to cope with the effects of "climate change" simply by acting on local policies. Mitigation implies the need to remove the causes that generate the risks: it is a question on a global scale that must be dealt with through global policies and network interventions, as demonstrated by the intergovernmental convergences from Kyoto to Paris COP21. Cities, however, must also deal with the effects of climate change in an adaptive way, closer to the space of the city and its fluxes, which means the construction of solutions capable of responding to risk in relation to specific contexts and realities. A space that takes shape as a "soil design ", not only as a sectorial approach to the topographical condition "security", but as a principle that generates new forms of urban public space, flexible and multifunctional, with a high vocation for landscape, as for example in the projects for the new resilient "waterfront" of Manhattan (re-conceived on the basis of a profound demand for innovation as a result of the damage caused by Hurricane Sandy in 2012)[7], or in the system of public spaces in the city of Copenhagen, where the squares become - if necessary - lamination tanks to protect public spaces from flooding.
The search for variable structures capable of guaranteeing the safety and livability of the city, are part of a more general notion of metabolism, understood as the balance of the territorial organism and the balance of the flows of different nature that cross the structure, such as the balance between the production of waste and its disposal, or the balance within energy budgets. The knowledge and control of the territorial metabolism, as control of the residual footprint that the flows of matter and energy leave as they pass through, is a new form of project, essential to ensure the development of the territory's potential and at the same time to outline the conditions for its resilience: in this sense, metabolism can be understood as a delicate point of territorial balance that the different forms of territorial government are responsible for maintaining. Metab-

7. "Working Together to Build a More Resilient Region" was the motto of "Rebuild by design", an initiative to promote a competition developed in New York after Hurricane Sandy in 2014-15. This initiative was launched by the Hurricane Sandy Rebuilding Task Force and the HUD, in collaboration with the Institute for Public Knowledge of the New York University, the Municipal Art Society of New York and the Van Alen Institute, and with the support of the Rockefeller Foundation.

olism can become the objective of territorial policies and take the form of the project, thus improving the liveability of spaces whilst also giving them new meanings. Mobility, approached from the viewpoint of an ecology of the city (reduction of emissions and a new relationship with road space and its various crossings) becomes a way to give shape to the space, and therefore also the perception and livability of the city; it becomes a sort of lens through which it is possible to see the city and its spaces, to polarize the nodes of the urban region with new functions and new green and blue infrastructures (watercourses and green areas), pursuing urban and environmental effects - i.e. the design of the system of functions and equipment and their relationship, and at the same time the mitigation of environmental risk - as in the case of the Green Network Plan of Hamburg (2010).
These relations between space, activities and functions generate economies and at the same time derive from economies: circular economy is, for example, a different and innovative way of considering resources, their reproducibility and their use, which is based on the conversion of the very principle of raw material. There are no longer "raw materials" in a balanced metabolism between production, consumption and waste, but " second-hand" raw materials, as a result of a necessary conversion of the linear chain of production processes: recycling is not a way to look back and imagine what to do with waste, on the contrary it represents a stimulus to rethink the ways of using and producing what most easily becomes waste; what has an improperly short life cycle, what can be "reduced" in consumption and "reused" by others and with other purposes. Ultimately, designing according to the logic of recycling allows us to influence the way we live/traverse space, and shapes our everyday lifestyles. Recycling, well beyond the mere creative reinvention of objects, is a perspective that defines the ambition of building a broader vision linked to the metabolism of the territory, the balance of environmental cycles, the territory and the landscape. Precisely because "landscape" is a value of both identity and belonging, it is the life of communities but it is also continuity and environmental integrity, balance and protection of biodiversity; it is both "landscape" and "environment"; perhaps because of the plurality of its meanings and the potential of its transformation, for its centrality in the discourse on contemporary space, the landscape has acquired in recent years an increasingly structural dimension in urbanism.[8]. The landscape is an interpretative lens, a cognitive strategy, a construction device for contemporary design, but it also includes "waste", which can be built and reconstructed through careful consideration of what is "existing" first of all, and then of what is residual, through forms of restoration and regeneration of the territory components that are to be considered in relation to their cycles.

In these pages, pictures of "Ambito 43", Naples, by A. Formisano.

8. «Across a range of disciplines, landscape has become a lens through which the contemporary city has represented and a medium through which is constructed», cit. in C. Waldheim, "A reference Manifesto", in C. Waldheim (editor), "The Landscape Urbanism Reader", Princeton Architectural Press, New York, 2006, pag. 15

In this sense, the landscape pemits - against any conventional vision - the combination of principles and materials that differ in nature, form and state of use. It is an approach that links nature and artefact and has as its continuous reference the perception of space, its visibility, its communication, its sharing; it recalls the knowledge of the territory, its image, its description, the memory of the palimpsest as an interpretative and cognitive practice. The combination of different concepts that come into play when thinking about the future of the city and the project, such as

"resilience and recycling, growth and value, waste and landscape", brings problematic relationships into conflict with the networks of ecological values of the urban environment, thus permitting a radical reflection on new methodologies and new materials of contemporary urban design. These solicit 6 points for reflection:

1. Multiscalarity[9], as a distinctive feature of the ecological dimension of environmental phenomena, is the interpretative category for understanding and treating the complexity of urban and territorial phenomena in contemporary times, especially when the ecological theme and sensitivity to the landscape represent a substantial foundation of the territorial project. Thinking about the territory in a multi-scalar perspective represents a way to constantly cross the physical, material and visible components of that space with the immaterial dimension of the subjects who are its active protagonists; it also allows us to understand the production of the space as the result of a process in which it is essential to define a point of mediation between multiple instances - economic, social and cultural - which, precisely in the name of this complexity, can no longer be segmented by separate rationality and sectorial action. The theme of multiscalarity, therefore, allows us to distinguish and recognize the morphologies of the contemporary territory, in the forms of their transformation, in a sense of adaptivity proper to contemporary design.
2. Visibility is based on images and their communicative capacity, as a means of knowledge for the project, which is always a form of interpretation. The construction of an image "of change" for the territory represents an exercise in the collective definition of its identity; it represents the history and sense of belonging of its inhabitants, its housing traditions and construction. The image is a representation of the project and is its ability to envisage a possible future, while making it a communicative, plural and collective objective. Urban planning reclaims the image to re-launch its capacity for prefiguring space by combining social sciences with architecture.
3. Participation and inclusion, for a development based on an idea of open source urban planning, and on the ability to interact with citizens, to create an alternative economy, assigning value to local action and to the empowerment of the people, making them capable of doing, giving, sharing and saving energy, services, goods, knowledge and skills. The sharing dimension also becomes a new creative form of city economy: rather than consuming the city, residents share its production.[10].
4. Time. If it is true that the project, exploring the nature of new urban phenomena, must identify possible strategies to

On the opposite page, above, Copenhagen, masterplan Nordhavn, Spazi Pubblici, Cobe studio, 2013.

Below, Istanbul, Tactics of post-urban development for TOKI mass housing, Superpool & Atelier d'Architecture Autogérée, 2014 (Uneven Growth, Moma, 2014).

9. M. Russo, "Multiscalarità. Dimensioni e spazi della contemporaneità", in "Archivio di Studi Urbani e Regionali", n. 113, 2015, pp. 5-22

10. «Instead of consuming the city, residents will now resiliently coproduce it», Superpool, Atelier d'Architecture Autogèrèe, in P. Gadanho (editor), "Uneven Growth. Tactical Urbanism for Expanding Megacities", The Museum of Modern Art, New York, 2014, pag. 84

Infrastructural landscape

Natural matrix

Open spaces landscape

Ecological matrix

Garden matrix

Ecological landscape

rationalize resources and to "DIY" with what is available, to recycle and optimize the use and reuse of waste, then the time of the project becomes a usable tool. It shortens, has a pragmatic horizon related to what can be done immediately in the short term. It faces the uncertainty of the future and often requires small, temporary, molecular actions, taken by means of a tactical rationality rather than a strategic, variable over time, capable of directing the very nature of the uses, their temporality. Considering the inadequacy of "long time" as a form of control conforming to a model, to a fixed and superordinate design, to rules and programs as an exclusive reference for every minimal "modification" of the territory, it can be said that time is a fractionable, and always modular, unit. A renewed notion of "temporality" of urban planning as a multiscalar context requires instead the dimension of "long duration" as a reference framework for issues of "general interest", such as climate change, energy emergency, the redistribution of population and production on the planet. The long term requires a structural dimension of project thinking, requires forms of knowledge based on multifaceted know-how and skills, requires a relationship between the scales of the project, to connect the strategic dimension with the tactical one. That is, it claims the need to relate the "visions" in a timeline that gives shape to the space. Time can be broken down into a multiple "sequence of life cycles"[11] with which the project must interact on the basis of knowledge of territorial morphologies different from the past, that can subvert traditional interpretative paradigms. Knowing the territory means building a topography of deposited values; but it can also mean building a cartography of the life cycles of the city divided into parts, subjects and objects, for forms of use and life that define new intentional geographies of the city.

5. Ecology as a project material. Vulnerability, mitigation, safety, resilience, eco-innovation and efficient use of resources, energy regeneration and reduction of CO_2 emissions, can no longer be considered background issues for the design of contemporary territory. Instead, they are goals essential for building visions and for directing the future. They are the real new materials of urban planning on every scale, that can redefine the form, the techniques, and language. Materials that urge for the mutation of cognitive models, techniques of investigation and forms of "description" that induce strategies of valorization of public goods and production of "welfare" (including, for example, the production of public space, energy from renewable sources, etc.). Ecology is the new palimpsest of the relational dynamics that provide logic and forms of continuity to design and constitute a dense potential for the reinvention of language,

Residential clusters and open spaces, "Ambito 43" Naples, Uap Studio, 2017.

On the opposite page, below, "Ambito 43" Naples, Masterplan, Uap Studio, 2017.

11. W. McDonough, M. Braungart, "Dalla culla alla culla. Come conciliare tutela dell'ambiente, equità sociale e sviluppo", Torino, Blu Edizioni, 2003; ed. or. "Cradle to Cradle: Remaking the Way We Make Things", North Point Press, New York, 2002

Oudolf's garden, "Ambito 43" Naples, picture by D. Capasso.

of common space, of soil design. New ecologies meet the need to develop the awareness of a dynamic - and stable - concept of "balance"[12] (resilience, for example, means constantly redefining and guaranteeing the achievement of new and different balances), of limits, of the functioning of the territory as a complex and mutant organism, as an ecosystem, with its constitutive mechanisms.

6. Territorial mapping represents the forms and values that define new project maps, indispensable for managing the uncertainty of future impacts on the city and society, without neglecting the interactions between urban models, energy consumption, micro-climatic conditions within the city and the impact of extreme weather events on people's health. An innovative cartography invents a new way to thematise space, for example in forms of obsolescence, intensity and temporality of life cycles, forms and intensity of customs, etc.; such cartography can also cover relationships and behavioural aspects in cases of risk, in a view of the conditions of impoverishment, segregation and migration, as well as the effects of extreme events on the conditions of habitability of the urban and peri-urban space.

In short, "re-cycle" is not a punctual action or one specifically aimed at a final treatment of the territory or some of its partial components. Recycling represents an idea of the project and its role - above all social - as an action that looks to the future with a strong consideration for the reasons of the present: it is a form of thought that resorts to an innovative way of knowing and selecting materials indicative of the reality in which we should act, of reconsidering the value of what must be preserved or recovered, modified and rethought with forms and meanings different from the original, but especially with a subtle attention to the forms of balance that this potential modification establishes with other life cycles, not only those of objects, materials and buildings, but those of the environment with which they interact. It means conceiving the city as an organism endowed with its own metabolism, where interference causes the loss of that dynamic and changing equilibrium - evolutionary in the sense of new, unconventional, unprecedented[13] - which ultimately represents the degree of resilience of the city and its ability to respond, to second, to produce change.
"Recycling" means considering this balance as a creative space for innovation, as an invention of new structures and shared values, therefore as the centre of the project, as a mediation between transformation, development and continuity / preservation of environmental values.
This means, first and foremost, thinking and designing the territory in terms of habitability and security (and therefore resilience), understood as a precondition for any possible

On the opposite page, above, Residential clusters with water tanks, "Ambito 43" Naples, Uap Studio, 2017.

On the opposite page, below, Interior gardens, "Ambito 43" Naples, Uap Studio, 2017.

12. S. Davoudi, "Resilience: A Bridging Concept or a Dead End?", in "Planning Theory & Practice", Vol. 13, No. 2, June, pp. 299–333, 2012

13. S.R. Carpenter, F. Westley, G. Turner, "Surrogates for resilience of social–ecological systems", in "Ecosystems", 8(8), pp. 941–944, 2005

"welfare" policy: in this sense, the reproducibility or elimination of resources primarily concerns the city as a substantive and relational value. Its reproduction is increasingly complex over time in less expansive and conventional forms, and functions as a new way of conceiving contemporary urban planning.

Clues: Roots of an evidential paradigm

Roberto Serino

Implosion is a process generated by the cessation of the use of buildings and spaces, and is manifested not only in peripheral areas but also, and above all, within urban agglomerations.
In the architecture of a city, it represents an indisputable possibility both in terms of growth and in terms of the more recent possibility of "de-growth" - or rather, contraction. One of the problems that this new condition poses to architects and urban planners who are used to expressing themselves by designing and intervening in the building environment, is that of often having to contemplate processes of transformation that develop without significant physical intervention. Architecture is thus confronted with a sort of shift in meaning that makes its transition from a physical fact to a more social one, shifting from "hard tools" to "soft tools" and making the latter a key factor in planning.
Rather than dealing with housing typologies and the location of buildings on the territory, the new generation urban architect increasingly collaborates on projects aimed at training new types of actors, associations and consortia that promote construction, and they are also involved in the definition of new tools for planning, law and financing. Such a process was foretold in the 1960s by sociologists such as Turner[14] and Burckhardt or by architects linked to participatory movements such as Knoll in Belgium and De Carlo in Italy.
In short, it is a question of considering the trite axiom "relationship between planning institutions and users" from an innovative point of view in the direction of a new awareness of the need to give shape to an "architectural culture of recycling" that is quite distinct from that of restoration. "An overwriting" -as Ciorra affirms- "that puts recycling in relation to what happens at the same time in all fields of art, music, visual arts, cinema". Many examples, in this sense, show how one can intervene without adding anything while restoring a different meaning and use to the materials that are already available. In the wake of a consequent mobility and unpredictability, it is evident that we are faced with a process and no longer with one of the many "isms", and/or inconsistent concreteness.
The phenomenon has deep roots. Individual and environment, no longer separated from one another, as subject, within the other, as container, grow together - they "con-grow" - and only their inextricable unity constitutes the "con-crete". This statement,

14. V. Turner, "Il processo rituale. Struttura e anti-struttura" (Original title *The Ritual Process: Structure and Anti-Structure*, 1969), introduction by Dario Zadra, Brescia, Morcelliana, 1972.

of a philosophical nature, leads back to Jaspers[15], who wrote in 1919 that the image of the world, the emotionally experienced environment felt through the objects that are represented, has grown intimately with us. This position appears clear today, as a result of a new widespread sensitivity, which leads to the affirmation that everyone is what he/she cultivates and cares for. The contemporary development of modern anthropology has by now liquidated the overused Cartesian scheme of the two substances, irreducible to each other, "the thinking" and "the extended" and, even if this is found in most areas of knowledge, it is not yet completely so in the field of architectural science due to an objective difficulty in grasping the deepest and most vital desires hidden under the thick layer of experiences and "active debris" that characterizes the urban landscape and, in particular, the Italian one. It is inevitable and also the only possible condition, that architects find themselves working within the living body of existing "worlds" and that their task is to transform them.

For architecture, the contemporary challenge is to transform buildings, cities and sites from crowded multi-ethnic environments into creative inter-ethnic worlds. An intersection that is not only functional, but also meaningful and that can only take place within the context of the modern city.

The modern city is, in fact, a social product and this is determined by the social relationship based on production, which assigns certain roles to the city. In the wake of changes in the form and organisation of work brought about by technological innovations, changes in lifestyle and so on, the city confirms and preserves its functional, social and political-ideological role. What changes, however, is the way in which it is observed, experienced and used.

Through the flow of young people who move freely in urban structures, drawing significant traces to create chains of fragments, the image of the city in the mirror reveals new scenarios where centrality, and seemingly indestructible symbols fade, "sympathetic nuclei are visited as in a bored and schizophrenic via crucis (the Way of the Cross) with the conviction that the city is beautiful because [...] it is nothing more than a place produced from nothing and destined to play", a very young Neapolitan reports. A singular interpretation that is also a sign of an evidential paradigm, as Ginzburg[16] calls it. If such

15. K. Jaspers, "Psychologie der Weltanschauungen", 1919; translated by V. Loriga, "Psicologia delle visioni del mondo", Astrolabio, Roma, 1950.

16. C. Ginzburg, "Spie. Radici di un paradigma indiziario" (original title *Clues: Roots of an Evidential Paradigm*), in "Crisi della ragione", A. Gargani (edited by), Einaudi, 1979, pp. 57–106.

a paradigm is an overall conceptual construction capable of determining a particular and coherent tradition of research, the evidential paradigm would seem to identify, then, a familiarity between the fields of knowledge, even though very distant, but united by a similar methodological path. Paying attention to the marginal, to the anomaly, to the revealing detail, is a fundamental opportunity to discover new aspects of the past that would certainly be missed in the absence of these occasions. The art of deciphering seems to have accompanied the work of Paola Galante, rigorously and stubbornly centred on the observation of fragile details revealing sensitive changes in the fabric that has historically been responsible for production in the eastern area of Naples, highlighting its most important urban aspect, which is that of a progressive branching and connotation within an "inter-zone" between mono-functional areas and the city. In fact, after analysing the new productive realities of excellence, established in disused production sites, the author points out that the operating methods are more similar to those of the "artistic districts" than to the industrial realities. Reinterpreted "serendipity" is a sought-after asset, inasmuch as it is an unparalleled supplier of prolific stimuli. The result is a renewed relationship between production models and the idea of the city. In these spaces, which we can define as transition spaces, tertiary functions for collective use have been accepted through minimal transformations that relate to the sphere of urban "recycle": "[...] by accepting functions for collective use, these spaces no longer concern exclusively the private sector or the work space but rather the entire city, and are configured as promoters of rebirths that are often indispensable, and are not feasible when traditional urban planning tools are used [...]".
In my opinion, an intriguing interpretation opens up on the mutations of the public space concept, and even more so in the case of private space for public use. In the background, the question always arises: if the city is seen from its constituent data, is there still continuity in its transformations?
New spaces for public use, therefore, that re-propose - although concentrated and with greater guarantees - some of the features and functions of the city. However, what emerges is the different organization of elements that can be ascribed to the historical city, often importing models that are foreign to our tradition, as a response to the absence and, even more so, to the widespread degradation of public space. Therefore, the role and function of the space for public use is renewed, albeit in dimensions that conform to the characteristics of these settlements, indicating an antagonistic position to any idea of fragmentation, disharmony, or occasional, individualistic opportunity. A sort of city plan appears, one which has no ambition of improbable unity and harmony, but which confirms the idea of the city as a vital territory of contradictions.

Its nature is to make possible the emergence of processes, but it is also the one on which these processes are poured. (In this interdependence, the analogy to natural processes is disconcerting). It is like the "tom-tom" and all the eccentricities of communication: they work where there is connectivity. What makes communication possible and determines the processes of mutation, is the existence of a superstructure, of a network, and this is in itself the greatest productive activity. To exist, it needs a "constellation" of points. We could say that we are witnessing the passage from the "stellar point" to the "satellite point". The city as a place of exchange exists as a possibility to make the traces of information and material exchanges evident. The territory, traversed, is a place of slowing down of the route. Line/time is a line that is interrupted to create new visible and invisible paths. If this is the fundamental nature of these modifications, that are cyclical only in appearance because they are influenced by multiple variables, what is most interesting is the induced effect, as it generates further and unpredictable (or at least only partly predictable) mutations.
What happens when everything stops or slows down?
What has happened, for example, in those communities where productive settlement has disappeared and how can a community be equipped to replace interrupted processes?
Drought dries up the tracks of communication unless we have reserves to use as happened with the peasant culture that provided for "storage" and the spaces allocated to it. Part of the (global) production goes into (local) reserves and this can be valid not only for the finished process, but also for the energy that made its formation possible. How is this energy storage transformed? This is a very interesting subject. It is not so much the place that changes, as the way in which we cross it and therefore use it. The "objects" that we unload on it testify to our transit.
Such objects are similar to the ruins of a suddenly decayed empire. The attention to the ruins, to their fascination, is a phenomenon attributable to the all-western phenomenon of the recurring obsession for some kind of "classic", always considered dead and always reborn. A phenomenon that crosses the whole of European culture and that marks its difference from that of the East. It is European culture's ability to strive for a singular balance between continuity and the discontinuity that follows a sudden interruption and that is embodied to the highest degree in ruins as presence and absence. It extends to languages, gestures, institutions, organizational methods of work and all traces of ancient/classic culture re-birth as an essential condition of tradition and memory ... and this also applies when reorganizing East Naples - we try to balance the Sprint project and the Chinese siege of Prato - we appear as "cluster" and "castrum". In the end, we consolidate architectural codes in

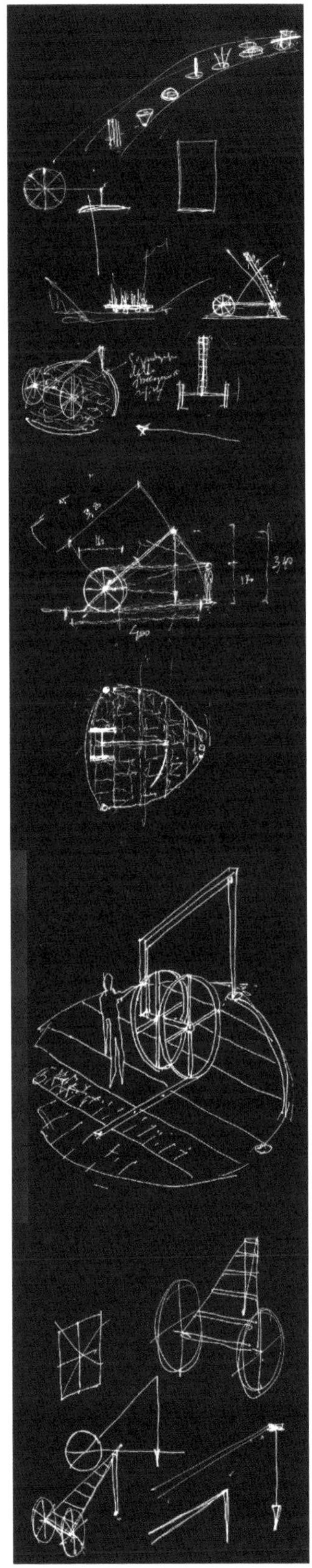

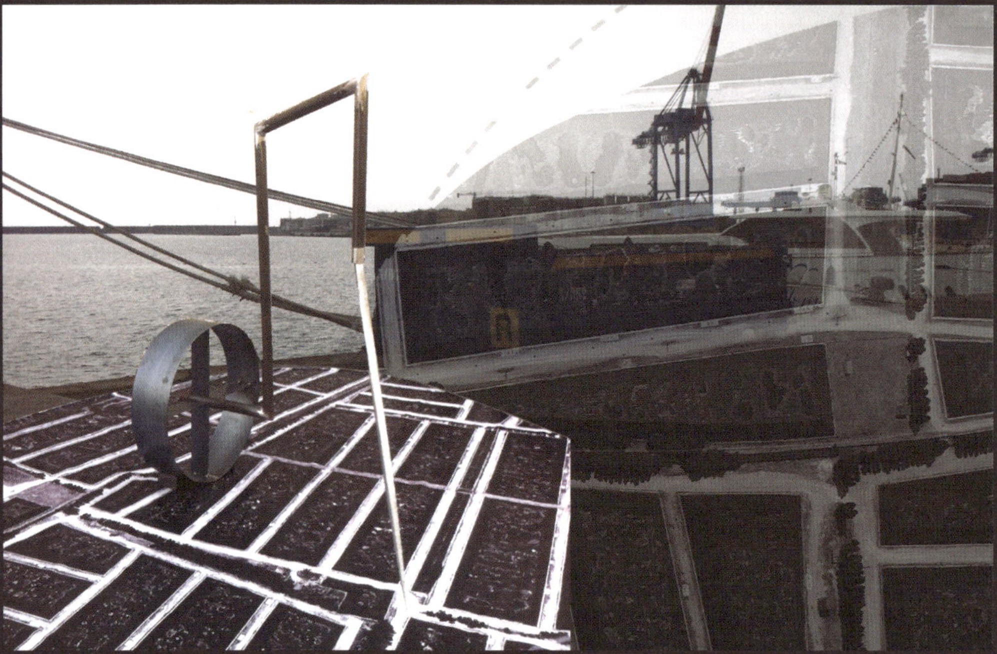

"automatic" innovation when we try to prefigure the production sites as possible solutions to contradictions between the need for expansion and the need for rooting; incredible bridges that manifestly presuppose a desire for lost harmonies and, as for the ruins, require attention to the traces of interrupted stories in order to clear the present.

Images in this contribution are by Roberto Serino

17. R.Serino. Intervention carried out at "Foot Print SUPERELEVATA", curated by Re-Cycle Research Unity of Genova, on 21st September 2014.

Corollary
40°57'39,2" N - 14°04'19,6" E[17]

It's an archaic, rural metal structure, like a mobile vector overhanging a territory that has been stratified by use and abuse. It embodies a tension, perhaps the only one possible, to engrave, recycle and regenerate from the depths a history of fertility: Campania Felix. The intervention is proposed in the visible contradiction between the "seductive disgust" of a trace of so-called ecological fuel pellets (*ecoballe*, in Italian) and the "reassuring ambiguity" of a machine that dominates it. A tool that moves like a drone over the geometry of the city of Leonia, exploring its corrugated surface with the tip of the needle, appearing therefore as a metaphor of a territory not abandoned to itself in its cycle of "percolation". On the contrary, it is controlled and prepared to wait - not for a solution, but for intertwined solutions. "The "*ecoballe*" pyramids of rubbish have been there for over ten years. They hide from the public eye between 6 and 7 million cubic metres of rubbish. They are waiting to be burned in a waste-to-energy incinerator that will have to be specially built with an investment of 450 million euros. However, the process takes a long time and the result is uncertain, all the more so because the machinery would be built on fertile ground that cannot be waterproofed. There is someone, with the evocative name "*Rete Campania per la civiltà del sole e della biodiversità*" (literally translated, *the Campania Network for the civilization of the sun and biodiversity*), who proposes the project "Dalle ecoballe alle piramidi del sole" (lit: From fuel pellets to the pyramids of the sun), encouraging the transition from the protest (by now depleted), to real action. Using the sun to produce energy is essentially envisaged through the installation of photovoltaic panels on the pyramids of Taverna del Re, covering them from sight and protecting them from meteoric water. A plan in line with the "M2RES" project of the European Union, which, with funding for degraded areas, authorises the installation of plants for the production of renewable energy. It is proposed, then, to provide, at half the cost, a plant capable of producing clean energy at no cost. An initiative that fascinates ecological groups and that, if implemented, would contribute to the elimination of the much-disputed CIP6 without conflicting with the final objective, which is always that of disposal".

From landfills to art sites

Vito
Cappiello

On examination of countermeasures that are globally undertaken with regard to the waste problem, the comparison between what has already been done worldwide and what has never been done in the Campania region is truly disconcerting. In a constant state of "continuous emergency" that lasted several decades, while millions of euros were wasted in transportation abroad and advice-seeking actions, no one bothered to waste their precious time to learn about what had been taking place elsewhere over decades. Not only about the treatment of the waste itself, but also with regard to the precious materials and the artistic outcomes that could originate from it. The inability to come up with possible solutions by presenting what had already been achieved as new forms of urban and landscape equipment, together with a far more serious inability to manage the urban waste cycle, and to lead appropriate consultation and dialogue with the local communities that were supposed to host landfills, has led to a climate of suspicion; a fracture between institutions and citizens, a conflict where, in fact, nobody really knows, although everyone tries to impose their own blind point of view, unable to show actual perspectives based on concrete examples.
Massimo Fagnano's essay on the new techniques of "re-cycle" of the landscapes to be reconverted also opens a new "vision" of the procedures to be followed in order to rehabilitate contaminated landscapes. Many of these techniques refer to vegetation-related practices, indicating how the planting of trees and/or shrub species can be used to clean up these areas. The above mentioned technique highlights some important issues which, directly or indirectly, refer to a landscape architecture, or to what can be referred to as a "landscape project" which, passing through the functional aspects of the depollution, results in the "construction of new parts of the landscape" in both its intermediate and final phase.

It is clear that the combination of landscape engineering and work conducted on the themes of water and air resources can make a new form of landscape design possible. And on this basis it will be possible, in a new "timeline", to graft a definitive new landscape project in which technical requirements and artistic and scenic needs are also combined.
In fact, there are quite a few cases in which the landscape project is intertwined with the artistic project, even in non-degraded areas (just think of Niki de Sait Phalle's Pinocchio's park in Collodi, the park of the battle of Varo, the forest of Teutoburg in Germany, the Art Park in Verzegnis, the Hortus conclusus - Court of San Domenico in Benevento, by R. Serino and M. Paladino, the Ortus Artis event in the Certosa di Padula, etc.), determining such qualities that can now be ascribed to the history of international landscape architecture. But what needs to be investigated more closely, in our case, is the consistency of the creation of new and important park areas, not only with landscape features but also with strong artistic contributions in areas that were previously used as landfills.
One example is **Danehy Park** in Cambridge, Massachusetts, a 55 acre site, relatively close to the urban area. Litter dumping was stopped here in 1970, following a process of redevelopment and landscaping, that was started by landscape architect John Kissida. In 1990, it was converted into an urban park which increased the city's greenery by about 20%.
Beyond the problem of continuous monitoring of the levels of use compatible with the site, it should be noted that, since 1990, the municipality has identified a major project of artistic transformation of the park by Mierle Laderman Ukeles[18].
A second example is **Fresh Kills** (Staten Islands, New York), one of the world's largest former landfills, covering about 2,200 hectares (i.e. two and a half times the surface of Central Park). The dumping of waste ceased in 2001 but the

18. Mierle Laderman Ukeles, a well-known artist who exhibited and presented installations wordwide, was described by the Art Review as one of the main artwork makers in public spaces. She was awarded very important prizes by the National Endowment of the Arts.

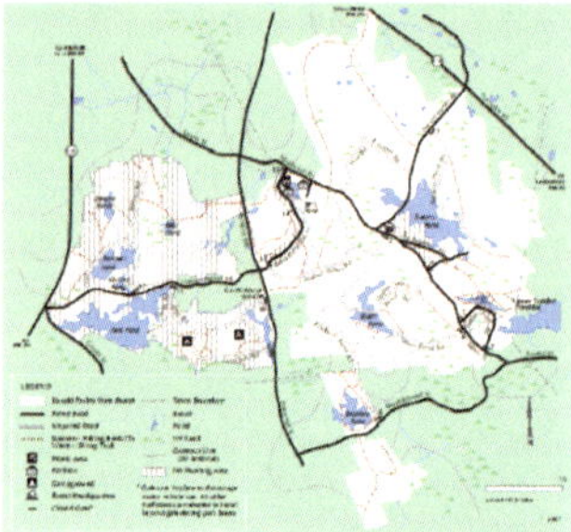

Danhey Park by John Kissida, Cambridge, Massachussets, USA

clean-up process had already started in 1980 and continued for the following thirty years.
In 2001, the landfill had been reopened to accommodate both the Twin Towers debris and that part of dust containing unidentified human remains (this section occupies a special area of about 44 hectares, which for many years will continue to be inaccessible). At the same time, however, through the stratification of the various stages that created it, the site was reclaimed and thus plans have been made for its transformation into a future landscape, with a considerable part of Fresh Kills becoming accessible to the public as a large urban park in the coming years. In this case, also, Mierle Laderman Ukeles was entrusted with the task of adding her artistic installations as a contribution to the masterplan.
Let's now switch to another nation, namely Israel. Here we find the **Hiriya** site, an area of about 70 hectares, used for decades as a landfill to dump undifferentiated urban waste from 17 municipalities between Jerusalem and Tel Haviv, which formed a hill almost a hundred meters high. The landfill was closed in 1998 but it is still possible to find undifferentiated household waste, stratified at various stages corresponding to various regulations and sources. The area, besides being very close to the urban centres, is situated at the crossroads between the two motorways (the Highway 1 between Jerusalem and Tel Haviv, and Highway 4 between Beersheba and Haifa), and is strongly visible due to the extremely flat local landscape. For decades there has been strong pressure from environmental groups to turn this site into a park. Although Israel most probably had many other problems to solve apart from this one, administrators and citizens have been working for decades to carry out the project to transform the landfill into one of the most interesting contemporary parks, also known as "Recycling Park".
The planned disposal began in 2000; Hiriya currently receives 2,700 tons of undifferentiated waste per day from 800 trucks, and has become one of the largest transit stations in the world, as well as Israel's largest and most advanced environmental centre.
Here everything is separated and recycled: plastic, wood, metals, wet waste. In the bio-treatment plant operated by the public company Environmental Services Company (ESC), light materials (which float) are separated from heavy materials (which settle on the bottom). After the inorganic substances have been eliminated, the biological treatment material remains, from which water, compost and methane are obtained. Methane is used to produce electricity with a 1.5 megawatts turbine. The plant can handle 200 tons per day, the same as the one that receives and separates building materials. A pilot gasification plant produces closed loop syngas. A structure for the treatment of tyres is also being planned. Then there is a wetland

On the opposite page, Fresh Kills, by Field Operations, Staten Island, New York, USA

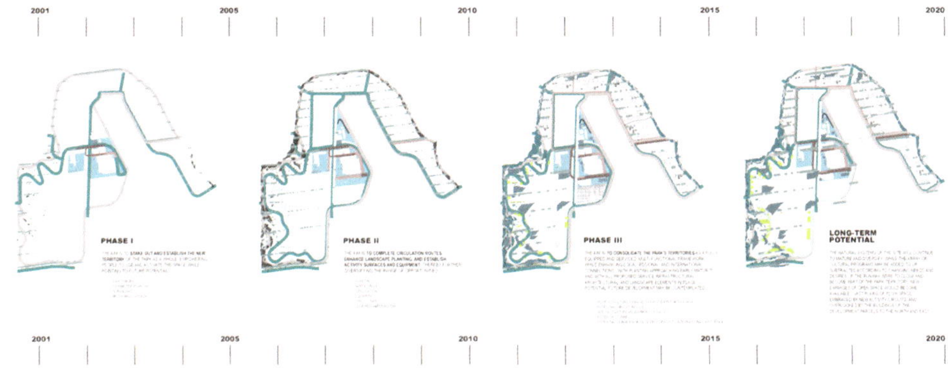

2001
2005
2010
2015
2020
PHASE I
PHASE II
PHASE III
LONG-TERM POTENTIAL

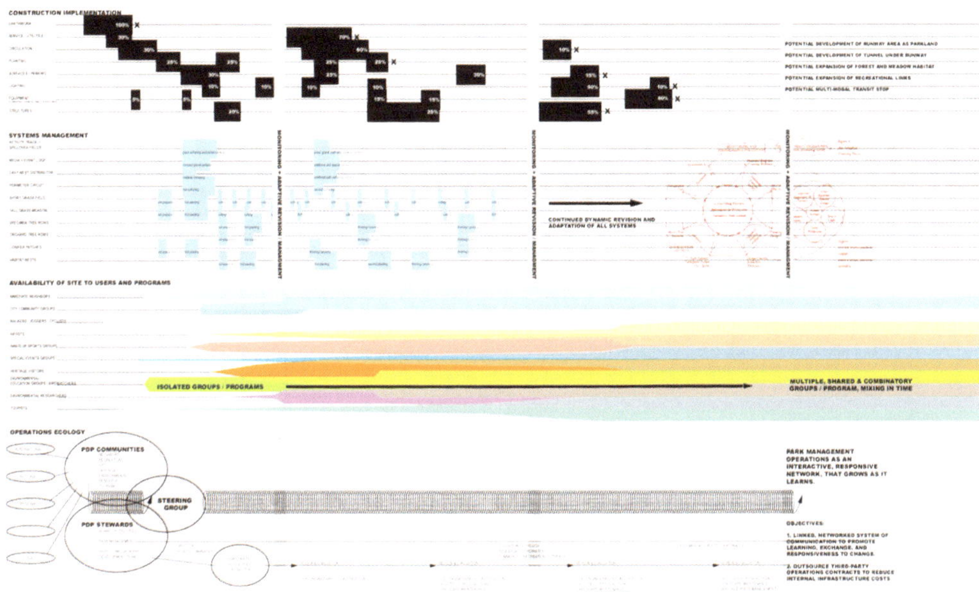

CONSTRUCTION IMPLEMENTATION
POTENTIAL DEVELOPMENT OF RUNWAY AREA AS PARKLAND
POTENTIAL DEVELOPMENT OF TUNNEL UNDER RUNWAY
POTENTIAL EXPANSION OF FOREST AND MEADOW HABITAT
POTENTIAL EXPANSION OF RECREATIONAL LINKS
POTENTIAL MULTI-MODAL TRANSIT STOP
SYSTEMS MANAGEMENT
MONITORING + ADAPTIVE REVISION / MANAGEMENT
CONTINUED DYNAMIC REVISION AND ADAPTATION OF ALL SYSTEMS
AVAILABILITY OF SITE TO USERS AND PROGRAMS
ISOLATED GROUPS / PROGRAMS
MULTIPLE, SHARED & COMBINATORY GROUPS / PROGRAM, MIXING IN TIME
OPERATIONS ECOLOGY
PDP COMMUNITIES
STEERING GROUP
PDP STEWARDS
PARK MANAGEMENT OPERATIONS AS AN INTERACTIVE, RESPONSIVE NETWORK, THAT GROWS AS IT LEARNS.
OBJECTIVES:
1. LINKED, NETWORKED SYSTEM OF COMMUNICATION TO PROMOTE LEARNING, EXCHANGE, AND RESPONSIVENESS TO CHANGE.
2. OUTSOURCE THIRD-PARTY OPERATIONS CONTRACTS TO REDUCE INTERNAL INFRASTRUCTURE COSTS

143

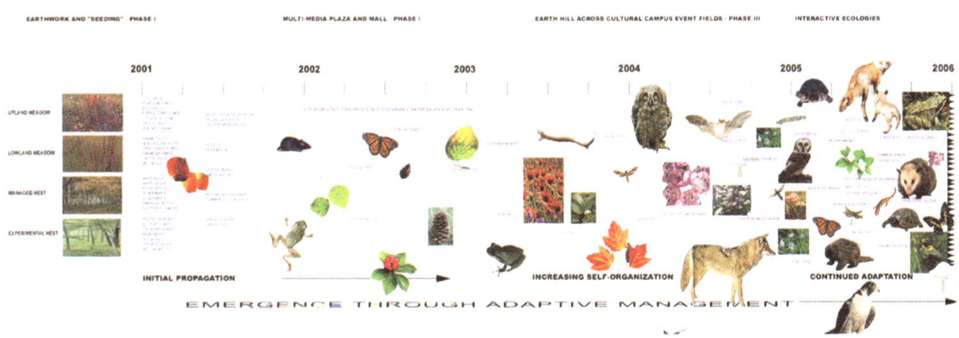

EARTHWORK AND "SEEDING" · PHASE I
MULTI-MEDIA PLAZA AND MALL · PHASE I
EARTH HILL ACROSS CULTURAL CAMPUS EVENT FIELDS · PHASE III
INTERACTIVE ECOLOGIES
2001
2002
2003
2004
2005
2006
UPLAND MEADOW
LOWLAND MEADOW
MANAGED WEST
EXPERIMENTAL WEST
INITIAL PROPAGATION
INCREASING SELF-ORGANIZATION
CONTINUED ADAPTATION
EMERGENCE THROUGH ADAPTIVE MANAGEMENT

Hiryia Park by Latz+Partner,
Tel Aviv, Israel

Norman Levy Park,
Merrick, New York, USA

plant - five basins where different species of flowers have their roots, with the task of purifying waste - and a wastewater treatment plant for the wastewater from the disposal process. Finally, with 63 wells scattered throughout the area, the biogas produced by underground materials is collected, producing 4 megawatts of energy that power a textile company a few kilometres away[19].

Ayalon Park was a huge construction site but, since spring 2004, ten kilometres of bicycle paths have been opened to the public, connecting Hiriya to the Begin Park in Tel Aviv, while the eastern part of the park was inaugurated in 2010. Following an international tender, the project was entrusted to the German landscape architect Peter Latz. To showcase this jewel of environmental management, the deputy mayor of Tel Aviv Doron Sapir organized a banquet at the top of the garbage mountain that he proudly showed to his guests.

The **Norman J. Levy Park** (Merrick, New York) is another innovative example of reuse of an old landfill whose success was, to say the least, overwhelming; due to an average intake of more than 6,000 visitors per month, a dedicated Park Rangers Corps was set up to assist them.

The main landscape features are: 3 kilometres of hiking trails (many of which are suitable for disabled people), with beautiful panoramic views on the Manhattan skyline and the surrounding bays; the creation of "wetlands" (about 2 acres); two rainwater ponds; a long pier - belvedere, also suitable for fishing; a docking area for kayaks; and so on.

The pavements were created by recycling the building materials that had accumulated in the former landfill. Furthermore, systems for biogas recovery and water purification have been adopted.

The **Fonsanta park**, created by landscape architect Manuel Ruisanchez in Barcelona, is situated on an unexpected geographical depression that, in some places, reaches 30 metres and is therefore hardly usable without planning intervention. For many years, it was actually used as the city's landfill. The minimalist intervention restores the pedestrian path in the lower part of the valley through a zigzagging pattern, mostly made of wood and stone: an obligatory route that does not allow alternatives and stopovers but offers unexpected glimpses and perspectives. **Al Azhar Park** (Cairo) was designed by Sites International and opened in 2005. Also in this case, the project targets a degraded area: a mountain of waste and debris in the heart of the city, transformed into a thirty-hectare green oasis with more than 655,000 plants, three new water reserves, and an artificial lake.

The many heaps of debris and rubbish were replaced by citrus orchards and rows of palm trees, exotic gardens and fountains, in the district where the world's most important Islamic

19. See "Le Scienze", February 2008

monuments are located and a long marble path leads to the splendid view of the citadel of Cairo and the Mohammed Alì mosque (the most beautiful and important in the city), which stands in front of the hill. The park also contains a playground, large sports areas, an amphitheatre and a restaurant. Thanks to the redevelopment of the area, it has been possible to discover and restore a mile of the ancient walls surrounding Cairo.
West 8's **Schelpenproject** (East Scheldt, Holland) is one of the most innovative "creative ecology" projects. It is also called "The mussels and clams park". Because of a lack of funds, when the barrier to the East Scheldt was completed, the rubble accumulated during the construction was not removed from the island. The problem was aggravated by the need to dispose of tons of mussel and clam shells accumulated by the canning industries. At that time, West 8 received a commission for a project for the island's development. The idea was to cover the surface in white and black shells, arranged as elongated strips or in a chess-like structure. The solution stemmed from the observation of dark plumage birds, which tend to be attracted by dark shells, while light plumage birds tend to be attracted by light shells due to camouflage issues.
Passing over the East Scheldt barrier by car, motorists can enjoy a great landscape design in black and white, in contrast with the colours of nature and the sea. The contrast leads to a better understanding of the naturalistic and landscape values of the place. In addition, when white and black birds rise up in the sky, the structure becomes a large geometric white and black pattern in the sky, thus duplicating the abstract Zen garden designed for the ground.
If knowledge can be considered the first step in planning and implementation, I would say that there is sufficient material available. It would be enough just to inquire.
However, I believe that the time has also come for the second step: to start doing things, or at least to make proposals. A good starting point could be the opening of operational workshops on sample areas involving the Landscape specialists and the Departments of Architecture, Engineering and Agriculture as tutors. The participants in the workshops would be young architects eager to tackle the challenge of transforming the "*munnezza*" (Italian slang for rubbish) into art, energy (recovery) and, why not, economic and social benefits.
Let us not forget, among other things, that, in the case of Hiriya, 30% of the employees of the Municipal Refuse Collection company are graduates, while 42% hold a diploma.
Knowing how to use it, the "*munnezza*" can be gold, and not only for the camorra and other criminals.

Fonsanta Park by Manuela Ruisanchez, Barcelona, Spain

Al Azhar Park by Sites International, Cairo, Egypt

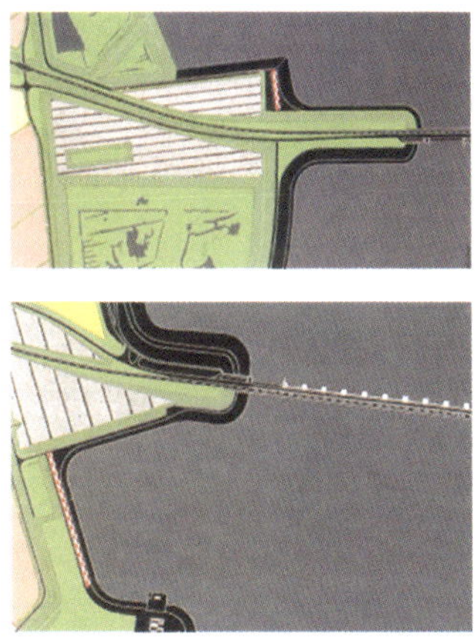

Schenplonproject by West 8, East Scheldt, The Netherlands

Drossland, abuse, disguises

Enrico
Formato

Individuality and inertia

Architectures, infrastructures and cities have ground-laying relationships with the territory; their conformation is stable in the structure and their relationships feature elements of high inertia. This "fixity" can deeply affect the notion of recycled material used in the industrial treatment of solid urban waste (referred to as Rifiuti solidi Urbani, R.S.U.).

Plastic and aluminum, paper and cardboard, glass, organic, undifferentiated: in the R.S.U. cycle, objects are placed in the above categories by sorting the waste according to the characteristics of the material. In this case, the recycling project aims at maximizing economic convenience and environmental sustainability. In the collection cycle of the solid urban waste, however, spatialization is missing. In fact waste, being mobile, can be aggregated in infinite ways, is removed by mechanical means and undergoes post-selection, screening and classification processes that foster assembly chains.

Conversely, in the case of the recycling project applied to the territory, the "one-dimensional space" of waste becomes more complicated and intertwined with the topological and contextual dimension, and with the soil and its corrugations (buildings included) . In this sense, drosscapes seems to only take on real meaning when set in a context of drossland/waste. The act of waste sorting becomes thus a reflexive operation, carried out in a multi-dimensional space, where the combinations of matter and object disposition are virtually endless. An operation that, unlike the space-deprived sorting of urban waste, may also have no solution. On the other hand, when it does have a solution, it can offer more "equivalent" solutions (it is in this dimension of equivalence that the role of the author can be fulfilled and, as we shall discuss here below, the time deconstruction of the project can take place).

The issue can be seen from many points of view.

Looking at the objects (and the urban fragments), it takes on linguistic significance, that is, the distance that is needed between the present condition and that of the expected life cycle. In the industrial recycling process of the solid urban waste, the process leads to a substantial transformation of the

starting object: it is the material that is recycled, not the object. The fixity of the architectural works and their condition of belonging to the territory, to the site where they are built, makes this hypothesis impracticable, unless one assumes that their recycling is reduced to a mere selection of demolition materials. This is not the ambition of Re-Cycle; it's the transfiguration of the function and meaning of useful items, selected according to a reference model (defined by topological and systemic conditions). Moreover, it is precisely the "difference" between the 'before' and 'after' on the object level that takes into account the particular "individuality" established from time to time by similar and repeated forms, with an ever-changing substructure.
As far as the "relational" plan is concerned, the recycling project is based on the study of the belonging of a given object to a system of relationships, affinities of use and / or topological continuity. The space tends to be articulated into flow, centrality, and voids, and turns into a porous surface, made up of moving parts, full of nooks and crannies, torn apart by sudden distances. Such porousity is, on a general level, an endless overlapping of folds: a multiplicity that is continuously thickening and stretching.
Therefore, a movement is generated and makes it possible to identify "rhythms", that is, recurrent spatial modifications and their interruptions. The focus is then shifted on the way in which conformations evolve, and / or how similar topological or material / conformational conditions repeat over time. The project of transformation projects the life of objects - and their relationships with the context - in an evolving scenario, with reference to the temporal development of relationships and divergences, and of affinities and conflicts that are latent to different degrees.

About the model and the context

In urban planning, the potential for endless combinations of architecture and the city is traditionally reduced and conveyed by using "models"; typical and recurring conformations recognized through abstraction by observing particular cases and

grouping them by similar characteristics, of a relational, functional and localizational kind. One way in which urban architecture and urban planning usually work is, in fact, the approach based on the comparison between reality and expected models (prototypes), in an attempt to bring reality and project scenarios into convergence.
The background against which the Re-Cycle project is set is different from that of the twentieth-century urban planning, with the cold and rational modernity that resulted in a "*tabula rasa*" of the pre-existence. In a recycling project, this relationship is even reversed because pre-existence, or the object or system of objects to be transposed into another life cycle, plays a key role in determining the field of possible transformations. In this condition, the expected models - spatial and use pattern (typology), invariant in the relationship between empty and full (morphology), due to their material conformation, and the constancy of relationships between conforming parts - are not intended as well-defined urban facts; they are not ideal architectures or cities. Their nature is "multi-scalar", partly blurry, their conformation is defined over time. The prefiguration process is based on a dialectic between a systemic analysis of reality and the "political" figures of orientation. The first motion is based on pure analytical rationality: the reconstruction of biographies, their superimposition, the scientific recomposition of broken relations. The direction through which this clutter of signs is extricated makes use of the "political" orientation of new models. The latter, no longer determined and "finite", seem to assume a "figural" characterization instead, with the synthetic reduction of a complex world of references and perspectives. While the classic model, type-morphological, is implemented in a given set of invariant relationships, the "figure" defines an open formal field: that is, it represents the possible difference in the repetition of the same principle, accentuating its conceptual characterization and promoting its communicability. Some recurring figures of Re-Cycle Italy are, for example:
1. the resilient and sustainable dimension of the environmental project (i. e. the importance given to 'blue' and 'green' networks) and soil research as a topological continuity plan between the parties, with strong public significance;
2. the symbolic importance given to the permanence of elements typical of the life cycles of origin. Therefore, the choice of what has to be recycled and what has to be disposed of is made on the basis of a double reasoning:"logical-deductive", through systemic analysis and evaluation of the opportunities of a functional, economic, localising type; "strategic-inductive," on the basis of a modelling of the case study in the light of a certain future perspective.
Transformational scenarios are generally long-term. Individual, even minute, transformation actions are directed by macro-strategies (and their related figures), defined as structural scenarios, which can be specified over time.

Continuous project: ephemeral disguises and abuses.

The Re-Cycle code is structurally analogous to that of architecture, in oscillation between construction and conception. In fact, the work of construction is primarily the transport of an (originally natural) element from one life cycle to another: the tree that becomes a column, the fronds woven into the roof of a hut. In the context of the transposition between life cycles, profound changes are made in the meaning of the single object that transcends from the figural analogy (for which a column remains the idealization of a tree) and makes it possible to identify the epiphany of a distance between a "before" and an "after": the introduction of the design process and its concretization in the real world of time, which characterizes the space of architecture at least as much as the material of the sculpture. The design process is not a game - that is, it needs not only rules (it has, in fact, the characteristics of a regulation that is applicable everywhere but is serving as a precursor of the contextual project) - but a continuous and elusive, as well as always responsible, succession of choices. Among the infinite possible configurations in the world of ideas, only one takes place in the reality in which ideas are shaped and

implemented: a multitude of parallel realities is not possible, but nevertheless it is typical of architecture to proceed by juxtaposition and selection of alternative solutions, even during the construction process. It is not easy to understand how the options are defined, but of course the choice is a responsibility for which the project has developed specific techniques.
The first consideration is that the multiplication of solutions is possible considering an interweaving between conception, use and construction. In other words: thinking of the recycling project as a continuous, cyclical operation. It is not possible that there are more solutions for the same problem, but it is certainly possible (and desirable) that they evolve over time, in relation to changes in the context and sense of the constituent elements (each of these changes needs time as it needs to break up grevious inertias, amplified by significant periods of abandonment and/or underuse). Therefore, the temporal factor introduces an additional dimension in the recycling differentiation process: interest is not aimed at the comparison between modified objects in fixed scenarios (such as reuse or recovery) but rather at the difference between different conformations, the "difference" (in algebraic sense) between terms. So if the object before the transformation is identified by number 1 and the object after the transformation by number 10, Re-Cycle understands the difference as oscillation between 2 and 9, infinite combination of irrational numbers. In this operation, the beginning (1) and end (10) are clear and invariant, as is the direction, from 1 to 10. In truth, with Foucault[20], this differential space is based on the beginning, or rather on the iteration of the act of beginning. Moreover, the peculiarity of irrational numbers, as we know, is their indeterminacy, i. e. the impossibility to represent algebraically the complete entity (e. g. 10/3 = 3,333, the 3 being a periodical, infinite number). In this perspective, the project becomes an adaptive flow, a repetitive and sometimes cyclical activity, whose time of deployment tends to be unfinished, regardless of the scale in which it operates. It makes increased use of the temporal and indeterminate characteristics of landscape design[21]. As previously mentioned, in Re-Cycle there is no "temporary use" except in the sense of infinite variation of uses; there is no closed form but ambition to form, rendered through changing configurations, or aspirations of form, and also made possible through temporary disguises and alterations of use ("ab-use").

20. M. Foucault, "The Order of Things", original edition, Paris, 1966

21. M. Desvigne, "Il paesaggio come punto di partenza" (Landscape as Prerequisite), in "Lotus" n. 150, 2011, page 20-27

From an open project to a blurred project

Paola Galante

For decades, urban and peri-urban areas have witnessed the growth of building interventions aimed at the recovery or re-functionalisation of the existing building stock. In homogeneous critical fabrics, such as that of the Domitian Coast, or Gianturco's "left-landscape", recurring events that concern the substance of artefacts, and the needs of new users, provoke similar and often transferable interventions. These transformations, often neglected and rarely analysed, imply a lively planning activity, albeit silent and often unaware of the urban and territorial consequences it may trigger. Due to financial reasons, uncertainties about the future, or even to objectives and methods of implementation of regulatory measures that may make long-term replacement or planning interventions unsustainable or impossible, this planning activity features particular aspects still unknown to the past generation of professionals, both in the case of reconstructionist architects or big 'archi-star' studios.

They encourage new relationships with the client and new ways of working, which are declined in precise temporal intervals of the project which elude the traditional partition into preliminary, final, and executive; they also require new, close relationships with complementary professional figures (structural experts, installers, etc...), and solicit transversal readings of the regulations at different scales.

They act on existing structures, often with no particular merits other than that of the existing volume and the inherent capacity to increase floor areas.

The load bearing structure rather than the often neglected spatiality, becomes the protagonist of the project because it is almost always considered inviolable because the regulatory plans prevent its demolition, or because a replacement intervention is considered unprofitable. Considered the main obstacle to change, the structure (together with the environmental constraints, if they exist) is in fact the element that confirms the intended use, and lays the foundations for the transformation. The skeleton of the building must therefore be identified, redesigned, studied, verified in its suitability to withstand its own current load, and the future loads it will carry in the near or distant future when, as in the case of

industrial sheds, intermediate floors will cross wide spaces or when new vertical connection elements will subtract portions of the floors. In any case, the structure must be adapted to the regulations in force regarding the safety of the building structure and of the people who use it.
For each critical factory it is possible to identify the recurring structural types.
For example, for the production complexes that inhabit "left-landscapes", it is possible to draw up a list of the different load-bearing structures that are found in similar sectors in Europe. These range from the load-bearing walls of small buildings to framed steel structures or, more often, reinforced concrete made on site or using semi-prefabricated systems. The bays of these structures range from 5, 6, 7 metres for load-bearing masonry buildings to over 15 metres allowed by the use of steel arched structures. As far as heights are concerned, they range from 3-metre high reinforced concrete framed structures on several levels, up to the 12 to 20 metres that can be safely reached in large production or storage areas.
A set of knowledge of this kind would prove useful to the study of any critical material, because the history and the technology behind the construction of the structures would become an indispensable tool to approach the project. Through this tool, combined with a map of the locations of the structural types in the study areas, there would be potential for large-scale changes in use if - as expected, at least in the Campania area - building and urban restructuring is not planned: one can imagine a table in which each structural type is associated with its possible uses.
Another peculiar feature of contemporary design is the absence of a detailed functional program. At the beginning of the planning process, the degree of structural resistance to be ensured is established (ratio between structure and use) and, when foreseen, the type of reclamation to be carried out (ratio between health-safety on site and its use). In these cases, the value of the transformation (the difference between the value of the transformed structure and the cost of the transformation), calculated as short-term and low-risk investments,

represents the parameter that has the greatest impact on the choice of the range of possible uses, the relevant criteria to be met and, therefore, the projects to be drawn up.
The use of the available space is defined only to a large extent. On the basis of parameters that relate the floor area to the primary equipment required, only the location and the number of rooms for toilets and the spaces allocated to the facilities are specified.
The latter are non-starring players, supporting the structure in the design process. The need to adapt to the rules on energy containment, the possibility of accessing funds and incentives, as well as the market competitiveness of buildings with innovative systems, make it essential to design accurate and forward-looking projects with respect to the range of uses identified.
As a whole, the design activity consists of ensuring as many uses as possible of the available spaces.
These needs, dictated by the already mentioned economic and regulatory contingencies, come together in the idea of an open project, considered a "sine qua non" condition by the promoters of the transformations, not only when they are promisors in search of promisees, but also when they are themselves the future tenants. The construction sites proceed in successive "steps", each of which is associated with a building practice that precisely identifies the limits of the intervention. As a general rule, the building is recovered through the elimination of excesses, the verification and, when necessary, the reinforcement of the load-bearing structure, the restoration of all the elements of protection from atmospheric agents: waterproofing of the roof and systems for the outflow of water, fixtures, external facades.
Then we move on to a phase of enhancement and construction opportunities. This phase determines the future lives of the building. The systems are prepared, the methods to increase floor levels and the consequent vertical connections are set up, working on the existing structure or programming with precision the spaces of a new structure.

In a subsequent phase, sometimes distant from the previous ones, the internal partitions and finishes are defined and are generally included in one of the previously programmed options.
In this context, there are many situations in which architectural design ceases to be a synthesis capable of orchestrating different needs and instead represses its own contribution.
The building remains a silent spectator, a blackboard for autonomous operations.
In these cases, the "non-project" affects the spatiality of the interiors: autonomous environments and repetitive sequences; the choice of technical alternatives: dry partitions, modular false ceilings, floating floors, recessed lights distributed with impartial repetitiveness, anonymous fixtures regardless of or unaware of the future usability of the spaces; the choice of finishes: neutral colours, washable paints, practical, low-cost and resistant materials, medium size formats: the technical specifications (when they exist) are homologated on what is considered to be the average taste. Even more subtle are the consequences of adaptation activities with regard to the exteriors: their regeneration is almost always the result of evaluations made with regard to energy containment rather than a study dedicated to their layout: the choices relating to plant engineering issues affect the sloping roofs covered with solar panels, on terraces occupied by technical premises, on facades often overburdened with double fixtures on their frames, divided into glass mirrors that resume subdivisions of the old iron window holders that have been removed or in other cases adapted to available formats; the protection of the buildings against atmospheric agents also intervenes to disrupt the rational facades of the artefacts created for production purposes, or the functional partitions of holiday villas.
The frequent splits through the multiplication of accesses cause a proportional increase in the number of canopies protecting against rainwater: the rays of the gutters and the number of downspouts increase exponentially, in some cases the facades are covered with metal panels that ensure ventilation; the multiplication of flooring levels is also proportional to the increase in sunscreen systems (opaque glass or *breisoleil*) that were formerly not essential. The need to

eliminate architectural barriers is the reason for the internal and external ramps, and anti-panic handles are scattered here and there to ensure safety against fire. The personalization of the spaces is entrusted to furnishings and accessories chosen by the tenants once they settle in. The sad circumstances of such buildings and open spaces, deprived of a spatial idea, are improved by the addition of heterogeneous languages.
What sense do these themes have with respect to the reasons and times of the "re-cycle"? The "re-cycle" project as a whole is a "relief" (in any form) of tensions in progress, and at the same time it allows for the construction of possible scenarios oriented to an idea of territory in which the form (as an organization of ordered sequences) and position of its components do make a difference.
Critical materials are transformed by an "accumulation" of small transformations. These transformations, affecting the way in which artefacts are used, have repercussions on the spaces of transition (inside and outside; public/private; closed/open) and therefore reverberate on the urban structure, introducing new forms of spaces that are public - if not yet collective - that show the relevance of well-known themes of architecture, first of all the threshold.
This reasoning is not intended to bind the future of our areas to architectural projects, but to stimulate a reflection on the 'bark' of the elements and its thickness, or rather on the elements that form the transition.
The need for an open project rather than an invitation to disengagement from the designers has been grasped by some as the possibility of a new type of project that Jordi Bellmunt likes to call a 'blurred' project. "Not clear in its forms but precisely identified in the substance of the relationships it expresses". The definition opens the field to experiments that, aware of the urban role of the inter-zones, deal with them as a key to interpreting design possibilities.
The 'bark', even if with variable meshes, must assume the role of a founding element that dictates further design instances. It is indeed important to organize the elements around an idea of space that can definitely guarantee maximum flexibility of the indoor private spaces but, first and foremost what is needed is rigorous design of all the parts that guide their relationships.
If there is a possibility of understanding the value of such transition - not only of space (inside/outside) - but also of time (in relation to the different compatible uses, to the different times of reclamation, etc.), then perhaps the architectural project cannot be the result of "inputs" coming from specific needs but should rather show awareness of the relationships it triggers.

Drosscape Urbanism. Incremental projects, strategies and tactics

Anna Terracciano

Emanuela De Marco

"Drosscapes" and re-cycle. Experiences and references

The panorama of multiple and diverse urban waste materials that settle between the either dense or rarefied meshes of urban sprawl is the result of physiological growth. Generally speaking, it results from mutations in the metabolic dynamics of cities. A critical condition emerges, that makes it necessary to rethink the functioning of urban structure, both from a spatial, environmental and relational point of view. Most of all, however, there is a need to trigger new mechanisms of "recycling" and new "resilient" strategies for urban planning which have as their premise the recognition of the potential regenerative value of "drosscapes" as a new and renewable resource for contemporary city design.
As often happens, in the very structure of these "waste areas" there is a latent demand for recycling. It needs to be identified, interpreted and encouraged through a progressive strategy of re-appropriation and inclusion in the social, economic and ecological life cycles of the city, turning them from places of expectation to new territories of urban design within a new collective dimension.
Contemporary urbanism increasingly focuses on the need to intercept, among the immaterial components, precisely those dynamics of re-conquest and reuse. These are more or less spontaneous; and are able to involve the space affected by disposal and waste, all this by virtue of the relational potential that they express in the contexts into which they are inserted. Through this approach, the Strategic Framework Plan of Detroit[22] demonstrates how the drafting of a strategic plan - which starts from a condition of dissipation of urban components - can introduce a methodology of action capable of raising the fate of a city that has experienced a decisive decline due to the social and economic changes it has undergone during the twentieth century. Such plan can be implemented within gaps and residual spaces, using waste materials and available resources, as well as short, medium and long-term strategies. The Plan's priorities objectives are:

- to reconcile the need to rediscover a sustainable dimension with an innovative economic impulse;
- to reduce the very large social and economic gap;
- to prepare an urban structure capable of accommodating a suitable transformation of the territory, so that a new solid structure and identity may be traced;

22. Source: http://detroitfuturecity.com/framework/

- the systemic interaction between long-term structuring strategies and land use as medium/short term design tactics.

When applied to grey areas and abandoned spaces, this interaction contributes to the creation of an innovative, complex and ecologically-oriented framework within which the city may be rethought.

These reflections are made possible and desirable by the structures of these places in Detroit and, more generally, in contemporary territories. Such a structure is increasingly configured as a reticular system in which relations are made up of material and immaterial flows, as well as the different values of use and reuse which together contribute to determine the different phases of the materials' life cycles.

In fact, the differentiation of the "drosscapes" and the definition of their taxonomy (which, although it has a general character, is able to decline in the specificity of the contexts) is not only linked to their morphology and their peculiarities or to the spatial configuration of the contexts of reference. It is also connected to the different degree and type of relationships that they establish with the materials and communities with which they are inevitably involved.

The complexity of these dynamics, interactions and reciprocal influences necessarily lead to a rethinking of the traditional interpretive tools. It also contributes to the development of a new vision that allows a level of knowledge able to detect, above all, the potential value of the "drosscape".

Such recognition is already a first step within the project, in which disciplinary convergences - that are nowadays denser in contamination - contribute to trace new urban stories and future visions.

In this sense, the ways in which Paris[23] has been reread, as well as the new "diagnostic" tools put in place in the research project for Le Grand Paris by Bernardo Secchi and Paola Viganò, begin from the assumption that it is necessary to reinterpret the existing city in order to establish a new cognitive framework, shared and transferable, able to support the regeneration project.

In fact, reading and describing the territory as has been done in previous decades, according to pre-established schemes that are unsuitable for investigating a condition of reactivity and continuous change of the systemic and related aspects, makes it necessary to adopt a multidisciplinary approach, as well as new

In the following pages, images taken from "Detroit Future City" - Strategic Framework Plan.

23. In 2008, the Elysée invited ten multidisciplinary and international équipes (coordinated, among others, by Jean Nouvel, Christian de Portzamparc, Richard Rogers, MVRDV) to research the so-called "Grand Paris", an area of 12,000 square kilometers and about 12 million inhabitants; each group had 9 months to define the future of the Paris after Kyoto. The team coordinated by Secchi and Viganò presented the project "Ville poreuse", a proposal that attempts to respond to the problems - common to all world metropolises in the coming years - related to the environment, mobility and social inequality. Source: http://www.grand-paris.jll.fr/en/grand-paris-project/

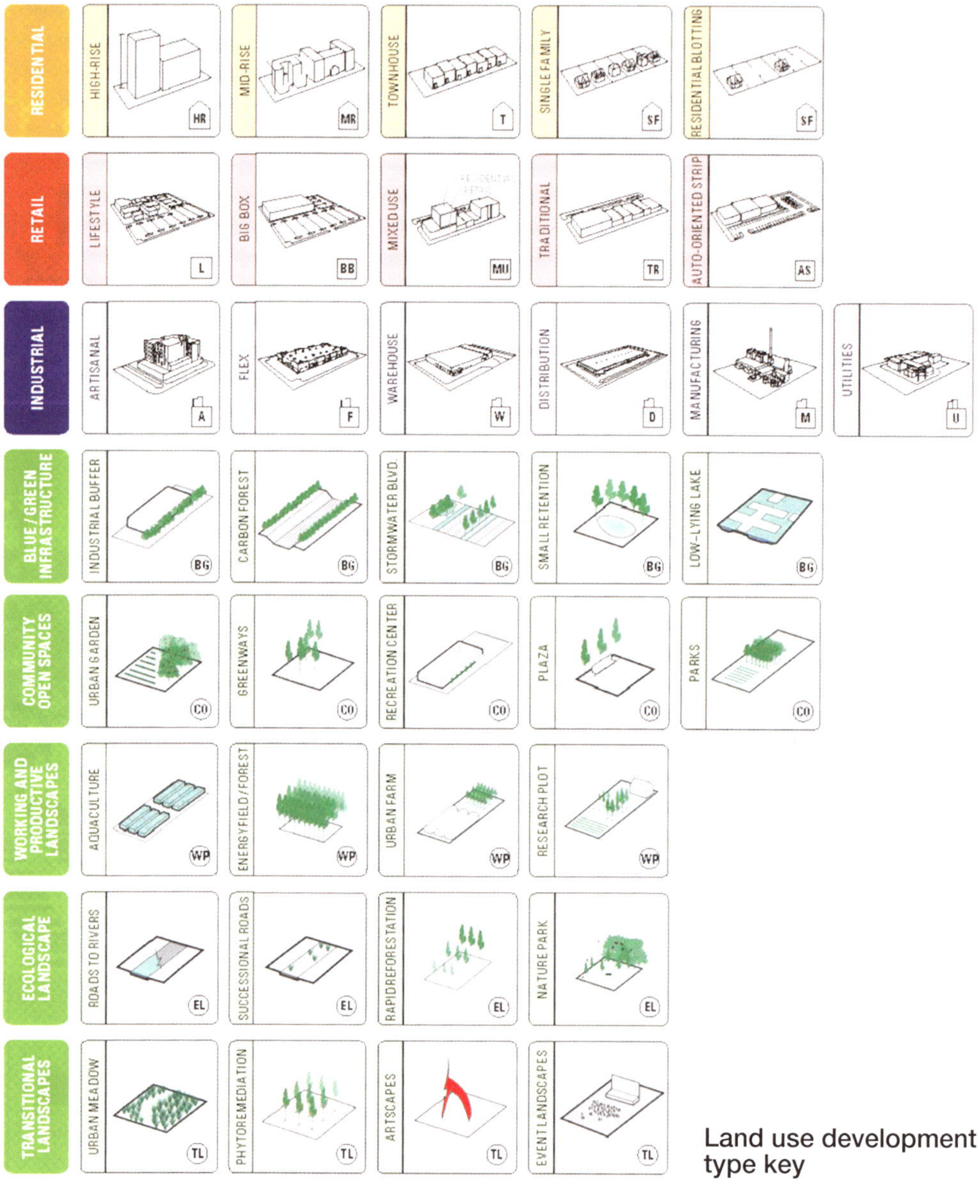

Land use development type key

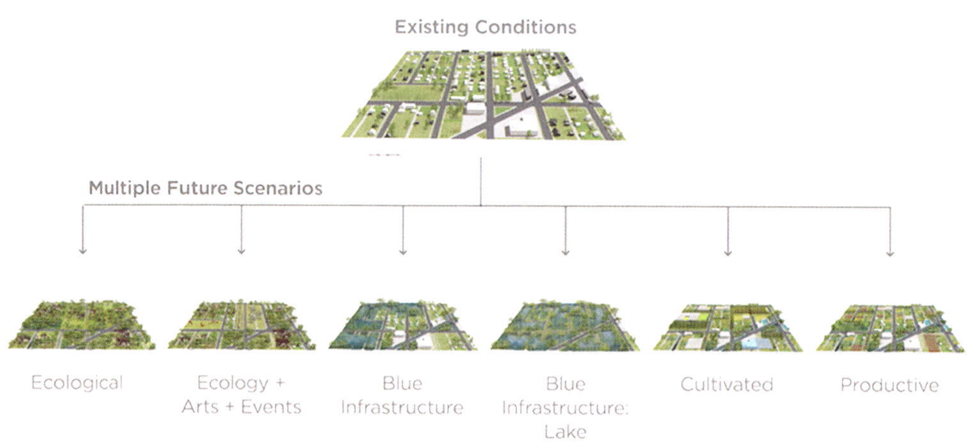

156

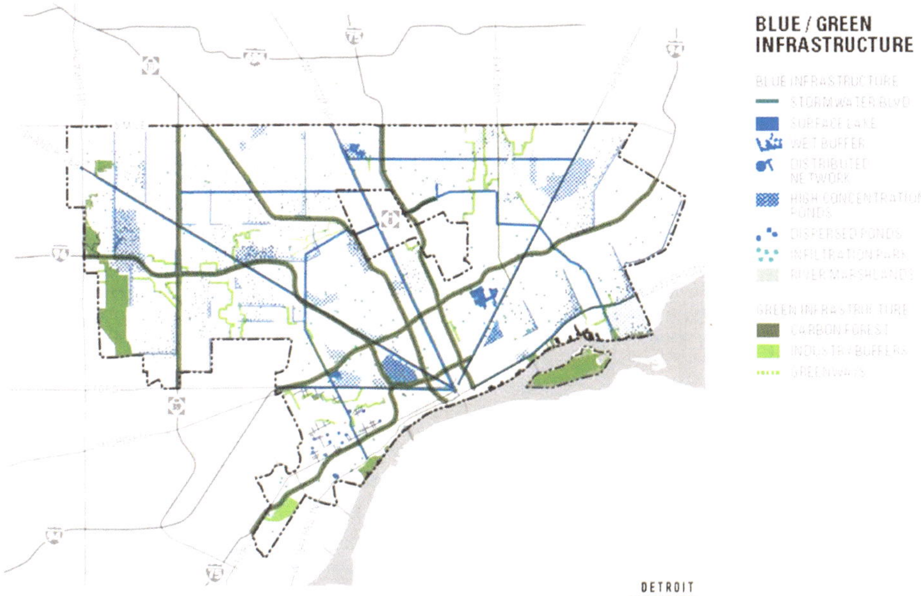

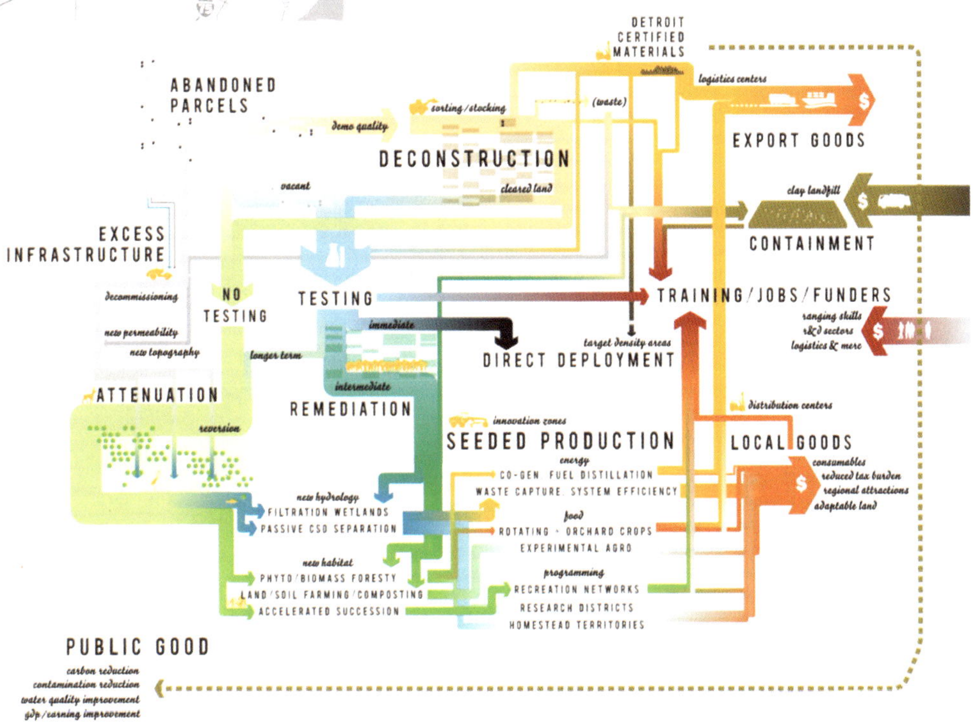

NEIGHBORHOODS: MANY TYPES OF DISTRICTS

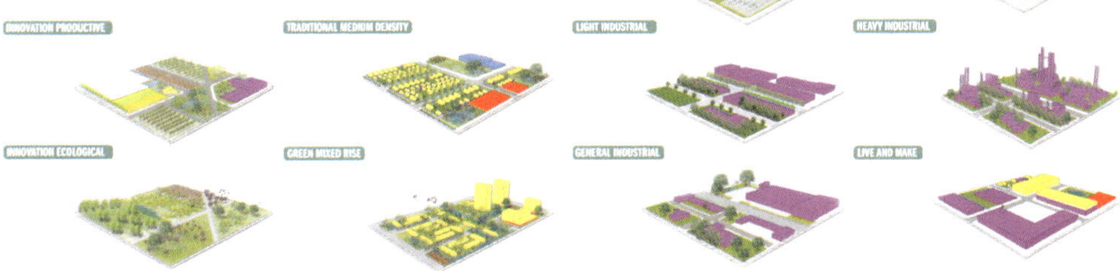

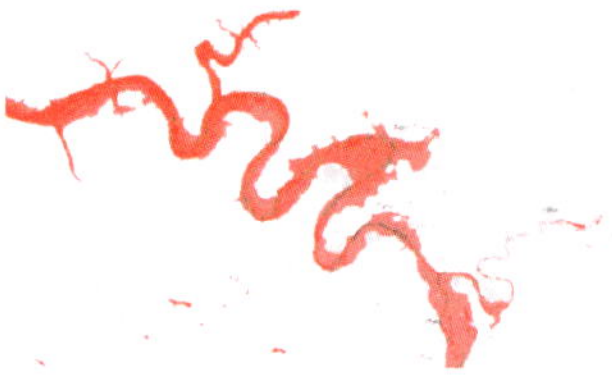

The river Seine and its flooding areas

Map of the drosscapes in Paris

To live with water: risks and biodiversity in a strategic project.

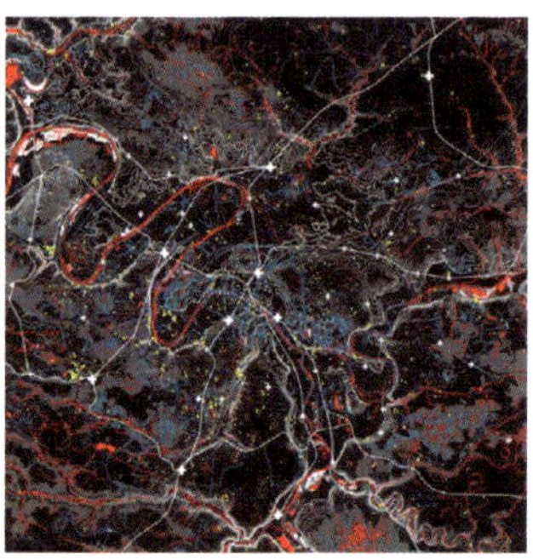

A vision of the project for a "porous city"

The images above are from "Equipe Studio 08-09" by Secchi Viganò, International Consultation for Le Grand Paris, 2009.

techniques and forms of representation. Since recycling is no longer intended as an isolated gesture, the landscape dimension plays a fundamental role. Recycling is now intended as a new process, capable of building new physical and relational networks, whilst regenerating existing ones, in order to connect different interventions within the multiple city fragments.

The London Thames Gateway[24], project is one of the largest urban regeneration and landscape redevelopment projects in Europe in recent decades, and forms part of this perspective. It aims at rebalancing the development and growth of the London metropolitan area. The project covers most of the flooding areas of the River Thames, extends for about 70 km east of London and includes the largest "brownfield" in the south of England. This "brownfield" is destined to become a significant example of "eco-region", where recycling is an integral component of a larger project of regeneration, based on the differential reactivation of areas. Intervention of this type can also create new economic activity, new uses and new supply chains.

The stratigraphic-relational dimension of the "drosscapes" recycling project

The complexity of these "waste areas" requires not only a change in perspective and in the ways in which knowledge is acquired but, also and above all, the construction of new plans and projects capable of dealing more effectively with the issues that are being raised. The above mentioned, as well as many others that are beginning to populate the international scene, are a useful reference to the design explorations developed during the time of the Re-cycle research.

Dross urbanism is therefore a paradigm shift that proposes, as we will see in the next chapter, a new vision of the territory. Its goal is to create incremental scenarios, starting from the recycling of "drosscapes" and the consequent effects on the geographical contexts in which they are inserted. This becomes an integral part of planning, precisely because they identify the most important priorities and strategic themes to start a comprehensive regeneration of the city. Also in Paris, the Secchi-Viganò[25] group proposes a future vision of a "ville poreuse" (or "horizontal metropolis") that attempts to address the problems - common to all world metropoli in the coming years - related to environment, mobility and social inequality. Such a vision is implemented through five strategies related to circulation within the great territorial "sponge":

1) the "monuments" as a project on significant places;
2) the water strategy;
3) the green and agricultural spaces as sociodiversity and biodiversity;
4) the recycling and energy efficiency of the existing building stock;
5) the policy of mobility and accessibility.

24. Source: https://www.gov.uk/government/organisations/london-thames-gateway-development-corporation. http://www.bbc.co.uk/kent/thamesgateway/maps_facts/thames_gateway_facts_whos_who_feature.shtml

Nevertheless, it is in the construction of incremental scenarios that "time" becomes the very material of the project, changing the geographical perspective with which the urban space is designed.
In Vincent Guallart's[25] City Protocol, together with the U.P.C. in Barcelona and the MIT in Boston, a "programme" has been set up that is no longer a spatial project in which what is imagined is a city made up of neighbourhoods with a human speed pace within a "hyper-connected", zero-emission metropolis. The time variable intervenes and modifies the design of urban space, thus becoming a "community asset" in various ways, in which the management of resources (soil, water, waste, etc.) and urban materials (isolated areas, roads, open and green spaces, etc.) interact organically with the relational components.
In this way, energy and digital networks contribute, together with environmental and infrastructure networks, to building the backbone of the city of the future. The idea that the future is not built through final or all-inclusive projects or images, but through the organization, on different levels, of a multiplicity of images and stories linked to this load-bearing structure and its networks, is becoming more and more evident.
This is also proposed in the already mentioned Detroit Strategic Framework Plan, in which the "blue" and "green" infrastructures define the supporting structure of the city's regeneration. The recovery of a necessary large-scale vision is thus marked by structuring and strategic priority actions, promoted mainly by public actors, which can be financed within a given time- They are oriented towards long-term objectives, in which the needs to secure and regenerate the territory converge. They also rely on a multiplicity of punctual and widespread actions and micro-actions, promoted by local participants, able to graft themselves within the framework of structuring and strategic choices in order to reconstitute the continuity of the "new public city" and of the "collective living", seen as the armature of the city itself.
What this research explores and proposes is also the overcoming of the sectorial meaning of reclamation in favour of a recycling project. This project can be considered as a dynamic process of design actions over time[26] as a result of the multiscalar interaction between "drosscape" and life cycles, within a "timeline strategy". This is a hyper cycle, that is, a differential reactivation of the various life cycles of the "drosscape" within three territorial frames/infrastructures, both spatial and temporal: the "blue", the "green" and the "innovative use network".
The experience of the recycling project is gained in the territorial areas of research. At the same time, it is stratigraphic/relational, since it intercepts the three dimensions of the exhaustion of soil/water life cycles. This is due to the effect of the exhaustion processes of industrial, agricultural, commercial, housing, extractive and metabolic cycles and can be placed

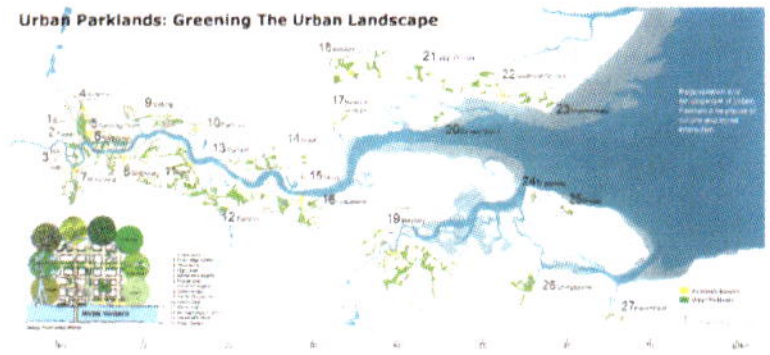

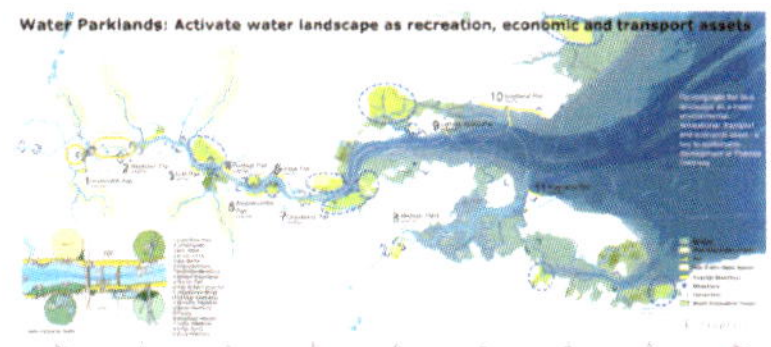

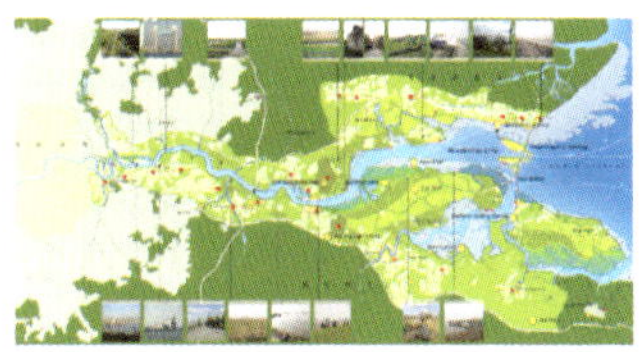

In the images above, The Thames Gateway, London, UK

25. Cf. video https://www.youtube.com/watch?v=oivMuQGZf30
26. J. Corner, "Recovering Landscape. Essays in Contemporary Landscape Theory", Princeton Architectural Press, New York, 1999

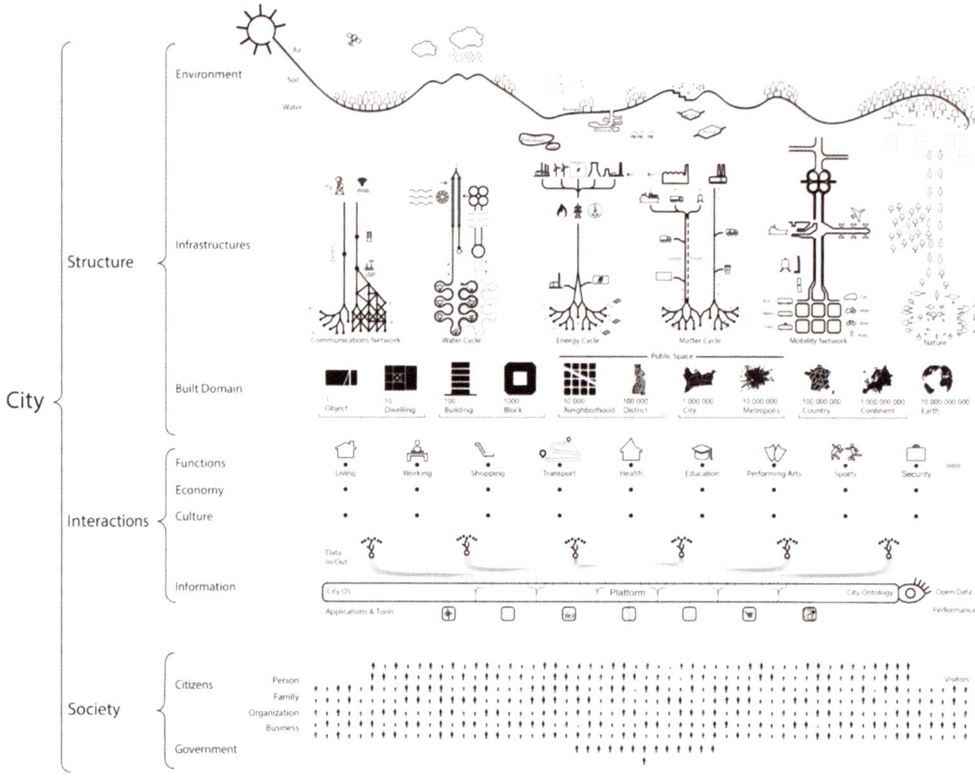

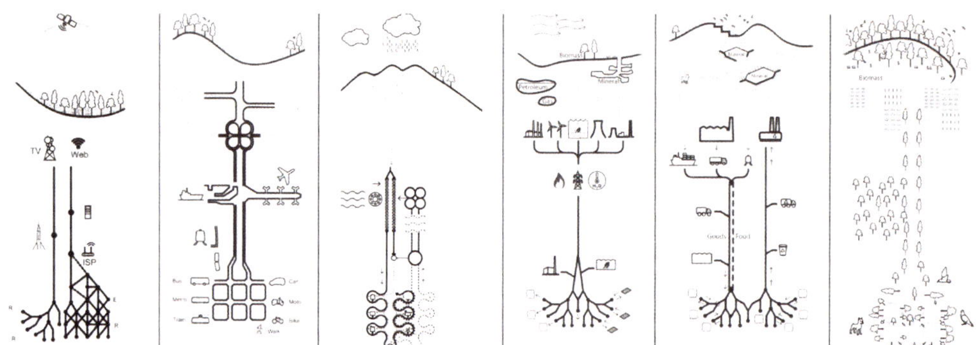

Network infrastructures | Cycle infrastructures

Class	Property	Value Restriction
Infrastructure	*owl:SubClassOf*	Structure_layer_component
	enable	***some*** Flow
	transports	***some*** TransportableThing
Network_infrastructure	*owl:SubClassOf*	Infrastructure
	isNodeOf	**some** Network_infrastructure

City Protocol – Anatomy of City Habitat
by Vincent Guallart

within the relational dimension of the dynamics of use and behaviour of the city; but the recycling project is also resilient/ adaptive, because it contextualizes and obliges the landscape and social practices to interact within a vision that recognizes which parts of the territory are presented as elements of "resistance". Instead, other elements can be susceptible to an evolutionary and "adaptive" model, which is able, over time, to return landscapes and scenarios that can link the design action to the methods and times of remediation.
Its double strategic/tactical level translates, in the Domitio-Flegrea town, into a "Land remediation network" capable of reinterpreting the historical machine of reclamation within a territorial infrastructure that innervates and reconciles the two main territorial systems, the coastal and the plain.
However, this system is also capable of intercepting the widespread regeneration of critical tissues and landfills, as well as the remediation of soils and water. Similarly, in the Eastern town where the water system is less clear and more compromised, a "hybrid machine" is oriented to the rethinking of the hydrographic network's management system and to the reclamation of surface and deep waters. It begins from a reworking of the design of the pre-unitary reclamation network and from the existing conditions of pollution and depletion of water and soil resources. It envisages independent scenarios, able to combine the control and reclamation devices of the aquifer with the widespread transformation according to the hydraulic resilience. These actions intercept the recycling of dismissed oil deposits for ecological uses, as well as the recycling of soils with ecological "*preverdissement*" along the plots and "patches" of abandoned and immediately usable agricultural soils.
The Sarnese town was historically based on a relationship of care and domestication between soil, urbanized structures and hydrographic network that is now greatly compromised. Therefore, a "Sustainable water management" works with the "green" and "blue" infrastructure to rebuild a virtuous cycle of surface and deep water and at the same time ensures the remediation. The recycling of the hydraulic devices in the system that regulates water from the Vesuvius, such as the Pianillo and Fornillo tanks, and also the recycling of the Conte Sarno Canal for sustainable water management, forms part of the new projects. New urban and peri-urban landscapes, practices and uses originate from the expression of the different forms of coexistence between decontamination, environmental transformation, and social re-appropriation. Their presence is progressively giving substance to the strategic dimension of a more comprehensive project for open public spaces, oftentimes very innovative, and capable of satisfying the current, largely unmet, needs.

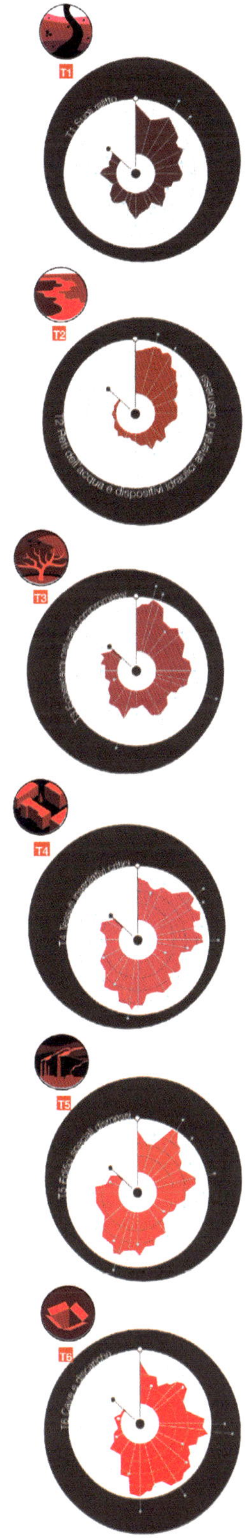

Drosscapes in the 3 areas studied by the research group in Naples: diagrams of lifecycles.

A Grammar of actions/relations

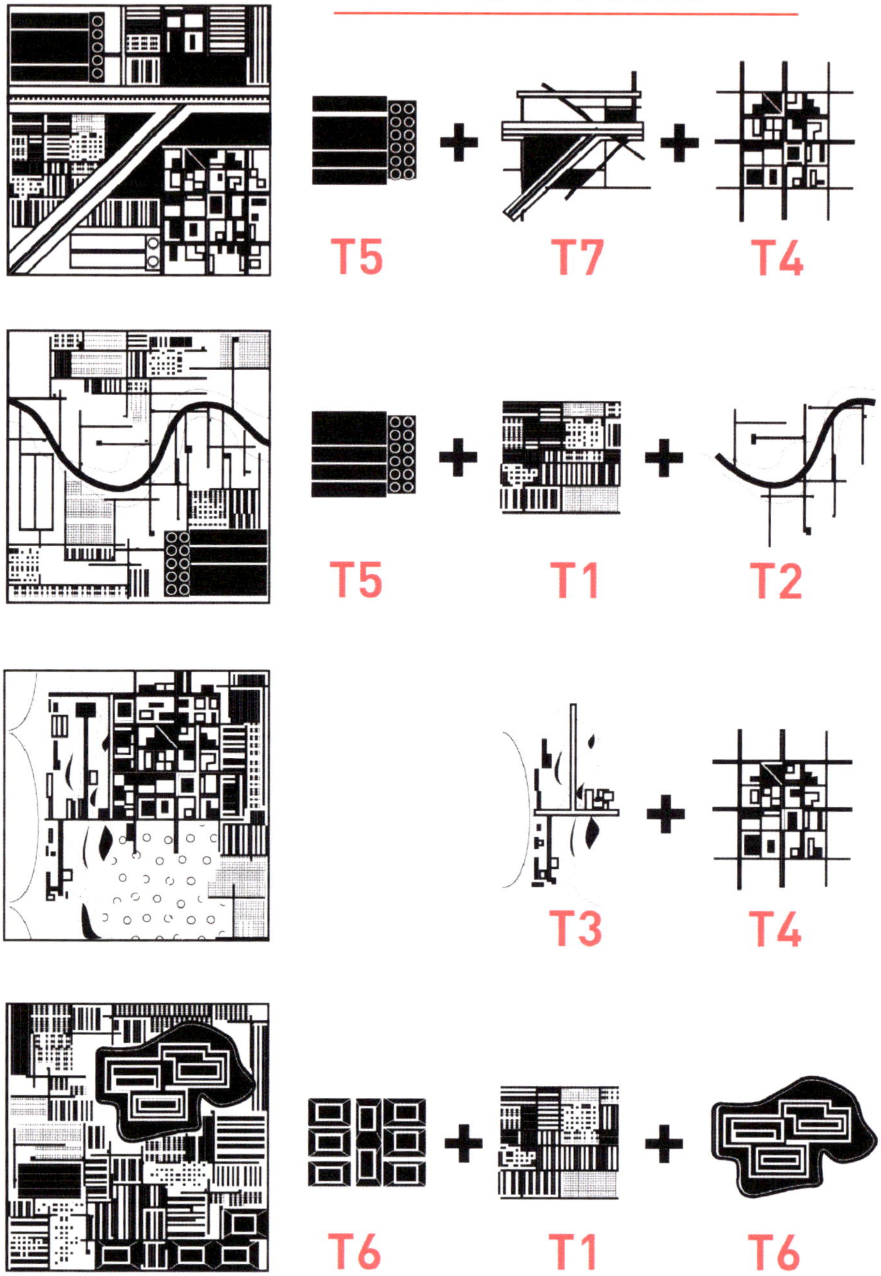

Above: Relational potentials and aggregative possibilities in the drosscapes studied by the group in Naples.

In the centre, grammar of actions/relations and recycling diagrams of the drosscapes.

Recycling of territorial contexts

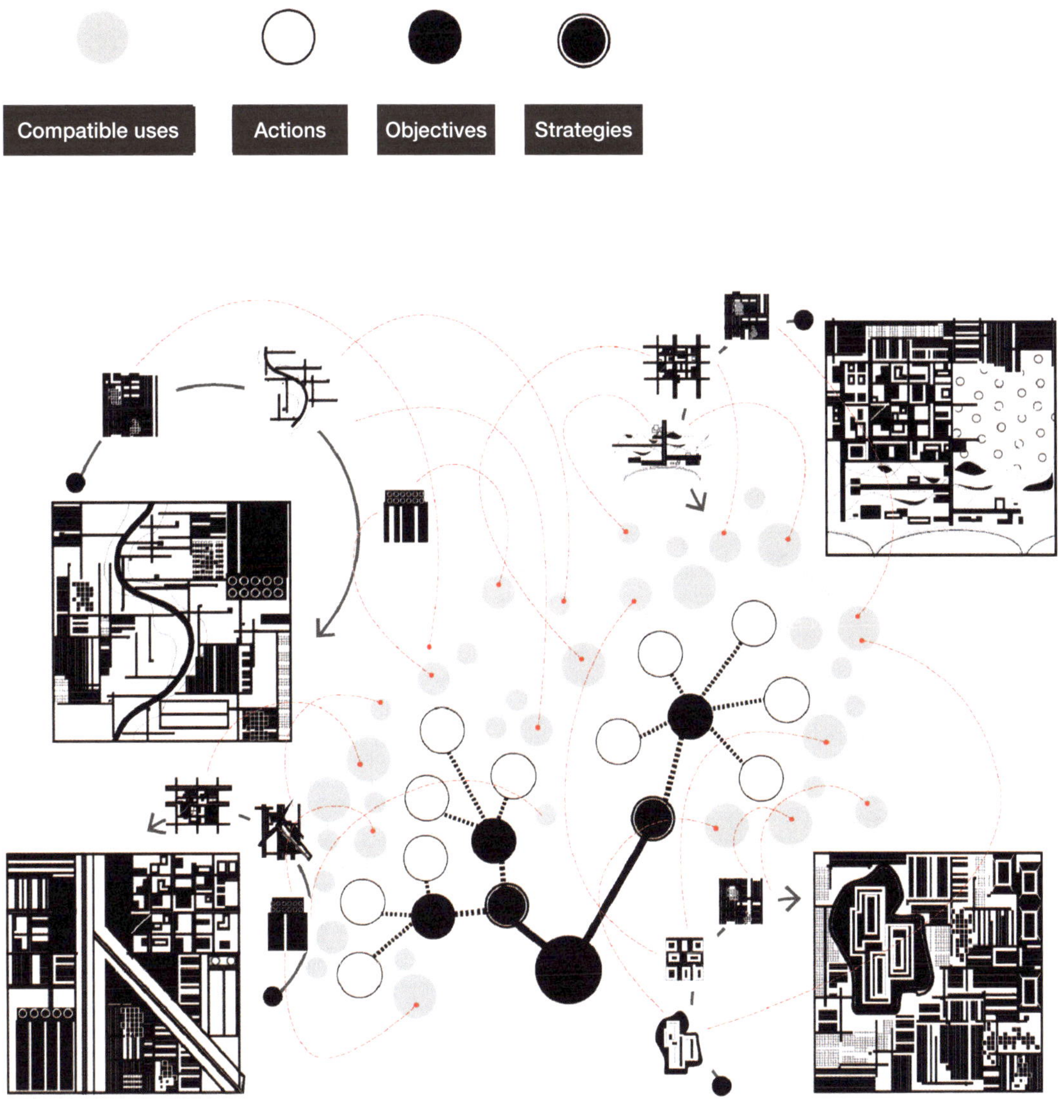

Hypercycle

Differential depletion
of life cycles

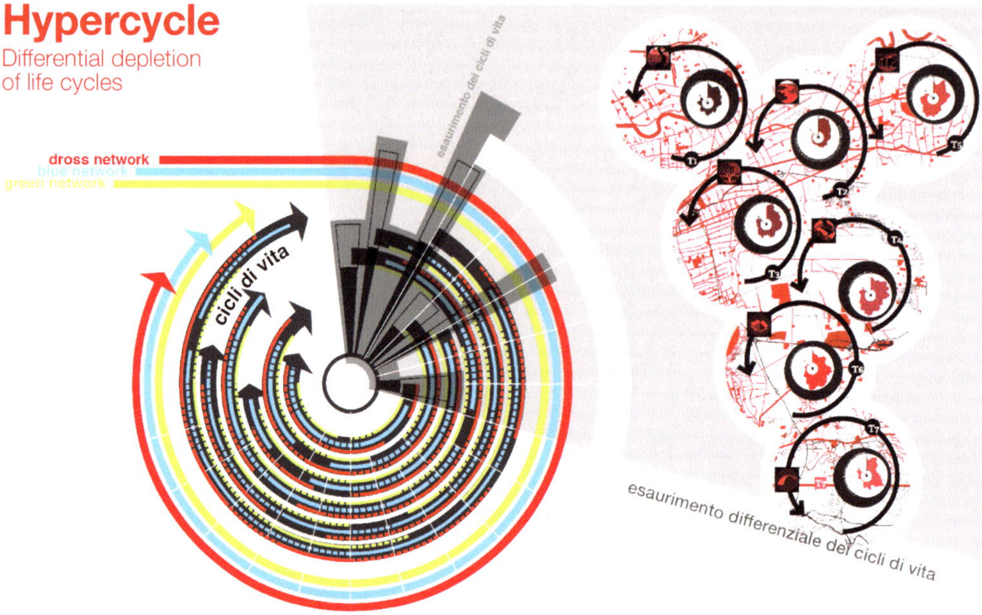

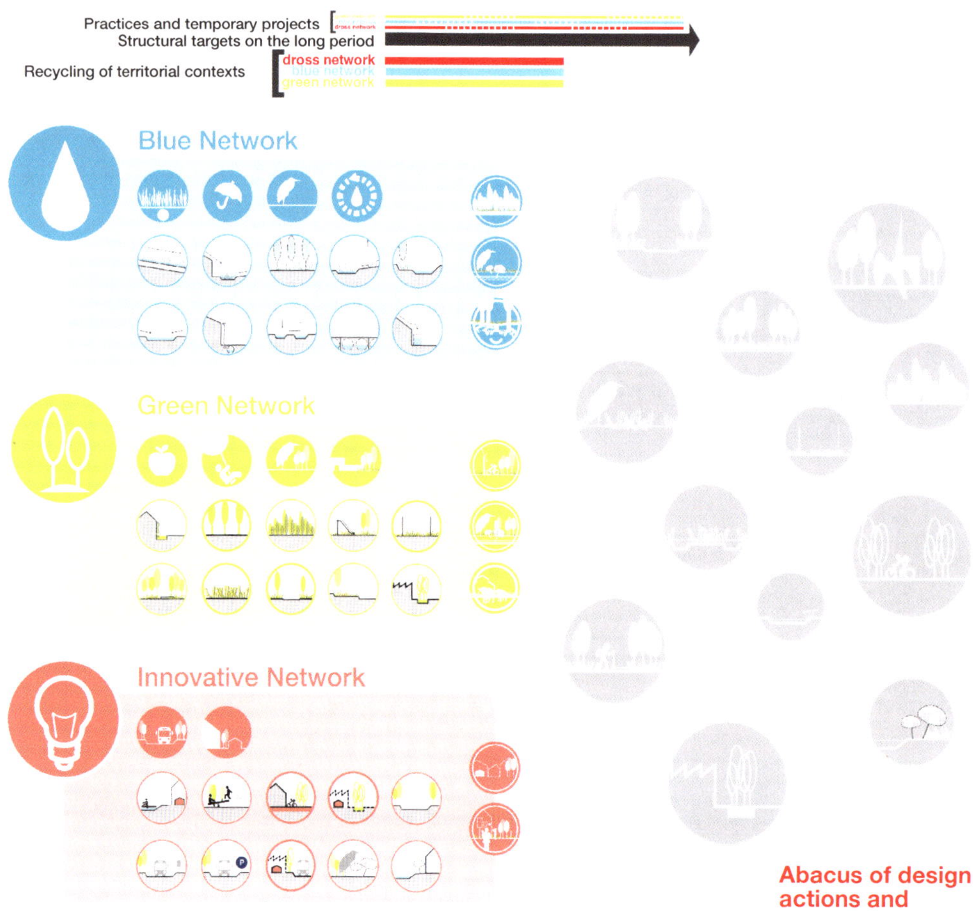

Abacus of design actions and compatible uses

Resilient/adaptive re-cycling

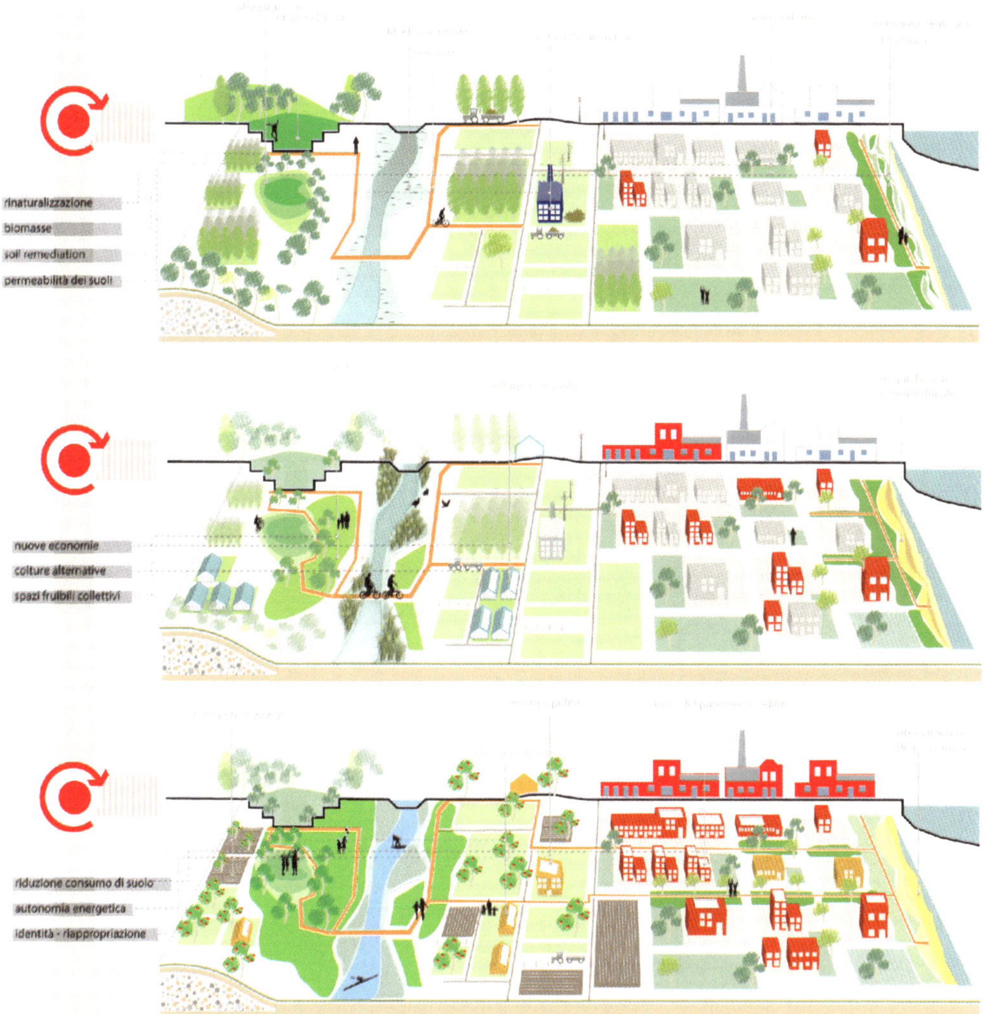

Stratigraphic/relational re-cycling

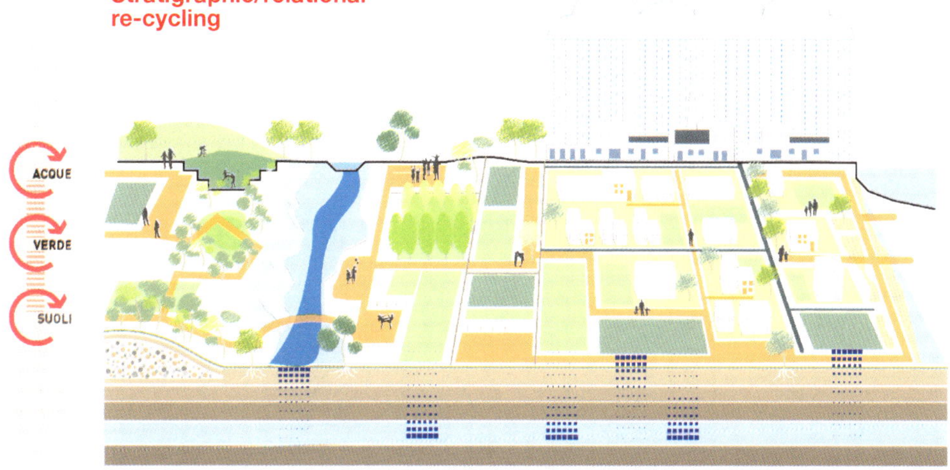

Part 4

Recycling projects, landscape projects, City planning

"In the city of Naples, research has explored the conditions and project trajectories especially in terms of multi-scale *hypercycle*, aiming to a different urban metabolism. These projects create quality landscape spaces, define virtuous combinations of practices, empathize with the construction of new infrastructure networks (slow mobility, micro-logistics, digital networks and power grids), and aspire to stimulate the *start-up* of new business activities in tune with the demand driven by the various waste industry supply chains." This approach is therefore a creative process built on a variable geometry method and borrows, in the dynamic movement between scales, several tools and practices typical of the landscape disciplines. It also introduces an idea of project that is at the same time *stratigraphic/relational* in space and *resilient/adaptable* over time. The stratigraphic interactions moving in the three-dimensions soil/subsoil/topsoil are in fact tightly bound to the relational interactions related to the "inverse city's" behavioural dynamics and practices (...) inspiring inductive and experience-based narrations that contribute to put into play new convincing life-cycles."

Carlo Gasparrini

Recycling of territorial frameworks

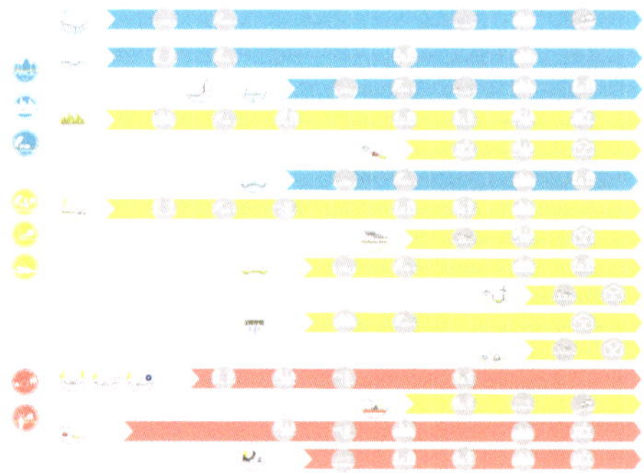

**Città
Domitio-Flegrea**

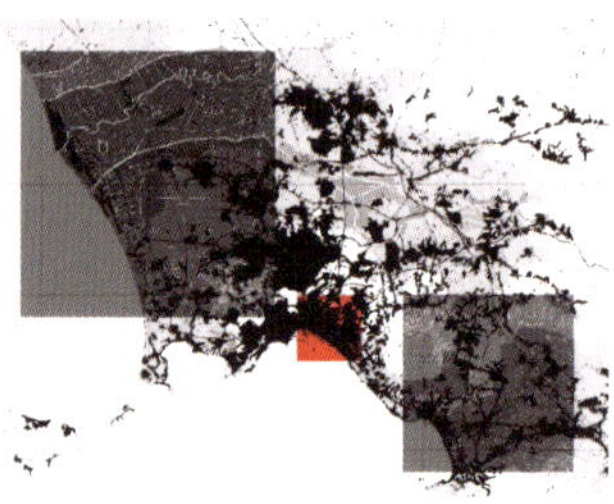

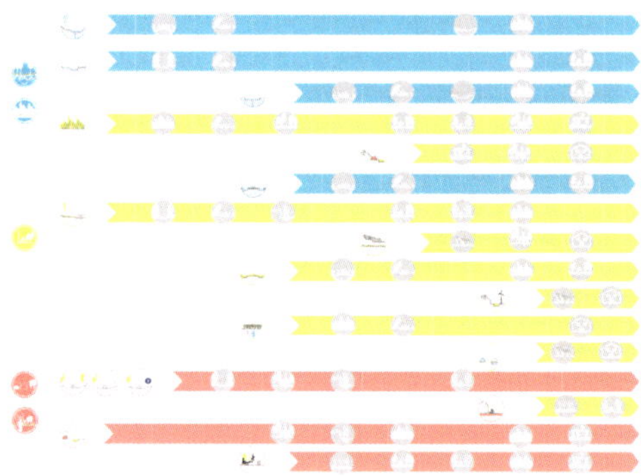

**Città
Orientale**

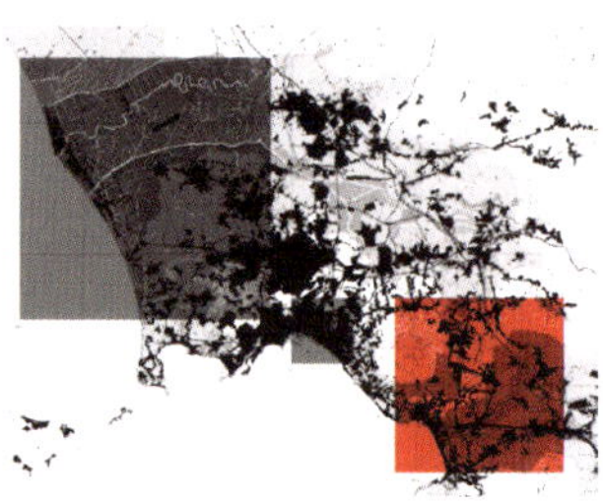

**Città
Sarnese**

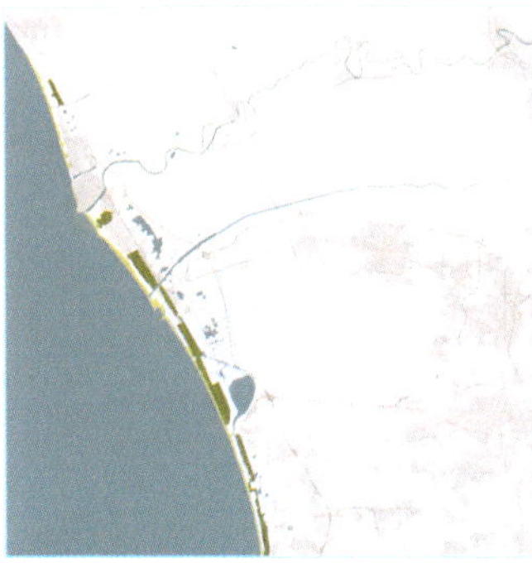

Waterfront park

Large park

New centrality park

Reclamation devices park

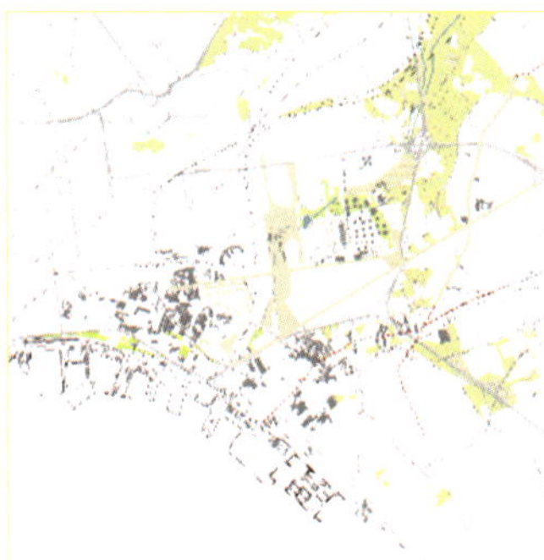

Hybrid park

Used city park

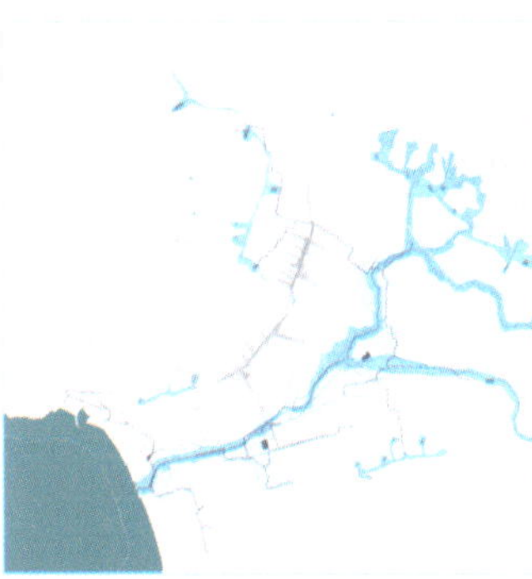

Pervasive park

Reticular park

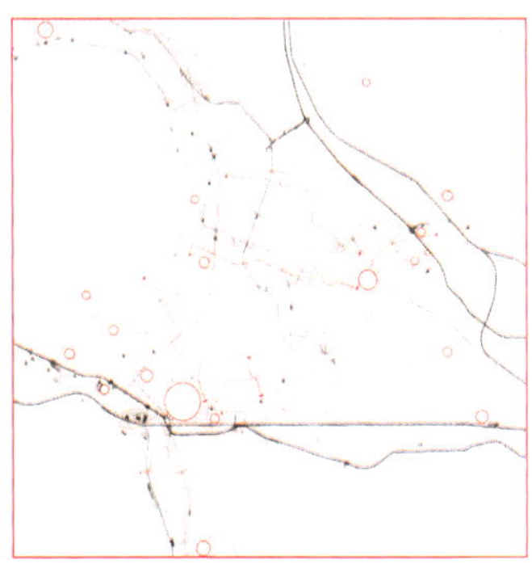

Archaeological park

LD/ LITORALE DOMITIO-FLEGREO

64 Comitati

- COMMISSARIO STRAORD. ALLE BONIFICHE
- ARPAC e APAT
- ASSESSORATO REGIONALE ALLA SANITÁ
- CORPO FORESTALE DI NAPOLI
- LEGAMBIENTE AFRAGOLA,CASAPESENNA
- WWF AGRO AVERSANO e WWF CAMPANIA
- IMPRENDITORI
- AGRICOLTORI

istituzioni e operatori di settore

82 inchieste

915 ordinanze di custodia cautelare

1.806 persone denunciate

443 aziende coinvolte

9.709.511 tonnellate di rifiuti smaltiti illegalmente

Traffici

11.02.1994	2001	2007	31.12.2009	2010
Stato d'emergenza rifiuti	Nuova emergenza	Picco emergenza	Fine stato d' emergenza rifiuti	Nuova crisi rifiuti

Emergenze

"Veleni, ogni giorno inquinati cinque centimetri di falda Il geologo Balestri: mai vista una situazione peggioredi quella Giugliano"

Corriere del Mezzogiorno, Napoi/Cronaca

le rotte dei traffici dei rifiuti in Campania
Legambiente [2013]

"L'emergenza nell'area tra le province di Caserta e Napoli dove l'inquinamento, i roghi e le morti collegate stanno dando vita a una vasta e profonda protesta popolare"

La Repubblica Napoli

The Domitio-Flegreo City

Anna Terracciano

Metropolitan archipelago

The great plain south of the Volturno river - a flat area[1] mainly composed of alluvial lands and coastal alluvial plains, innervated by a vast system of streams and drainage canals throughout the ancient wetlands until reaching the sea - has always been dedicated to agriculture, thanks to the fertility of its soil. Today, in the media, it is a symbol of abandonment and degradation.

This area, which had always been marshy, was reclaimed in the modern age. It has thus become a centre of agricultural and dairy products of the highest quality, especially thanks to the absolute uniqueness of this land - among the most fertiles of the planet - as documented by geochemical surveys and by the area's history.

Changes in the economic structures, which has usually coincided with a dynamic decommissioning of areas, buildings and infrastructure, here has also met the crisis of an entire real estate sector, that of tourists and accommodation facilities on the Domitio-Flegreo coast. Moreover, it has left us with a legacy of degraded landscapes, as well as urban areas that are socially and economically disadvantaged.

The settlements are organized into "filaments", "chessboards", and "villages" among which a slight and widespread "sprawl" infiltrates, inexorably depleting the areas; among these, the historical cities are still recognizable as dense and compact residential settlements, outposts of a past that resists through time. Nevertheless, in this particular case as well, the contemporary territory is not made by houses only. Garages, production and trade sites, greenhouses, quarries and dumps leave signs and marks of new morphologies, producing cuts and alterations in the structure and spatial continuity of places.

The urban explosion, while consuming space and soil, builds generic and equivalent landscapes all around the world. It homologates the area because it reduces the territory to an elementary grammar of "enclaves", one close to the other,

1. Historically defined as "Campania Felix" by Pliny the Elder in reference to its extraordinary fertility (see. A. Giordano, M. Natale, A. Caprio, "Terra di lavoro", Naples, Guida Editore, 2003), in the Middle Ages it took the name of "Terra Laboris" (see A. Gentile, "Da Leboriae (Terrae) a Terra di Lavoro, riflessi linguistici di storia, cultura e civiltà in Campania", in "Archivio storico di Terra di Lavoro", VI volume, 1978-79).

and this condition reflects a society in which individualism neglects collective space and fragments very different territories while making them all appear the same[2]. Among these, some "figures" are capable of capturing the dimensions and forms shaping up the the city today, evoking such condition through the design of its essential lines.
In the Domitian City, what is thus represented is a series of self-organized objects, juxtaposing each other in a vast territory deprived of any vision or perspective. The open space seems diluted and deeply unconcerned to what happens. An "archipelago" of materials, put together and mute, in which the large rural areas and consolidated urban neighbourhoods fade into an endless and pervasive dispersion of settlements[3], crossed by a multiplicity of overlapping reticular systems. This image can only be created by widening the horizon of our observation within a necessarily metropolitan dimension, in which the forms that these structures assume on the territory go beyond the administrative limits.
This area as we see it today - wedged like an enormous hinge between the provinces of Naples and Caserta - is the result of a long process of transformation and accumulation which has mainly occurred in the last fifty years, due to a rapid and widespread development of the territory, yet lacking a clear political and planning vision. The growth of new inequalities as well as multiple localisms in a setting of new identities, lifestyles, economies[4] and other dynamics, are not only in contrast with the growth - and the metropolitan dimension - of the city, but result also into different and contrasting practices of production, use and appropriation of spaces and places, such as the marginal and waste areas. The spatial consequences of previous economic crises are still clearly visible, and - here more than elsewhere - are also met by the waste cycle and land consumption crisis, leading to a deep change in the structures and metabolism of these places. It is here, in fact, that lies what is sadly known in the media as "La Terra dei Fuochi " (The Land of Fires),

2. S. Boeri, "L'Anticittà". Rome-Bari: Editori Laterza, 2010.

3. See, for example, the research (2008) carried out by Lorenzo Bellicini for CRESME, entitled "Le città italiane negli anni 2000. Passato recente e scenari di cambiamento" (Italian Cities in the Years 2000. Recent past and scenarios for change), commissioned by the General Commission for Landscape Quality and Protection, Architecture and Contemporary Art (PARC) of the Ministry for Cultural Heritage and Activities.

4. See the studies published in G. Amendola, "Le forme urbane della paura", "Urbanistica", 104. Rome: INU Edizioni, 1995; ID. (ed.), "Scenari della città nel futuro prossimo venturo", Rome-Bari: Laterza, 2000; ID, "La città postmoderna" Bari, Laterza, 2003.

Research surface area: 802,19 sq. km.

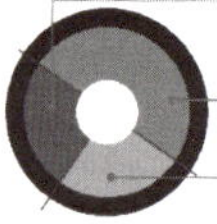

36 Municipalities

Inhabitants: 823.012

Density: 1.025,96 inh./sq.km

Critical situations

isolation

lack of accessibility

mono-functionality

scarcity of facilities

Figures of dispersion

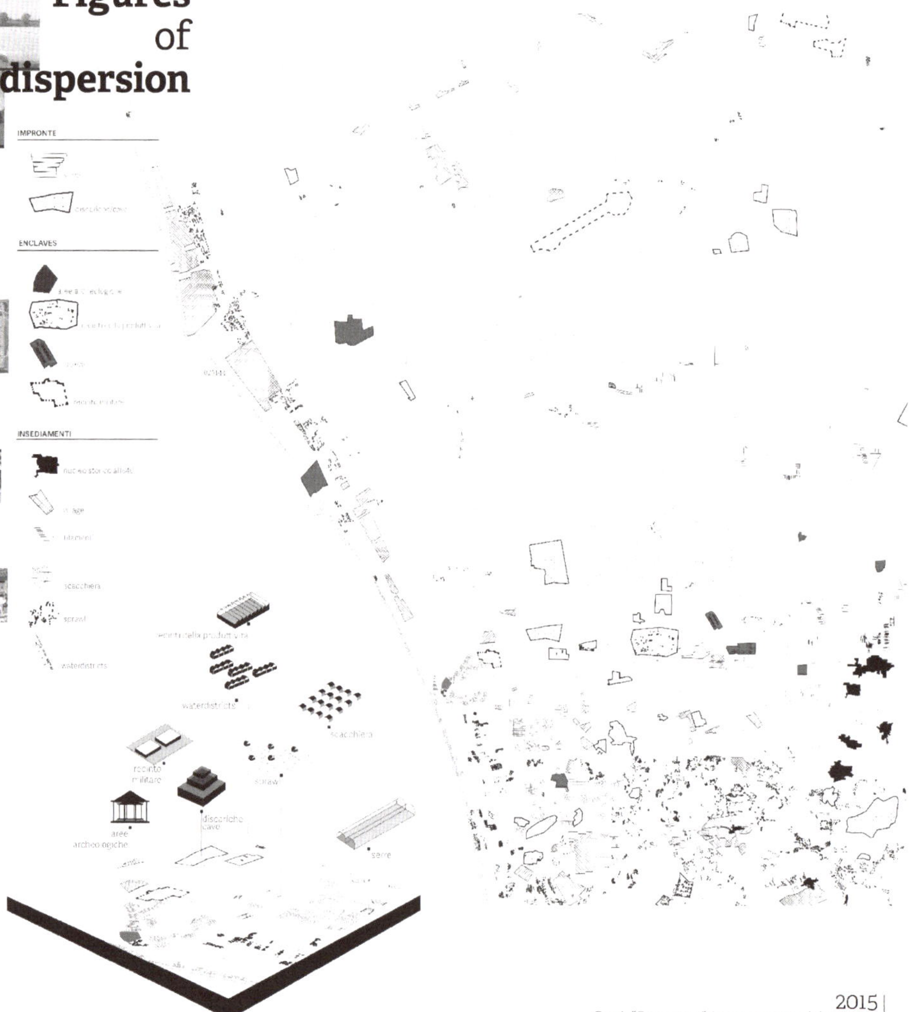

1990

1992
arresto Carmine Schiavone

1993
Primo Piano regionale di gestione dei rifiuti al fine di ridurre l'uso di discariche in Campania del 50%.

1994
Governo nazionale italiano dichiara prima emergenza in Campania. prefetto di Napoli 1'Commissario straordinario per l'emergenza rifiuti'.

2000
Costruzione 7 impianti CDR e inceneritori oltre 270 milioni di € finanziati dai fondi dell'UE,ma senza valutazioni di impatto

2001
"Emergenze nell' emergenza" siti di stoccaggio di CDR saturi, percentuali di arsenico oltre i limiti di legge, il Comune napoletano esporta due treni al giorno di 500/600t al giorno verso il Nord Reno-Westfalia al costo di 200.000 € al giorno.

2004
proteste violente contro la costruzione del termovalorizzatore di Acerra. Aumenti decessi triangolo della morte

2006
Scandolo ecoballe, inchesta Robero Saviano

2009
Acerra inizia afunzionare

2010
Corte europea condanna l'italia per una mancata gestione dei rifiuti in Campania

2011
Barriere di prottezione intorno la base americana militare a Gricignano di Aversa:rischi enormiper la salute

2014
progetto finanziato dall'UE Ecoremed inizia le prima prove in sito con tecniche di fitodepurazione.

2015
Oggi, "Cammora" è una società di business internazionale che lavora su scala globale. I tempi di uomini ben vestiti con revolver sono finiti, oggi lavorano con il computer organizzando milioni di transazioni i nternazionali. Sempre alla ricerca di modi semplici per guadagnare un sacco di soldi, commercio di armi, droga, prostituzione, eco-criminalità, lavori di costruzione, e affari immobiliari sono solo alcuni dei loro mercati attuali.

a vast area[5] in which tons of waste, mostly toxic, have been discharged for years. The name comes from the fires that, almost on a daily basis, are set among the heaps scattered throughout the countryside and towns, precisely for the lack of a legal disposal. For years, due to the waste cycle crisis, the state of emergency was declared, starting (in the Campania region) on 11th February 1994 and until 31st December 2009, making it through ten different special Commissioners.[6]

Reticular systems and compromised landscapes

The vastness of this territory includes a multiplicity of landscapes extending from the coastal strip to the urban centres of the hinterland, therefore crossing the huge agricultural plain, highly fragmented today: landscapes are profoundly transformed by those processes which need to be observed as a whole for the reciprocal consequences they have triggered on this territory and its water network and for the latter's proximity to the road network, the agricultural areas and the urban fabric.

In particular, the coastline strip draws a landscape with variable depths - a "waterscape" - which encompasses the area stretching from lake Patria to the river Volturno in which, from the 1950s and 1960s, following the definitive coastal reclamation, the delicate dune and back dune ecosystem has been consumed and compromised by the effect of a settlement process made of patchy areas and "islands" (purifiers, industrial enclosures, archaeological sites, tourist-residential clusters, etc.) which has provoked the erosion of the beaches[7].

Between the 70s and 80s, however, these conditions deteriorated further due to the ground water's excessive drainage which caused the salt wedge to contaminate coastal soils, compromising its quality. At the same time, the inadequate urban wastewater treatment system has compromised the quality of coastal marine waters, which have also been polluted as a result of illegal dumpings.

A problematic background[8] which is however also given by the still visible outcome of an extraordinary history of water domestication. This includes the construction of powerful hydraulic devices such as the Regi Lagni[9] which at the beginning of the 20th century - with the reclamation, defined as "a conscious act of landscape transformation aimed to a defined economic development plan"[10], redesigned the area through extensive cultivations and tree-lined canals, farms and eucalyptus windbreaks, within a dense network of drainage canals.

Landscapes like these, waterscapes in general, can play an essential role in the territory's governance and in the sustainable management of resources thanks to their particular ability to rethink, in terms of "resilience", both the effects induced by climate change and the historical relationship between cities and water, and the urban landscapes connected to it.

It is however during the years of the economic boom that a progressive aggression of the coastal strip begun due to

5. Included among the municipalities of Qualiano, Villaricca and Giugliano and 25 km far from Naples.

6. The state of emergency for the waste crisis in Campania was declared on 11/02/1994 and lasted until 31/12/2009, under the supervision of ten special Commissioners. In 1997, the Regional Plan for the disposal of solid urban waste was drawn up, which provided for the construction of 2 incinerators and 7 plants for the production of fuel derived from waste (RDF, or Refuse-Derived Fuel). However, the Plan was only partially implemented, causing the emergency to explode again in 2001, and 2007. On 31/12/2009 the state of emergency was declared to be over and the waste powers were returned to the competent local authorities. But, in 2010, the waste crisis re-emerged. (Source: Laboratoriocampano, 2012).

7. C. Gasparrini, "Waste, Drosscape and Project in the Reverse City", R. Pavia, R. Secchi, C. Gasparrini (ed.), "The territory of waste and refuse", Re-cycle Italy n.8, Aracne, Rome, 2014.

8. These areas are part of the S.I.N. (Site of National Interest) "Litorale Domizio-Flegreo ed Agro-aversano" which has an area of over 1500 square kilometres, established by Law 426/1998. Its perimeter has been established according to the following acts: D.M. 10/01/2000; D.M. 8/03/2001; D.M. 31/01/2006. Since this is a very extensive S.I.N., the sub-experimentation of individual polluted sites is being continuously updated.

9. The Regi Lagni, a system of canals started by the Spanish viceroy Pedro Fernández de Castro and completed between 1592 and 1616 under the direction of Giulio Cesare Fontana, have allowed the reclamation and recovery for agricultural use of tens of thousands of hectares of plain, putting an end to the frequent flooding of the river Clanio. This basin still covers an area of 1,095 square kilometers in 99 municipalities of the metropolitan city of Naples and the provinces of Caserta, Avellino and Benevento. The Campania Region, with resolution no. 1344 of 6/08/2009, has allocated 50 million euros for an action plan for interventions of closure or environmental remediation of the Regi Lagni, as a result of which, already in 2010, many of the channels have been closed or reclaimed.

10. E. Formato, M. Lanzi, ""Litorale Domitio: nuovi racconti nelle trame dell'acqua", R. Pavia, R. Secchi, C. Gasparrini (ed. by), "Il territorio degli scarti e dei rifiuti", Re-cycle Italy Nr.8 Series, Aracne, Rome, 2014.

3 Territorial networks

Accessibility& induced flows

attrattore principale
attrattore 2.000-3.000
attrattore 1.000-2.000
attrattore 0-1.000
flussi 0-100
flussi 100-300
flussi 300-500
flussi 500-600
flussi 600-800
flussi 800-900
flussi 20.000 ...

generatore principale
generatore 1.000-3.000
generatore 0-1.000
flussi 400-700
flussi 0-400
flussi 700-800
flussi 800-900
flussi 900-2.000
flussi 2.000-3.000

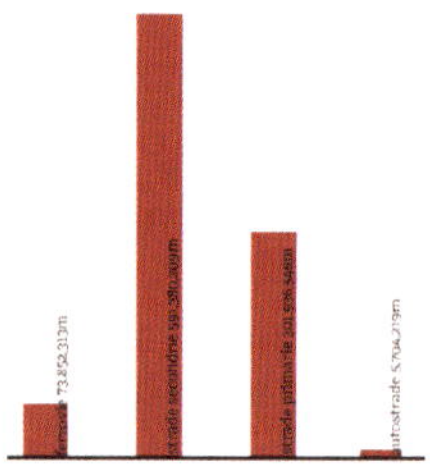

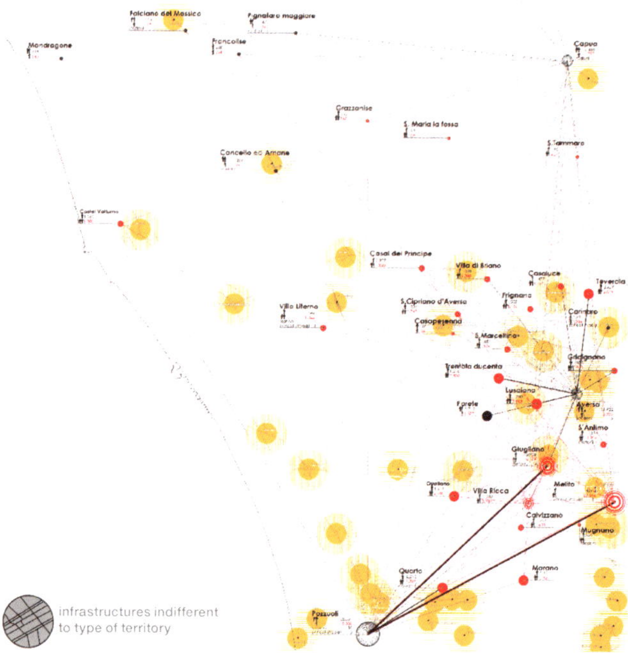

Critics

infrastructural compression of fabrics in abandoned areas

reduced accessibility

infrastructures indifferent to type of territory

strade primarie
strade secondarie
sistema trasporto su ferro
dross (aree di scarto/ cave/discariche ...)
aree/edifici industriali
svincoli stradali
stazioni ferroviarie
buffer 500m
buffer 1000m

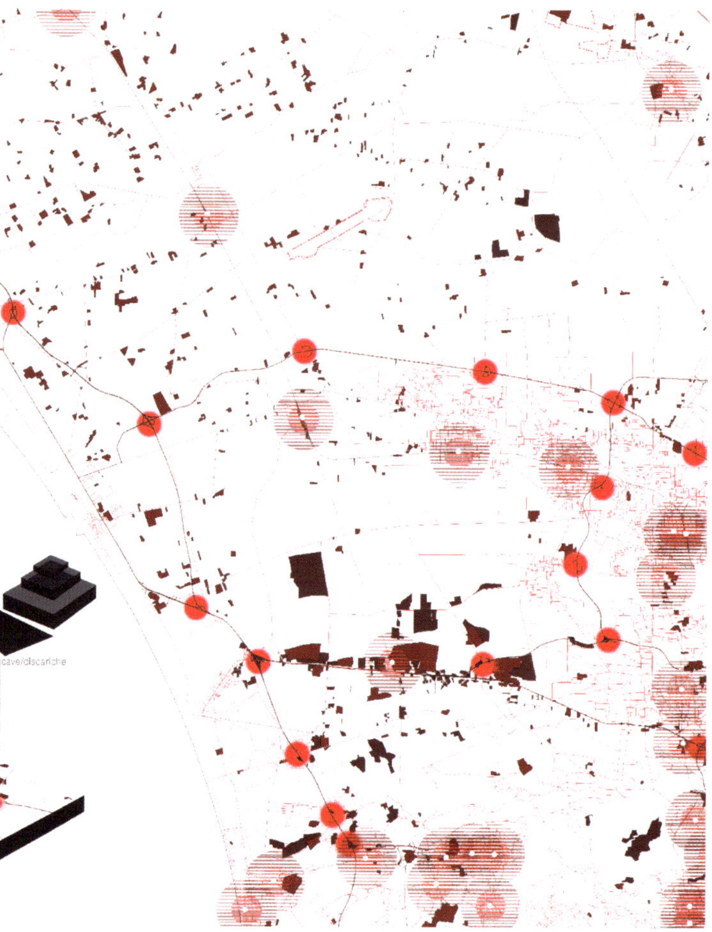

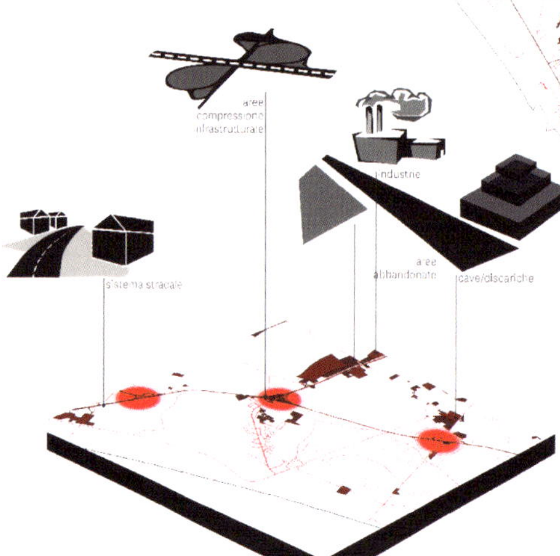

Water&hydraulic systems

Critics

pollution of surface water

ground water pollution

artificial banks

disfunctions of canals

coastal pollution

bathing prohibition

1817

1840

2016

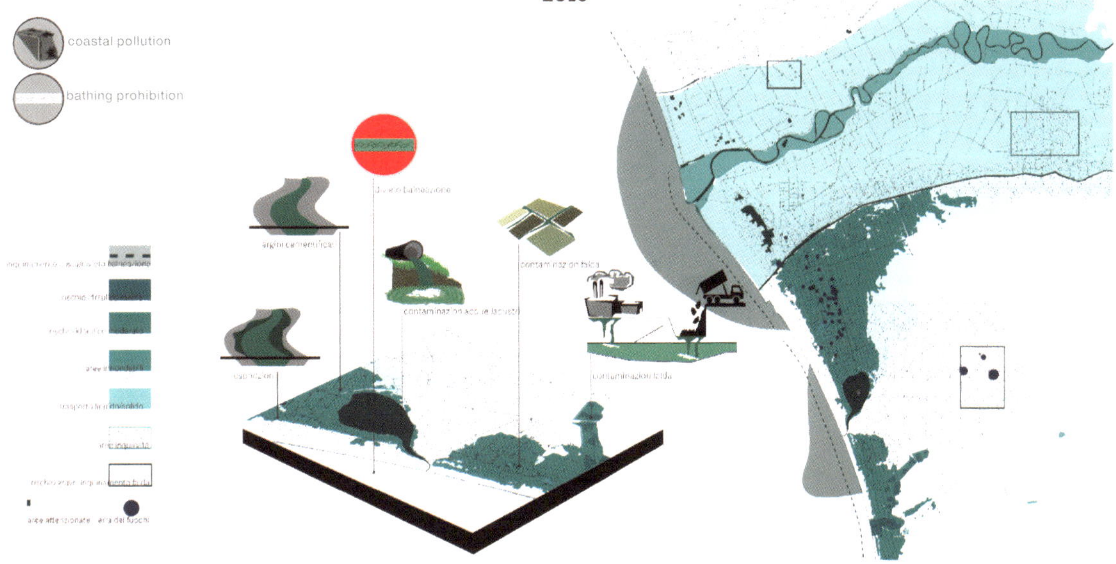

Green areas&Soil

Critics

land contamination

intensive farming

fragmental territory

discontinuity of coastal territory

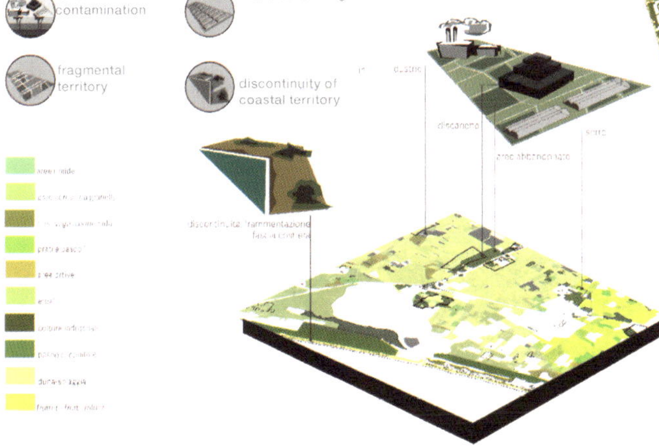

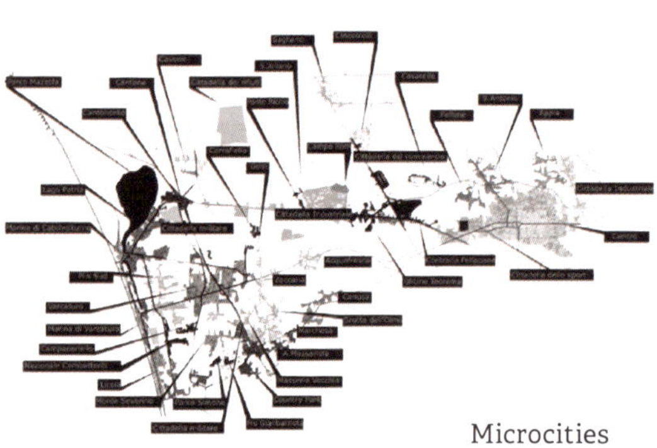

Microcities

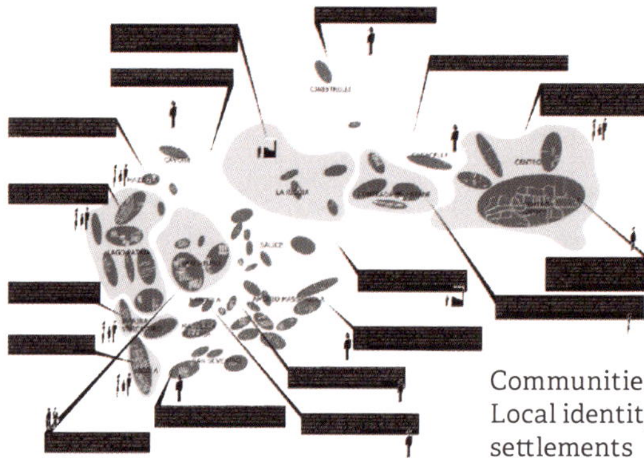

Communities&
Local identity
settlements

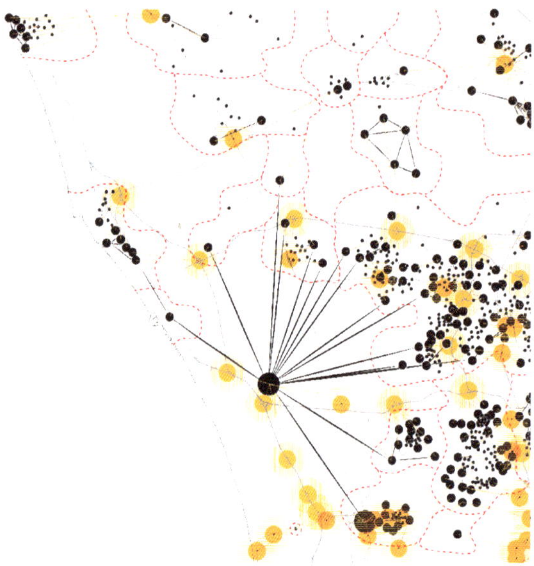

5 Centrality& services

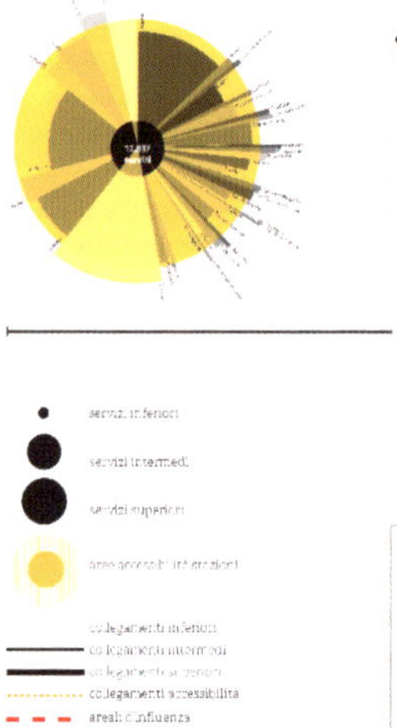

distribution
services

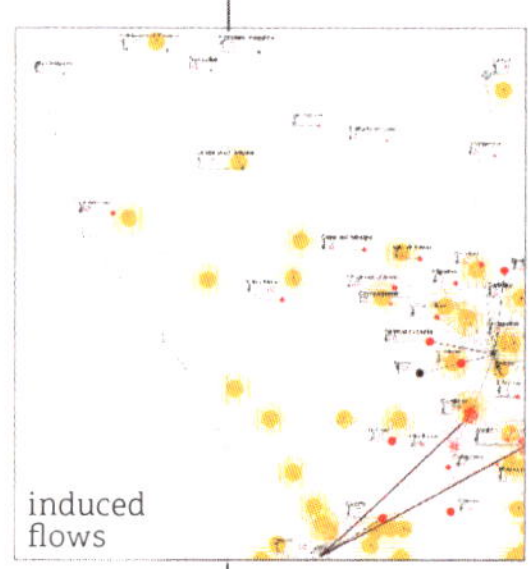

induced
flows

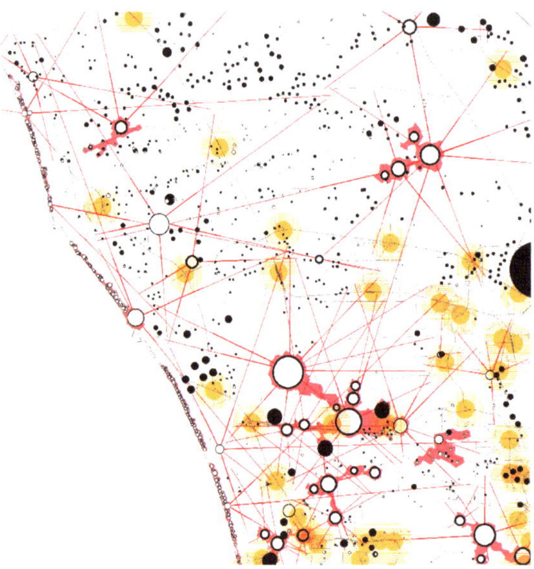

6 Economy& Production

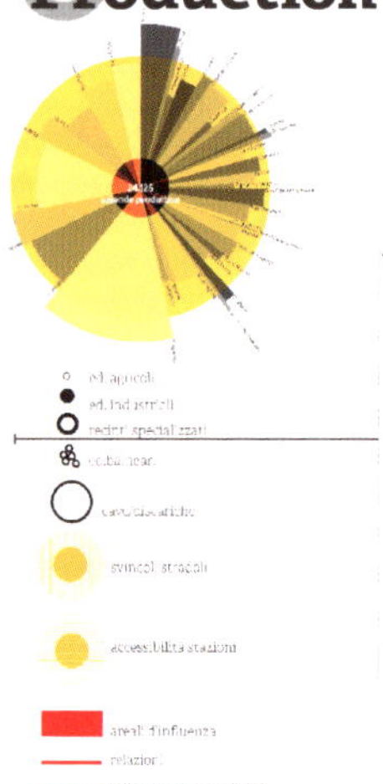

distribution
productive
activities

touristic settlements (of low quality and very often illegal constructions), industrial sites and to the highly-densed infrastructure present in the hinterland, largely financed by the Cassa per il Mezzogiorno[11] (National Funds for Southern Italy). At the same time, the progressive aggression of rural areas - "ruralscape" - by a firstly extensive and then widespread urban expansion ended up breaking that centuries-old balance between water landscapes, countryside and historical centres. The reclamation, from both a spatial and functional viewpoint, was intended as a territorial infrastructure that would have kept these elements together in a single, perfectly functioning organism. The contemporary degradation, on the other hand, directly affects the superficial hydrographic network and the agricultural soils, overlapping, in the marginal and interstitial areas, the growing increase of illegal practices of waste disposal. All this added tothe waste emergencies of the last twenty years, have caused the Domitio-Flegreo coastline to be the landfill[12] and temporary storage sites among the biggest in Italy, on the very edge of those soils already compromised by chemical pollution due to agricultural and livestock activities.

Even such deeply compromised areas can take part in the territory's regeneration, precisely for its still strong agricultural vocation which determines the protection, maintenance and enhancement of an agricultural landscape capable of infiltrating the fabrics of the historical centers, drawing their porosity and the system of green spaces.

Here more than elsewhere, the great patches of agricultural landscapes - "ruralscapes" - as well as the dense and complex network of surface and deep waters - "waterscapes" - together with the different hierarchies identified by the infrastructure system - "infrascapes" - overlap the variable granulometry of the settlement fabric, intercepting the myriad of wastes, degraded and marginal areas - "drosscapes" - produced by this area's metabolism, thus delivering a multiplicity of places and materials (new, hybrid or potential) "that can play a decisive role in the rethinking of the city itself, acquiring new meanings and establishing new relationships "[13].

The construction of knowledge

The complexity of these territories, and of contemporary territories in general, requires a necessary change in the perspective and in the ways in which knowledge is acquired. The construction of a problematic background in which we try to bring out information that is not obvious, or at least not immediately perceivable, inevitably passes through the intersection of different ways of investigation. The observation of places, the inspection through interviews, the acquisition of data through computer platforms and digital tools[14], all these elements help to provide articulated answers to the proposed research objectives.

The focus is on enhancing the different relationships between methods, tools, data and forms of the territory in order to build an updated, shared and transferable dimension of knowledge,

11. Abbreviated in Casmez, the Cassa per il Mezzogiorno was an Italian public body set up by Law No 646 of 10/08/1950 by the De Gasperi VI Government to finance industrial initiatives aimed at the economic development of southern Italy. From the beginning of its operations in 1951 until 1992 (the last known data), it has given the southern regions a total of 279,763 billion lire, while the average annual expenditure was 3.2 billion euro (G.A. Stella, "Lo Spreco", Dalai Editore, 1999).

12. In the non water-waterproofed "Resit" landfill in Giugliano alone, 806,509 tons of waste are landfilled, of which 30,000 tons come from the Ligurian chemical plant Acna in Cengio and 57,000 tons of leachate are produced. The contact between pollutants and relatively superficial water has already triggered processes of chemical transformation of the soil, gas fumes and deterioration of the well water. For this reason, about 200 agricultural wells in the area have been deactivated on an area of over 220 hectares. The peak of pollution will be reached in 2064, but will produce effects until 2080 (source: Antimafia prosecutor's office of Naples).

13. C. Gasparrini, "Città da riconoscere e reti eco-paesaggistiche", in R. Pavia (ed. by), "Eco-Logics", PPC, 25-26. Pescara: Listlab, 2011.

14. The main concept behind these interpretations - "liquid city" by Baumann (2006), the "city of bits" (Mitchell, 2005), or the "soft city" (Pickles, 2004) - is based on the possibility of reading it as a network of temporary data generated by the interaction between man and places, within cities seen and experienced through the media, which include new ways to create collective maps through digital tools, new ways to travel, to report travel, to build itineraries, to describe, interpret and share personal urban experiences.

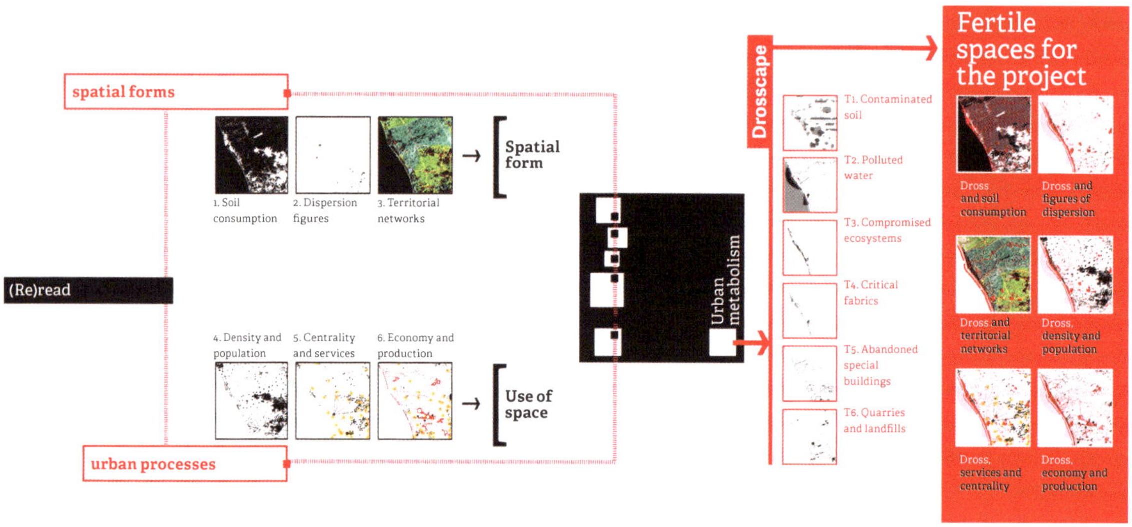
spatial forms
1. Soil consumption
2. Dispersion figures
3. Territorial networks
Spatial form
(Re)read
4. Density and population
5. Centrality and services
6. Economy and production
Use of space
urban processes
Urban metabolism
Drosscape
T1. Contaminated soil
T2. Polluted water
T3. Compromised ecosystems
T4. Critical fabrics
T5. Abandoned special buildings
T6. Quarries and landfills
Fertile spaces for the project
Dross and soil consumption
Dross and figures of dispersion
Dross and territorial networks
Dross, density and population
Dross, services and centrality
Dross, economy and production

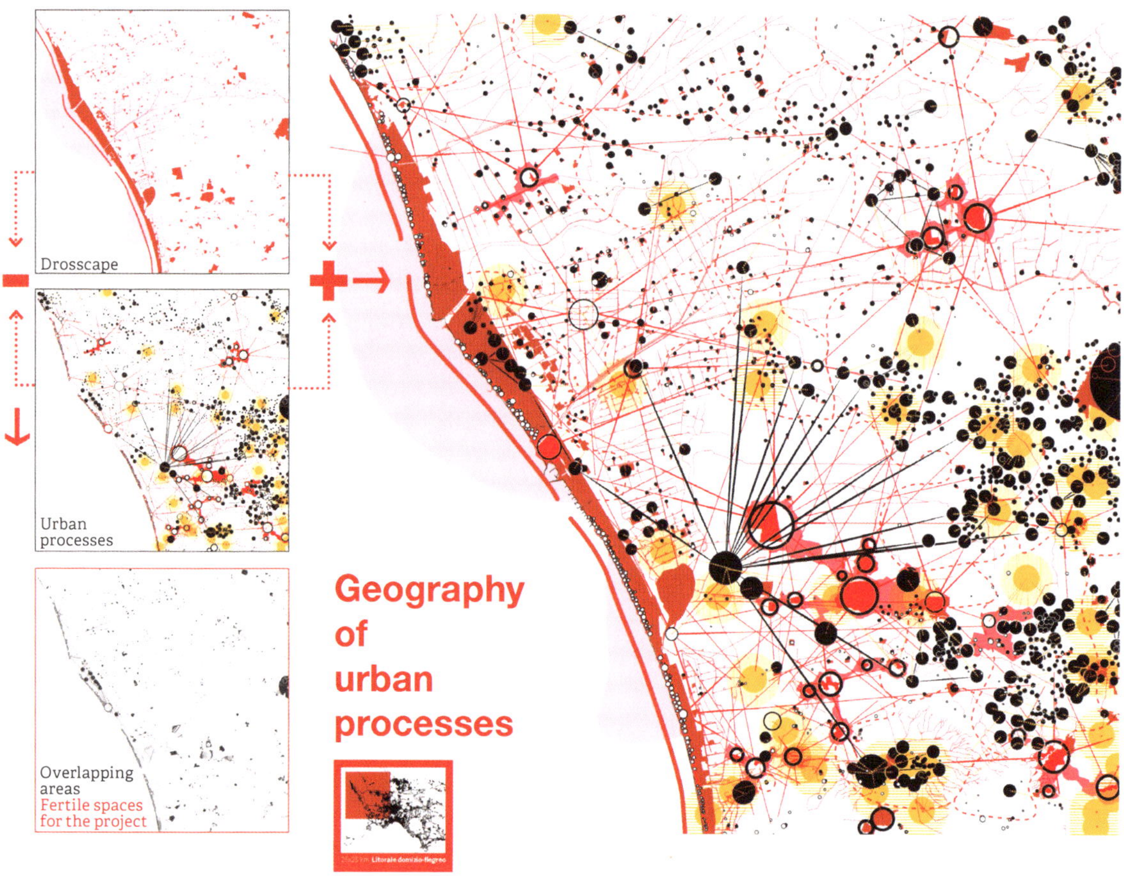
Drosscape
Urban processes
Overlapping areas
Fertile spaces for the project
Geography of urban processes
Litorale domizio-flegreo

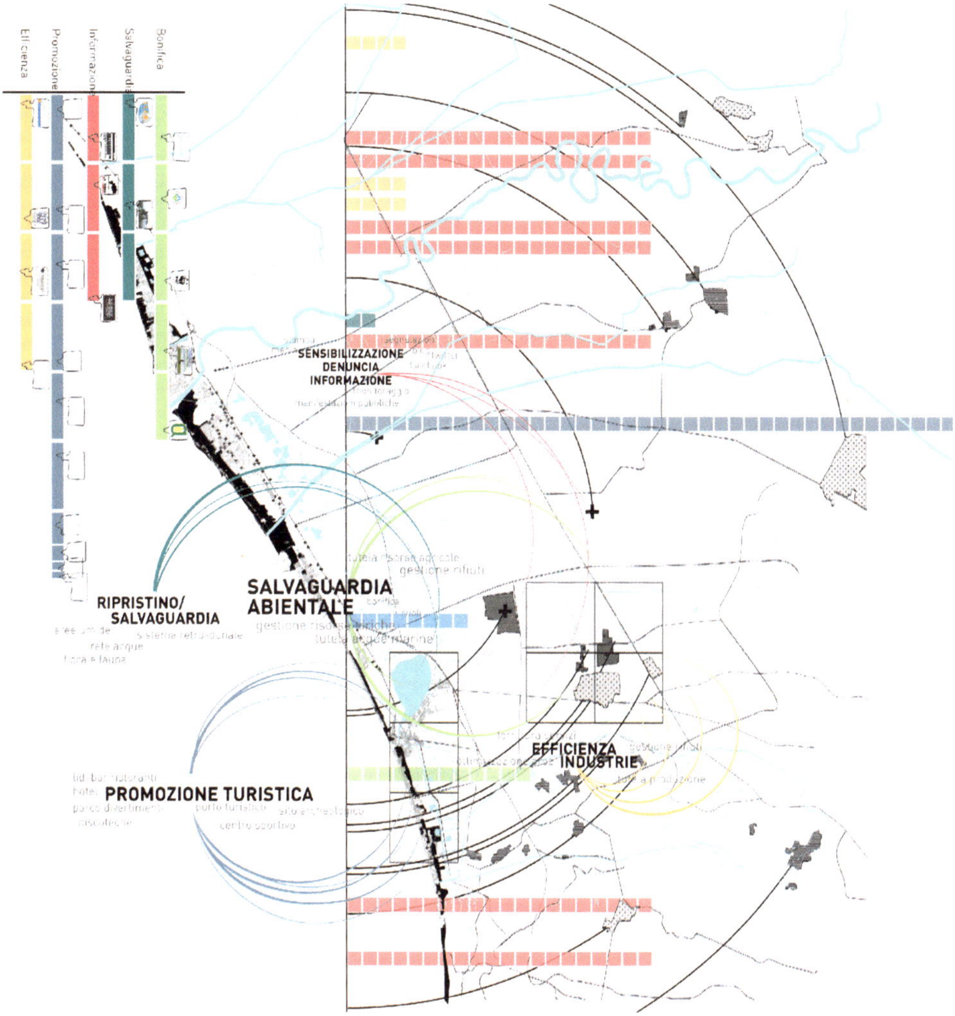
Bonifica
Salvaguardia
Informazione
Promozione
Efficienza
SENSIBILIZZAZIONE
DENUNCIA
INFORMAZIONE
SALVAGUARDIA ABIENTALE
RIPRISTINO/ SALVAGUARDIA
sistema retrodunale
rete acque
flora e fauna
gestione rifiuti
PROMOZIONE TURISTICA
parco divertimenti
discoteche
porto turistico
centro sportivo
EFFICIENZA INDUSTRIE

181

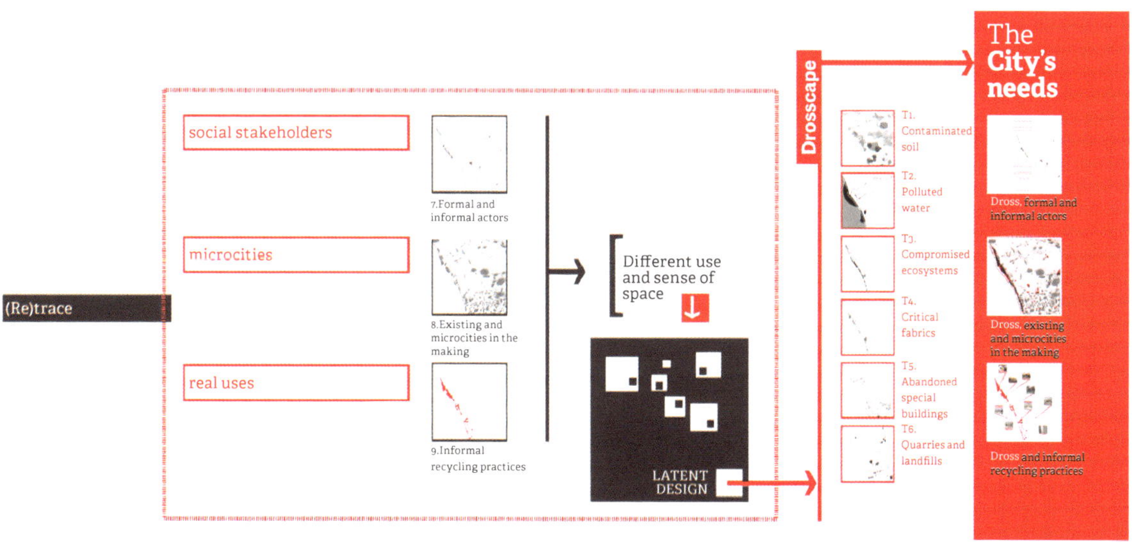
(Re)trace
social stakeholders
microcities
real uses
7.Formal and informal actors
8.Existing and microcities in the making
9.Informal recycling practices
Different use and sense of space
LATENT DESIGN
Drosscape
T1. Contaminated soil
T2. Polluted water
T3. Compromised ecosystems
T4. Critical fabrics
T5. Abandoned special buildings
T6. Quarries and landfills
The City's needs
Dross, formal and informal actors
Dross, existing and microcities in the making
Dross and informal recycling practices

while planning becomes both an instrument and a method of investigation and storytelling. A place for reflection in which we represent the questions and requests that the city submits to us. In this way, we try to get a hold of that apparently indecipherable, damaged script, which nevertheless exists and is "implicit "[15] in the structure of these places.
A sort of "latent drawing" we trace between the spaces and the questions that we try to describe and to recompose within new configurations, giving back new images in which the compositional methods of the drawings we refer to intentionally operate towards the breakdown/recomposition of the places inside the images, in order to deliver, beyond the figurative operations and recompositions of other contents, new senses and fertile spaces for future projects.
In the Domitio-Flegreo City, in fact, we find a variety of actors (institutional, sectorial stakeholders, associations and individuals). They often share competences and requests with a limited number of places and artefacts, most of which can be traced back to the waste cycle[16] and environmental emergencies[17]. Those same places are frequently, however, the object of informal practices of appropriation and reuse, often at the limit of legality, which define a complexity and variety of conditions that are difficult to spatialize and that end up designing a different city, almost superimposable to the physical one but perhaps realer than the latter.
Today more than ever, in fact, in most Italian urban areas people with different ideas, backgrounds, needs and lifestyles cohabit or just coexist, resulting in different and contrasting practices of production, use and appropriation of spaces and places[18]. This finding was achieved through the analysis of evidence in urban signs and the way of living in such places[19], recording their pace and rhythms in order to recreate that sample of actors and uses of space capable of designing new urban realities as relational structures, even before being referred to as defined spatial forms[20].
On the other hand, the relations between different poles of production are triggered by the economic-spatial flows governing the functioning of the city itself and therefore draw new "constellations" within the nebula of dispersion.
As we can see, the major players of industrial production, in parallel with a widespread system of supply chains linked to small and medium production mainly related to agriculture, end up creating structures comparable to "city networks "[21] in which the tendency is to find associations among similar realities, therefore according to "prevailing systems". Spatial realities arising from the aggregation of functional systems, often monothematic, that act as catalysts and end up generating, in the surrounding territory (to which they often remain spatially indifferent), strong relations of interdependence, if not subsistence.
Of course, all this has implications for the metropolitan dimension as well, since it describes the production system as a self-referential structure on the territory which produces spaces capable of building relations with the other poles of

15. R. Koolhaas, "Junkspace. Per un ripensamento radicale dello spazio urbano", Macerata, Quodlibet, 2006 .

16. See G.Tizian (5 September 2013), "Smaltimento rifiuti ed evasioni fiscali, per il nuovo business i clan cambiano volto", laRepubblica.it, source: http://inchieste.repubblica.it/it/repubblica/

17. See regulatory references: Regione Campania (2003), "Piano di Tutela delle Acque" (Plan for the preservation of the water resources), adaptation to Legislative Decree 152/2006 and subsequent amendments and additions, Convenzione Regione Campania - Sogesid S.p.A. Rep. n.13360Campania Region, Government Commissioner for the emergency reclamation and protection of water in the Campania Region, ARPAC (ed. by), (2005), "Regional Plan for the Reclamation of Polluted Sites of the Campania Region"; Campania Region, North-West Basin Authority, (2010), "Plan for the Protection of Soil and Water Resources"; Campania Region, Government Commissioner for the emergency reclamation and protection of water in the Campania Region (ed. by), "Updated Regional Plan for the Reclamation of Polluted Sites of the Campania Region", Naples, 2011.

18. A. Amin, N. Thrift, "Città. Ripensare la dimensione urbana", Bologna: Il Mulino, 2002.

19. F. Ippolito, "Tattiche", Pensare Architettura (series), Il Nuovo Melangolo, Genoa, 2012.

20. S. Sassen, "I 'senza potere', protagonisti del futuro", in AA.VV, V. Cobelli and G. Naletto (ed. by) "Atlante di un'altra economia. Politiche e and pratiche del cambiamento", Manifestolibri, 2005.

21. P.J.Taylor, G. Catalano, D.R.F. Walker, "Measurement of the World City Network", in "Urban Studies", no. 39(13), pp. 2367-2376, 2002; P.J. Taylor, "Leading World Cities: Empirical Evaluations of Urban Nodes", in "Multiple Networks", Urban Studies, no. 42(9), pp. 1593-1608, 2005.

production, even over long distances, but which are incapable of producing relations in the contexts in which they are inserted.
However, what is of interested to us in these readings is to understand how economic as well as social processes are able to lead to a different use of space[22]. And it is precisely in this process of crossing all the city's different scales (in terms of space and of the socio-economic context) that we experience the modes of production and transformation of its spaces, while we tirelessly try to understand, represent and narrate them.

The images in this contribution are by Antonella Senatore (see her contribution in the chapter).

22. See the action of some actors documented on the following websites: http://www.arpacampania.it/; http://www.autoritadibacino.it/; http://www.munnezza.info/; http://www.laterradeifuochi.it/; http://www.legambiente.it/; http://www.provincia.napoli.it/; http://www.argacampania.it/; http://www.regione.campania.it/; http://www.wwfcaserta.org/; http://web.archive.org/web/20120108020514/http://www.laboratoriocampano.org/

Villaggio Coppola, Castel Volturno (Ce)

The rehabilitation of compromised agricultural soils

Massimo Fagnano

Soil contamination: reality and perception

According to the current law (152/06), if there are violations of the Contamination Thresholds Concentrations or Fund Values, a site can only be defined as "potentially polluted". In order to define a site as "polluted" and therefore to plan possible reclamation or risk management interventions, it is necessary to perform a risk analysis and define the Risk Threshold Concentrations, which depend on the use of the site causing the possible health exposure of people and the risks for the environment.

In the case of agricultural soils, the issue is a bit more complicated, because in addition to the usual field or workplace analyses, one has to assess:

"Risks for workers" (by inhalation, ingestion or skin contact) in relation to emissions into the atmosphere and the hours spent at work.

"Risks for the environment" (due to the leaching of contaminants to the groundwater) in relation to the physical characteristics of the soil and the depth of the groundwater.

However, there are no methods to predict the "consumers' risks" associated with the absorption and accumulation of pollutants in the edible parts of agricultural products.

One could analyze the soil content of the bioavailable fraction of pollutants, but not always does the presence of a contaminant in available form automatically mean that the products of that site will be polluted, because plants are living organisms that have their own selectivity and, if a certain mineral does not contribute to its growth, the plant does not absorb it. "Lead" and "cadmium", which are often absorbed by plants through a mechanism of chemical mimesis, are an exception. In other words, plants absorb and process them, confusing them with calcium (which is an important nutrient in plant physiology and beyond).

And, even when the plant absorbs a certain pollutant, it does not necessarily move from the root to the aerial part. In fact, the greatest accumulations of heavy metals (e.g. lead) are always recorded in the roots - with concentrations of different orders of magnitude, higher than those of the aerial part (leaves, fruits, seeds, etc.).
Therefore, in order to define an agricultural site that is dangerous for health, it is necessary to analyse the products that have been obtained in that site, perhaps by using plants that accumulate a specific contaminant, which is present on that specific site.

Forms of soil deterioration in the countryside of Aversano Litorale Domitio

In 2013 in the Agro Aversano, for reasons not yet fully understandable, some confiscations and the related media campaign of criminalization of the agricultural sector began, exclusively on the basis of some values of "beryllium" and "stagnation" higher than the CSC.
In the wake of the media pressure, the Government (inter-ministerial directive of 23/12/2013 referred to in DL 136 of 10/12/2013) has set up a working group (GdL) with the aim of mapping the suitability for agricultural use of the soils of the so-called "Land of Fires". The GdL collected all the analytical data present on the territory and available from the University and the ARPAC. in particular, it used the data collected by the LIFE-Ecoremed project, which were published on the project's[23] website on December 11th.
The map with the distribution of lead is shown as an example. These data were then cross-referenced with the land use changes inferable from the historical aerial photos, to identify the presence of waste or excavation-covering sequences that could represent a risk of waste burial.

23. See www.ecoremed.it

On the opposite page, Tables on the Distribution of Lead in the agricultural soils in the "Agro Aversano". Source: http://www.ecoremed.it/

Table 1 (below on the left): Thresholds of concentrations of green public and residential soils in Europe, Carlon, 2007.
Table 2 (below on the right): Lead accumulation in different kinds of plants growing in contaminated soil.
Table 3 (above): Agricultural areas at risk in the "Agro Aversano".

Following this analysis, the territory has been divided into 5 risk categories as shown in Table 3, from which it can be seen that most of the sites have been classified as being at risk.
This is mainly due to the presence of waste, and not so much to a real chemical contamination. From this analysis, about 2% of the agricultural territory was declared at alleged risk.
This 2% of agricultural soils has been surveyed by teams of sample collectors, composed of staff from the Region, ARPAC of the ASL and the Forestry Corps, to carry out geomagnetometric surveys (to find underground waste) and to perform detailed chemical analyses of the soil, irrigation water and agricultural products.
In July 2016, only 21 hectares (out of 50,000 = 0.04%) were declared unsuitable for agricultural crops (class D) while awaiting detailed characterization, risk analysis and possible remediation. All data, as well as reports, were received by the Government through Ministerial Decree and published in the official gazette.
The report for sites at risk 5 and 4 was delivered to the Government on 29 January (Decree Law 12/02/2015, OJ no. 56 of 9/03/2015).
The report for sites at risk 3 was delivered in May (Decree Law 7/07/2015, OJ no.191 of 19/08/2015). In July 2016, the report on sites at risk 2a will be delivered, adding a further 5 hectares to the soils that are banned from agricultural production. As a result, the amount of polluted agricultural soils will rise to around 26 hectares (0.05%).
It should be noted that, on none of these sites, the plant products were declared to be contaminated. This confirms the above statements, as well as the thousands of analyses carried out by various independent bodies (zooprophylactic institutes, University of Naples, ASL, MIPAF Fraud Repression Institute), but - above all - it confirms that the consumers of our food products have never been found to be contaminated with agricultural products that may contain toxicologically relevant minerals.
On these soils, risk analysis must be carried out in order to define possible remediation objectives, establishing the most suitable techniques to avoid destroying the soil resource, and to restore the ecosystem of the soil itself.
In fact, the most used technique, and the one preferred by a certain type of business because it is more expensive, is the excavation and landfill disposal of the contaminated layer.
Setting aside economic considerations (5 million euros per hectare) and logistical considerations (in which region are there so large landfills that they can accommodate several hundred thousand square meters of polluted land?), this scenario is still gruesome, because it would exacerbate the wounds of our territory instead of healing them. Just imagine

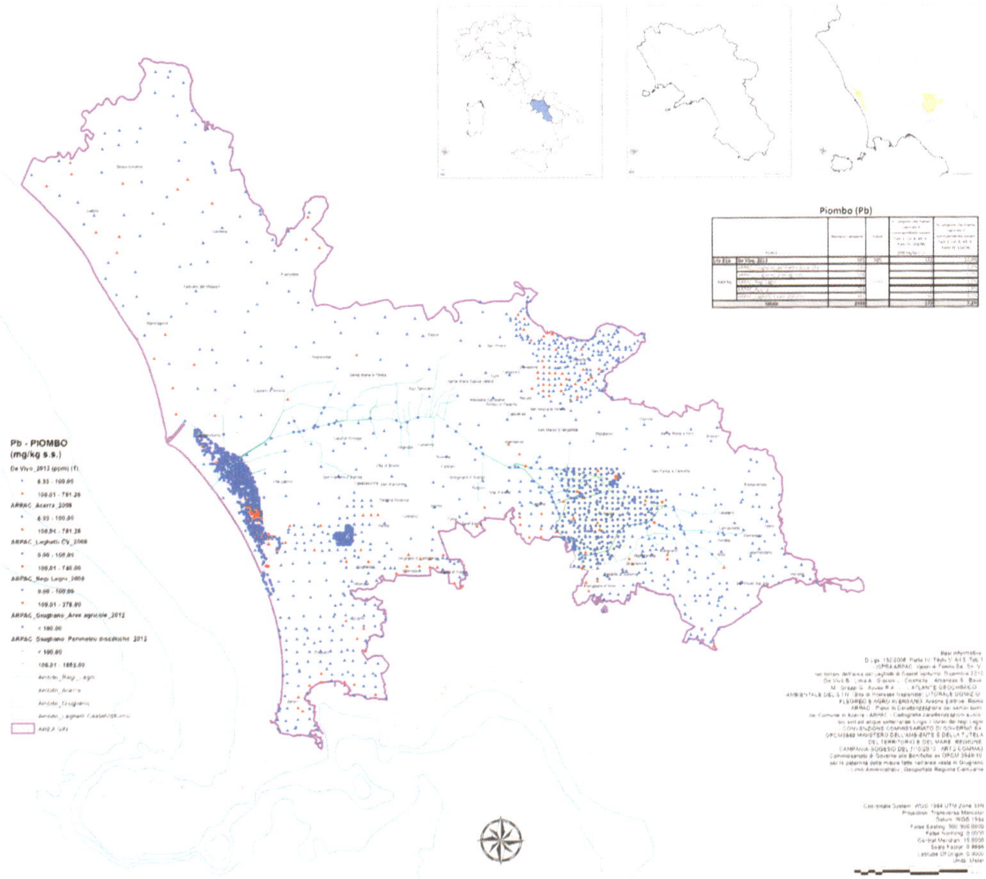

	AUT	BE(F)	BE(B)	BE(W)	CZE	FIN	ITA	LTU	NLD	POL	SVK	UK	DNK
As	50	110	110	300	70	50	20	10	55	22.5	50	20	20
Ba					1000			600	625	285	2000		
Be					20		2	10	30		30		
Cd	10	6	6	30	20	10	2	3	12	5.5	20	2	5
Co					300	100	20	30	240	45	300		
Cr	250		300	520	500	200	150	100	380	170	800	130	1000
Cu	600	400	400	290	600	150	120	100	190	100	500		1000
Hg	10	15	15	56	10	2	1	1.5	10	4	10	8	3
Pb	500	700	700	700	300	200	100	100	530	150	600	450	400
Mo					100			5	200	25	200		
Ni	140	470	470	300	250	100	120	75	210	75	500		30
Sb	5				40	10	10	10	15				
Se							3	5	100		20	35	
Sn					300		1	10	900	40	300		
Te										600			
Tl	10						1			15			
V					450	150	90	150	250		500		
Zn		1000	1000	710	2500	250	150	300	720	325	3000		1000

Crop	Pb Concentrarion (mg kg⁻¹)			
	Root	Stem	Leaf	Edible part
Carrot	2162,1 Aa	NA	25,4 Cb	54,6 Bb
Beet	NA	NA	40,3 Bb	109,3 Aa
Tomato	405,5 Ac	40,3 Bb	9,9 Cc	*
Sweet pepper	281,0 Ad	27,6 Bb	7,7 Cc	*
Green Collards	175,8 Ae	10,3 Bc	3,2 Bc	3,2 Bc
Cabbage	152,5 Ae	23,3 Bb	7,9 Cc	*
Okra	757,2 Ab	133,5 Ba	75,7 Ca	5,2 Dc
Eggplant	389,4 Ac	37,9 Bb	7,1 Cc	*

Livello di rischio presunto	Caratteristiche del sito	Indagini	Numero di siti	Specifiche agricole
5	Valori inquinanti > 10 x CSC (o VFN) e corrispondenza (entro 10 m) con siti a rischio da analisi foto aeree	analitiche e conoscitive (carotaggi, trincee, ecc) entro 90 gg	7	16,5
4	Valore inquinanti > 10 x CSC (o VFN)	analitiche entro 90 gg	40	40
3	Valore inquinanti = 3-10 x CSC (o VFN) e corrispondenza (entro 10 m) con siti a rischio da analisi foto aeree	analitiche e conoscitive (carotaggi, trincee, ecc) entro 90 gg	4	8,1
2a	Valore inquinanti = 2-10 x CSC (o VFN)	analitiche entro 180 gg	86	86
2b	Siti a rischio da analisi foto aeree	conoscitive (carotaggi, ecc) ed eventualmente analitiche	1249	820
2c	Aree agricole delle aree vaste Lo Utaro, Bortolotto-Sogeri e Masseria del Pozzo, aree agricole del PRB	analitiche entro 360 gg	da determinare entro 90 gg	da determinare entro 90 gg
2d	Aree agricole circostanti impianti smaltimenti di rifiuti, industriali, arterie di traffico, aste del sistema dei Regi Lagni, incendi di grande rilevanza, siti a rischio da analisi foto aeree (cl. 1)	analitiche entro 360 gg	da determinare entro 90 gg	da determinare entro 90 gg
1	Valore inquinanti = 1.2 x csc (o VFN)	analitiche oltre 360 gg	176	176

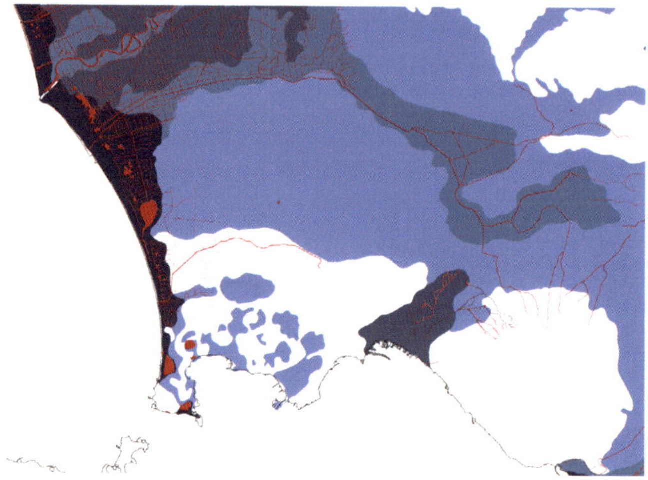

- aree relativamente rilevate delle pianure alluvionali
- pianura pedemontana
- aree relativamente rilevate delle pianure alluvionali (Piana Campana)
- aree morfologicamente depresse delle pianure alluvionali
- dune e depressioni retrodunari

Floodplain and coastline systems

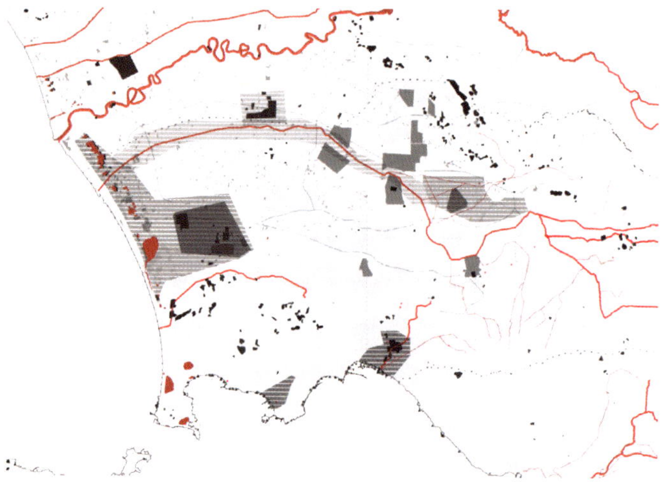

- Sistema dei territori inquinati
- Aree Vaste
- siti contaminati
- Aree industriali
- cave
- sistema idrografico

Contaminated soil systems

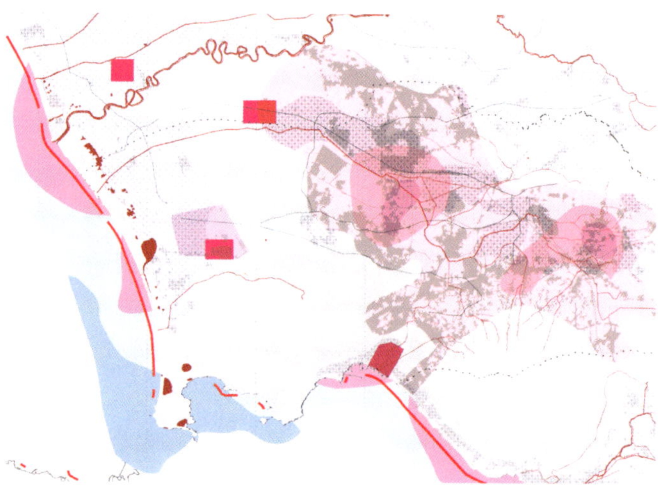

- rete idrografica
- inquinamento da Nitrati > 200ppm
- inquinamento da Nitrati > 150ppm
- inquinamento da Nitrati > 100ppm
- inquinamento diretto delle falde

Pollution and vulnerabilities of underground water

the fate of these areas if they are dug 1-2 metres deep (the foundations are already open, ready for new construction work!), the impact of these sites on the landscape, and the environmental impact linked to the passage of thousands of trucks that will have to carry this soil for years.

Environmentally friendly techniques for the rehabilitation of agricultural soils: the LIFE-Ecoremed protocol

The LIFE-Ecoremed project, funded by the EC in 2012, proposed an alternative approach to the remediation of contaminated soils, more respectful of soil resources and its ecosystem services and rural landscape.
The Ecoremed protocol, in the case of organic pollutants (e.g. hydrocarbons, dioxins, PAHs, etc.) provides for the collection of contaminated soil, the extraction and isolation of bacteria and fungi present in polluted soil, the selection of the most efficient strains in the biodegradation of contaminants of the site, their multiplication, and subsequent treatments of contaminated soil. This approach is based exclusively on the autochthonous microflora of the site, therefore it does not involve the use of strains from other parts of the world, nor the use of GMOs, thus avoiding the risks for biodiversity and the natural balance of the soil microflora. Subsequently, the soil is fertilized with large amounts of compost (20-30 t/ha), both to improve the physical and chemical fertility of the soil, and to enhance the metabolism of biodegradable bacteria. Such metabolism is also enhanced by the presence of the roots of plants that will be sown or transplanted to the site which, thanks to the substances released by the roots (rhizosphere effect), will further contribute to the biodegradation of organic pollutants. Furthermore, the choice of suitable species could also contribute to the extraction of the bioavailable share of heavy metals (e.g. lead) from the soil, in order to reduce the risks for the agri-food sector.
As far as the use of vegetation is concerned, the Ecoremed protocol has opted for the "multi-year" species (e.g. poplar, common cane, eucalyptus, etc.), because they allow to reach a series of important objectives for the management of contaminated sites:

1. They physically prevent the access and improper use of contaminated soils and the related health risks, thanks to the presence of the "non-pabular multi-yearly" species (e.g. eucalyptus and cane), and make it possible to grass, which limits the risks due to the lifting and dispersion of contaminated soil particles.
2. They limit the risks for agricultural operators, as they do not require annual tillage and sowing, which exposes farmers to the lift-off of contaminated soil particles.
3. They protect the soil for 12 months a year, not leaving it uncovered or unused for longer or shorter periods.

The figures on the opposite page are by Davide Di Martino

1. Extraction, isolation and classification of the biodegradable micro-flora

2. Multiplication

3. On-site inoculation of microbes that biodegrade organic pollutants (i.e. IPA, dioxins, hydrocarbon)

4. Microbial metabolism is strenghtened by organic fertilization

5. and by the interaction of the roots with the radical elements of the plants

6. that absorb the bio-available parts of the potentially toxic metals

Ecoremed protocol checklist for the remediation of contaminated soil

4. Some species (e.g. common cane and eucalyptus) also have water absorption in autumn-winter, thus protecting the aquifers from the leaching of soluble pollutants.
5. They allow the analysis of indirect risks of accumulation of toxic elements within agricultural products.
6. They improve the agricultural landscape and soil fertility by preserving the agricultural use (and function) of contaminated soils.
7. They increase the metabolism of the biodegradable micro-flora of organic pollutants, thanks to root esudation and the increase in organic matter resulting from the absence of soil tillage and the litter effect.
8. They help to extract the bioavailable share of potentially toxic metals from soils, in order to reduce the level of health risk.
9. They allow to return (in a variably short time) the soils to their traditional agricultural use (ex. fruit and vegetables), with better physical-chemical fertility.
10. They represent a much cheaper, more environmentally friendly technology if compared to traditional reclamations, which cost about 1-6 million euros/ha (against about 100,000 euros/ha under the Ecoremed protocol) and destroy soil fertility, making it unsuitable for agriculture (but ready for new concrete pours).

Former Trentola landfill

Case studies in the Aversano countryside
Former Trentola landfill site

The temporary waste storage site of the Municipality of Trentola-Ducenta has been freed from waste. Chemical pollution is only related to exceeding the CSCs for copper, which is obviously of agricultural origin (greenery as a pesticide). The concentrations of copper are not so high as to represent a risk to health, but the site presents significant problems of physical fertility. In particular, probably because of the passage of vehicles full of waste, as well as the presence of heavy machinery for the screening of waste, the site has a perspective at 30-40 cm depth, completely compacted and impermeable. The deep tillage of the soil, together with the development of the root system, could in a few years eliminate this sub-surface compaction, making the site suitable for any agricultural use.

Former Teverola landfill

Former Teverola landfill

A very similar situation was found in the temporary waste storage site of the Municipality of Teverola, where no violation of the CSCs was found. In other words, the site was devastated by the presence of waste, but was not chemically polluted. In this site, the physical degradation of the soil was so extreme (compaction, deconstruction and pulverization) that several of the tested species did not survive. The only species that successfully colonized the site was the common cane (A.

donax), which - after a first year of difficult growth - showed remarkable growth already in the second year, reaching a height of 5 metres.

Site of Giugliano, seized for illegal disposal of industrial waste

On the other hand, the site of the vast area of Giugliano, polluted by the spillage of industrial sludge containing chromium-zinc and hydrocarbons, represents a truly polluted site, due to the very high concentrations of the 3 contaminants.
In this case, a poplar grove has been planted, to prevent the improper use of the soil, and to limit the exposure of farmers to contact with contaminated dust. The vegetation will also allow the analysis of the risk that agricultural production may accumulate levels of chromium and zinc, at concentrations that are dangerous to human health.

Operational safety of soil in an industrial area

In industrial areas, contaminated by lead and other pollutants, the creation of a grassy poplar grove, in addition to having the obvious advantages on the quality of the landscape, is a real intervention of safety, as it prevents the uplift and dispersion of contaminated soil particles.

Conclusions

- There are very few polluted agricultural soils in the Aversano area and no one (EU, ISS, MIPAF, IZSM, UNINA, Government, foreign customers) has found contaminated agricultural products, so the risks to human health do not lie in the agro-food products produced in this area.
- The non-agricultural areas that are degraded (roads, waste fields, landfills, etc.) or contaminated, are many and well known (almost all of them have been confiscated already).
- Air and surface groundwater are polluted, as are all heavily man-made and industrialised European plains.

Does this mean that everything is fine for the agricultural sector? I would say no: our territory is dirty, and has become a disvalue for our products. The elimination of waste from roads would not simply reduce health risks due to its combustion, and the dispersion of toxic fumes. It would improve the quality of our landscapes, and could also have a positive impact on the value of our agri-food products.

Site of Giugliano, seized for illegal disposal of industrial waste

Securing of soil in an industrial area

Re-cycle perspectives in the Planning of the "Bacino and Distretto" catchment areas for the Integrated Management of water and water-related risks

Raffaella Nappi

Within the 'RE-Cycle Italy' PRIN Research program, the University "Federico II" of Naples promoted the involvement of administrations, agencies and institutions already active in the field to set up networks of cooperation among public stakeholders, in order to implement research activities and to expand the skills necessary for their development, with the aim of pursuing greater operational efficiency and efficient results.
With these premises, the cooperation immediately appeared to be a significant and useful experiment for all the institutional bodies involved in the exchange of knowledge, information, ideas and results, aimed at implementing research activities, expanding the skills needed to carry them out and, above all, sharing actions that through research could be disseminated and applied in the future. These collaborative experiments, aimed at creating transversality and sharing of information and training among projects and educational plans oriented to sustainable development, represent a challenge when facing the issue of the territory's future, of its vulnerability and potential. Therefore, of its urban reorganization and development in a strategic perspective of recovery, recycling, ecological re-generation, hydrogeological re-balancing and ecosystem-based social services.
For the Basin Authority of Liri Garigliano and Volturno, the interest in collaborating in this research activities is twofold. On the one hand, it is part of the participatory and dissemination actions of territorial knowledge in the scope of the "Bacino e Distretto" areas' planning implementation; on the other, the attention is for the focus of research, the "drosscape spaces", which is direct consequence of the territory's intensive modifications and subsequent abandonment and degradation. All these issues are a subject of constant study, updating and analysis. Moreover, they also fall within the assessment

and management of hydrogeological risks (landslides, floods, coastal erosion), the protection of water and soil resources; and the safeguard of the territory and the environment.
We must work together and reflect on the complex territorial situation, on the situations of deterioration (soil consumption, environmental detractors, SIN - Areas of National Interest and SIR - Areas of Regional Interest, and the polluting factors related to them), on the hydrogeological and seismic risks and on the fragmentation of biodiversity. In this way it is also possible to explore the contemporary complexity, diversity and criticalities of the territory and landscapes.
The work of exploration, knowledge and territorial proposition calls into play many different disciplines to describe, represent and imagine the territory as a new laboratory open to different views which help define a possible new vision capable of creating a new planning model that unifies and integrates development, re-balancing and regeneration.
"In this vision, project subsidiarity represents the synthesis of a methodology capable of combining the new paradigms, suitable for interpreting the necessary metamorphosis: identity and diversity, identity and risk, creative identity and sustainable, positive and conscious development".[24]
Following this logic, the Authority's collaboration, throughout the three phases of the research, has developed focusing on these aspects:
a. definition of the Bacino and Distretto planning and integrated approach for its management and governance;
b. SIR area of the Domitio Flegreo - Agro Aversano coastline;
c. environmental strategies in water and flood risk management;
d. internships at the Authority for students from the "Federico II" University of Naples; results of the exchange of information and ideas.

24. Persico P., "Subsidiarity and Non-linear Planning as a collective development experience", Quaderni del Medio Garigliano, Progetto Laboratorio Ambientale (Environmental Laboratory Project) - Autorita' di Bacino Nazionale (Basin authority) Liri-Garigliano e Volturno, p. 260-268, 2010. ISBN- 889553404-02

Environmental critical areas subject to floods;

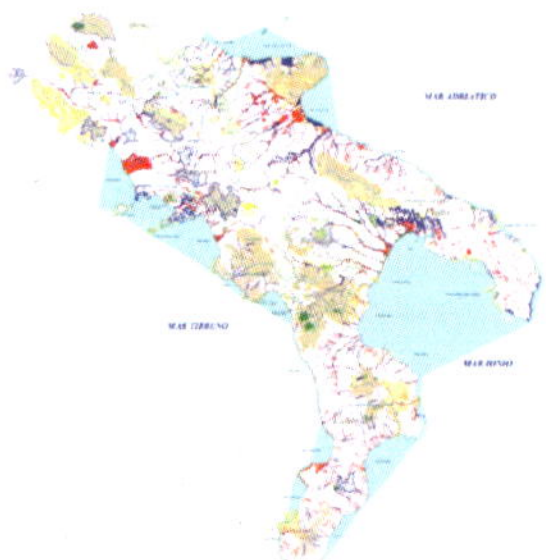

Natural protected areas and flood risks in the southern Appenines hydrographic district;

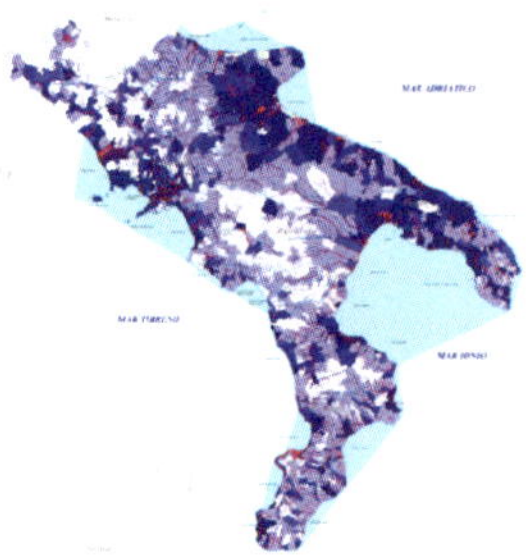

Population potentially at risk of floods.

The images and maps in these pages are by the Basin Authority of Liri Garigliano and Volturno rivers.

District Management Plans: integrated and unified management

The Bacino catchment area's planning was introduced by Law 183/89 "Regulations for the Organizational and Functional Rehabilitation of Soil Protection"; the planning of the river basin district referred to in the Water Framework Directive - WFD (2000/60/EC); the Legislative Decree 152/06 on "Environmental Regulations" by the Ministry of the Environment; Law 13/09 "Extraordinary measures in the field of water resources and environmental protection"; Framework Directive for the "Assessment and management of flood risks" (2007/60/CE); and Legislative Decree 49/2010 "Assessment and management of flood risks".
In fact, these tools address the complexity of the physical and environmental context according to a systemic approach that integrates the analysis and active policies of the catchment area Plans (Piani Stralcio: hydrogeological for landslide risks; hydraulic risks; protection of the underground and surface water resources for a quantitative and qualitative water analysis; protection from coastal erosion; and for the protection of the environment) while considering the river basin's complex ecosystem from a holistic point of view and by involving different experts: geologists, engineers, geotechnicians, urban planners, architects, pedologists, botanists, agronomists, foresters, biologists, economists and jurists.[25] Since 2006, to the planning of the Basin (L.183/89), the aforementioned Legislative Decree no. 152/2006 has added the one regarding the river basin area. Besides containing a comprehensive review of environmental legislation, the decree draws up the reorganisation of all territorial structures (Autorità di Distretto) responsible for planning and programming (District Management Plans) in the river basin district, in accordance with the provisions set by the Water Framework Directive (2000/60).
The District Management Plans are recently developed tools and include part of the Piani Stralcio di Bacino, revised and implemented on a broader level for a comprehensive and unitary management of water, flood and soil risks; and aimed at achieving a sustainable use of resources and at their preservation, through an impact assessment that also meets the need for economic growth. The Water Framework Directive has established a "Framework for Community Action in the field of Water", for the protection of inland surface waters, transitional waters, coastal and groundwater, in order to ensure the prevention and reduction of pollution, facilitate the usage of sustainable water, protect the environment, improve the conditions of the aquatic ecosystems and mitigate the effects of floods and droughts.
The Water Framework directive also identifies the Hydrographic Districts as physiographic units of reference and introduces the obligation to prepare "Hydrographic Water

25. Corbelli V., Nappi R., "Quaderni Piano di Bacino", Edizioni Graffiti, Naples, 1997.

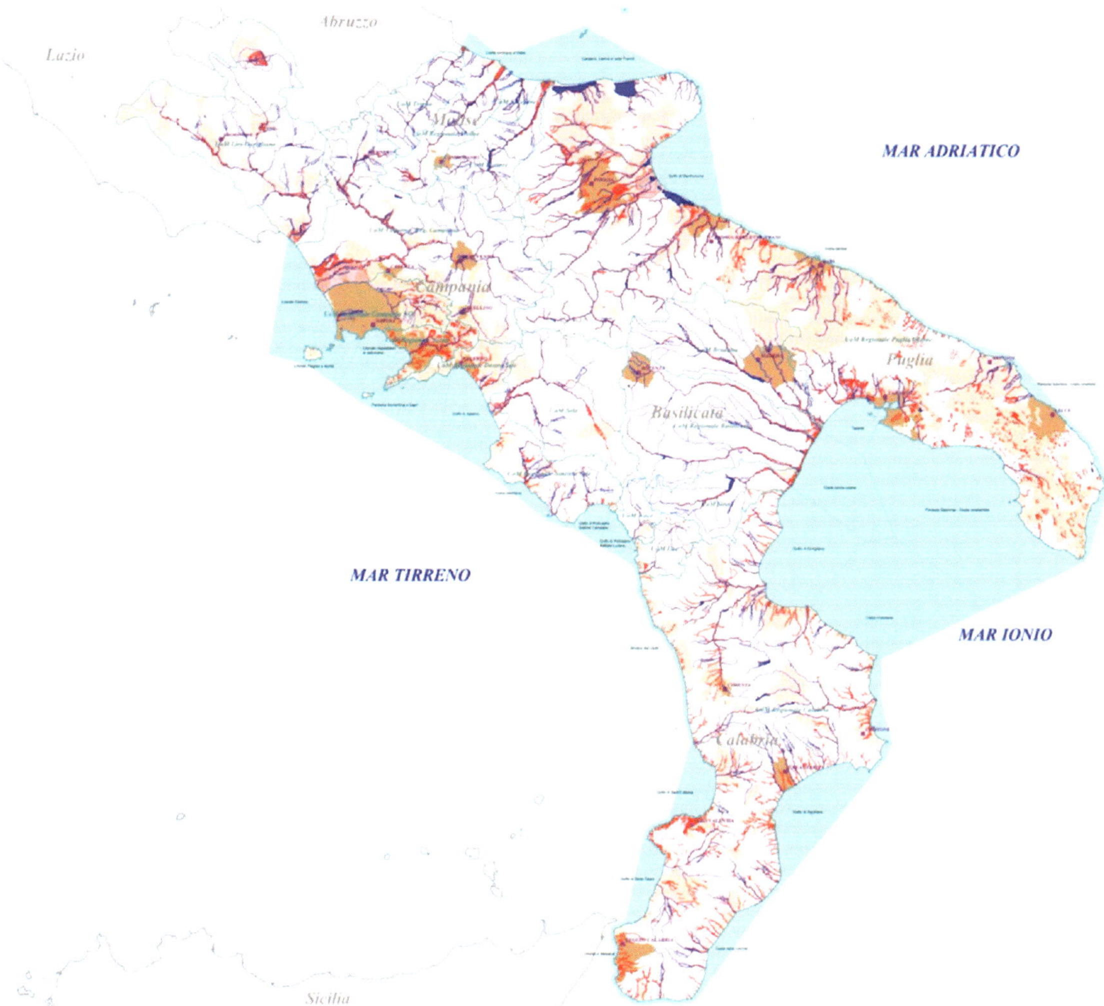

Management Plans" for all the involved districts. These plans are the operational tools for the implementation of the Directive's provisions. Framework Directive for the "Assessment and management of flood risks" (2007/60/CE) has established a coordinated programme in Europe mainly "aimed at reducing the negative consequences for human health, as well as reducing possible damage to the environment, to cultural

heritages and the economic activities concerned by the issues at stake"; in this sense, art. 7 of the Directive provides for the "Flood Risk Management Plan (PGRA) for all hydrographic districts". This Plan represents the operational and management tool for achieving any assessment analysis of flood risks. Both plans contribute to the "integrated management" of the hyrographic basins by maximizing synergies, with particular regard to interconnection in the development of measures (both non-structural and structural) and by promoting a coordinated implementation of such interventions. The two processes, therefore, enhance their mutual potential, synergies and common benefits, taking into account the environmental policy strategies set by the Water Framework Directive which ensure the efficient management and rational use of water resources, their sustainability and safeguard in quantitative and qualitative terms, and the implementation of subsequent strategies set by Directive 2007/60/EC that tend to establish a framework for the assessment and management of flood risks, thus integrating the Water Directive's action.
The two plans "may constitute a challenge, but also an opportunity, both for the environment - since they improve territorial and social security - and the economy - by improving the efficiency of production processes - with a particular focus on to the efficient use of natural resources, starting from land and water." It is in this logic that the PGRA is to be considered as a functional element to broaden the prospects of the European framework policy on water, as stated in the introductory notes of Directive 2007/60/EC which - by implementing the development of Flood Risk Management Plans within the districts - focuses on measures for the prevention, protection and management of emergencies in order to "reduce the risks of negative consequences arising from floods on human lives, the environment, cultural heritages, and the economic activities and infrastructure that are affected".

It is important to have a unified vision on flood risk management policies and water management, including:
- Convergence on the objectives, measures and actions set by the Flood Risk Management Plan (PGRA) with the environmental objectives described at paragraph 1, Article 9 of the Legislative Decree 49/2010 and the Water Management Plan (PGA).
- Interaction between risk mitigation and environmental protection in order to improve disaster resilience and, at the same time, preserve and strengthen the natural heritage.

The decision to act on the basis of a unified vision is, in fact, confirmed by the requirements of Directive 2007/60, which clearly states that the preparation of the Water Management

26. The first Water Management Plan of the Southern Apennines Hydrographic District (PGA DAM) was approved by the Prime Minister's Decree of 10/04/2013 and published in the Official Gazette no. 160 of 10/07/2013. The Second PGA DAM (2015/2021), drawn up in accordance with art. 13 of Directive 2000/60/EC and art. 17 of Legislative Decree 152/2006, was adopted pursuant to art. 66 of Legislative Decree 152/2006 by Resolution no. 2 of the Integrated Institutional Committee of 17/12/2015; approved pursuant to art. 4, paragraph 3 of Legislative Decree no. 219/2010 by Resolution no. 1 of the Integrated Institutional Committee of 3/3/2016. The process of verifying eligibility for SEA has been completed (Preliminary Report 06/ 2015; Public Consultation 07-09/2015; Supplementary Report and public re-consultation 12/2015 and 01/2016), which was concluded with the Administrative Provision protocol 0000060 - DVA of 25/02/2016 of exclusion from the SEA (Strategic Environmental Assessment) procedure. The first PGRA DAM (2016/2021), drawn up pursuant to art. 7 of Directive 2007/60/EC and art. 7 paragraph 8 of Legislative Decree 49/2010, was adopted pursuant to art. 66 of Legislative Decree 152/2006 by Resolution n.1 of the Integrated Institutional Committee of 17/12/2015; approved pursuant to art. 4, paragraph 3 of Legislative Decree no. 219/2010 by Resolution no. 2 of the Integrated Institutional Committee of 3/03/2016. The SEA procedure has been carried out in all its phases. In this regard, the documents presented within the SEA procedure - which was concluded by Motivated Opinion Decree no. 86 of 07/04/2016 of the Ministry of the Environment and Protection of Land and Sea in agreement with MIBACT - constitute an integral part of the approved Plan. The Ministerial Decree expresses a positive opinion on the PGRA DAM, the Environmental Report and the Monitoring Plan, with recommendations, suggestions and prescriptions referring to the contents of Assessments no. 1918 of 13 November 2015 by the Technical Commission for Environmental Impact Assessment - VIA and SEA and of the Assessments of the MIBACT Directorate General for Fine Arts and Landscape - Service III Landscape Protection prot. 3936 of 12/02/2016, which form an integral part of the Ministerial Decree.

Plans regarding the catchment areas (Directive 2000/60) and the preparation of PGRAs must be integrated. The measures, actions and interventions must be part of a systemic approach for district planning and be consistent with the guidelines provided by the Water Framework Directive and the Flood Directive. In this perspective, the elaboration of the PGRA, along with the review and updating of the Plans (for water and floods) and the related phases of active stakeholders participation must always be properly coordinated.
The two directives shared a common deadline (2015) for the updating of the Water Management Plan (already approved in the first version) and the first draft of the Flood Risk Management Plan. In Italy the deadline was met: in December 2015, in fact, both plans were adopted for the eight river basin districts in the country. They were subsequently approved in March 2016.[26] The unitary perspective for the management of flood risk and water policies is strenghtened, in the Southern Apennines district, by the convergence of objectives and actions in the Flood Risk Management Plan (PGRA) and the environmental objectives in accordance with article 9, paragraph 1 of Legislative Decree 49/2010 and as referred to in the Water Management Plan (PGA). The integrated management and perspective also takes shape from the participation and sharing of information and from raising awareness within local communities by promoting a series of participatory actions, which ease the implementation of the PGAs and PGRAs. Such actions are aimed on the one hand to mitigate and alleviate any conflicts of jurisdiction in the planning, while on the other hand to propose innovative, shared actions.

Environmental strategies, ecosystem services and green infrastructure in water management and flood risk control

Taking into account the environmental policy strategies of Directive 2000/60/EC, the protection of the environment and the safeguard of our territory's features are the core of this unitary management; and not only guarantee the level of quality of life, the preservation of acquatic ecosystems and flood risks prevention but also the preservation of the 'anthropic and cultural' environment's peculiarities deriving from our history and our culture. In particular, "the aim is to promote integrated Community policies with high standards of environmental protection and in accordance with the principle of sustainable development, as provided for in Article 37 of the Charter of Fundamental Rights of the European Union".[27] The strategy set by the PGRA consists in organizing an integrated management of the areas at risk of floods and coastal erosion by achieving the following priority targets: safeguarding of human life and health; protection of the environment; protection of the cultural heritage; safeguard of

27. Checcucci G., "La governance dell'acqua come pilastro dello sviluppo sostenibile" ("Water governance as a pillar of sustainable development"), EAI Energia, Ambiente e Innovazione, 2/2012.

economic activities. The measures identified with reference to the objectives are divided into: Prevention (M2 Actions), concerning all the activities aimed at avoiding or reducing the vulnerability of the above-mentioned assets' value (entity) - these are generally non-structural actions; Protection (M3 Actions), aimed at reducing the hazard (probability and intensity) of an event, the frequency of floods and their impact in specific locations - these are structural actions; Preparation (M4 Actions), which involve activities aimed at increasing the ability to react to events, to avoid or minimize the possibility of damages occurring as a result of events; and the Recovery of pre-event conditions (M5 Actions). It is important to point out that while identifying the actions, Prevention and Preparation measures have been favoured, as they are mainly "non-structural". Structural actions contributing to the territory's environmental safeguard - such as the District area - have also been favoured because they are most closely linked to the measures set by the Water Management Plan and Directive 2000/60.

The combined PGRA and PGA, which integrate the objectives of flood risk mitigation with those of safeguarding and improving the ecological functionality of water facilities, are called "win-win" measures, as well as the "green infrastructures".[28] The protection of natural resources and the search for sustainable development have become an important cross-sectional issue and are also present in management plans, in consideration of the fact that climate change, urban and infrastructural development make the areas already subject to hydrogeological risk even more vulnerable to natural disasters, such as floods, landslides, storms and swells. Such phenomena put the population and cities at risk, jeopardizing activities and damaging cultural and natural heritages, thus entailing considerable costs. The impact of these events on human society and the environment can, in many cases, be reduced by using solutions based on prevention, protection, preparedness and, above all, by considering the opportunities that arise from a good management of the natural heritage, of protected areas and of areas of high environmental, landscape, historical and cultural relevance. The opportunities arise from the eco-system's services, its known potentials and challenges, and from what the natural (protected areas and areas of high environmental importance), historical and cultural heritages may offer, especially when maintained, cared for, strengthened and enhanced. In particular for the protected areas, the sites of the Natura 2000 Network, the unprotected land strips and the degraded or disused areas, the Plan's purpose is twofold:

- on the one hand, to eliminate deterioration and improve the conservation status of animal species and river and coastal

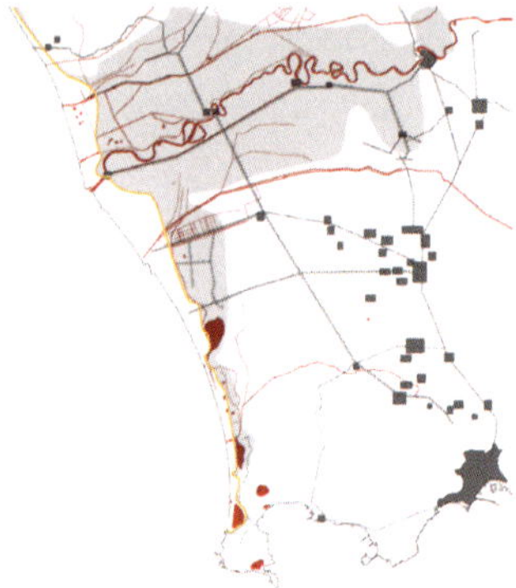

Re-design of the "general plan for the Bacino Inferiore del Volturno, with indications of the roads and canals built by the Royal Government until 1857". Overlapping of land systems. Source: Regno delle Due Sicilie, 1860.

Remodeling of the coastline by 1950, before the Piana's urban expansion and the coastal strip's colonization. Highlighted in blue are the rail lines. The extension of the canal and road system along the coast anticipate urbanization. Source: Campania Region, 2003.

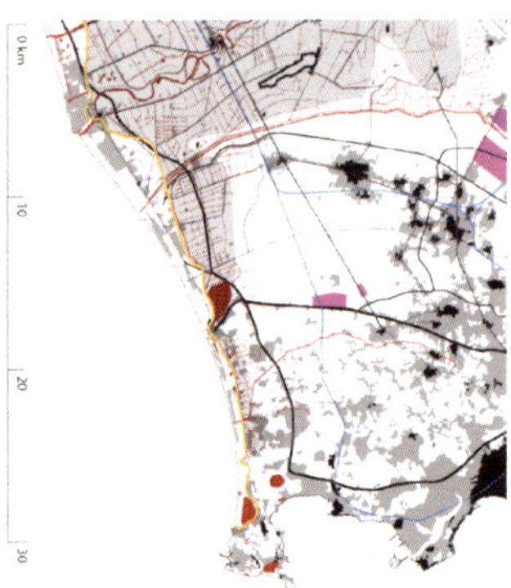

Urban development until 1990. The images in this page are by Davide Di Martino.

28. COM(2013) 249 - Green Infrastructures - Strengthening Europe's natural capital.

ecosystems of European relevance; to maintain and/or restore ecological connections between river, coastal and wet ecosystems;
- on the other hand, to focus on ecosystems that provide advantages to mankind, defined as "ecosystem goods and services", and that can contribute to fundamental aspects of the PGRA DAM, if preserved in good environmental and ecological conditions.
The relationships between the countless parts of an ecosystem constitute the processes generating the different ecological functions.[29] These functions are made by the ability of natural processes and components to provide goods and services that - directly or indirectly - satisfy human needs and guarantee the life of all species. Ecosystem-derived goods include, for example, food, water, fuel and timber; Services include, instead, water supply and air purification; flood and drought mitigation; biodiversity preservation; partial weather stabilization; the moderation of temperature peaks as well as wind and wave forces mitigation; natural waste recycling; land modeling and soil renewal; pollination and many other natural regulatory mechanisms.
A detailed knowledge and understanding of these areas' heritage, as well as the consideration of their intrinsic value and relative potential, is important for maintaining and strengthening their conservation, since their sustainable use is an essential element of human well-being and land balance. This is also the reason underlying the guidelines for biodiversity conservation of the Community and National Strategy: "Biodiversity and ecosystem services, our natural capital, are preserved, evaluated and, as far as possible, restored for their intrinsic value and so that they can continue to sustain economic prosperity and human well-being, despite the deep changes taking place at both global and local levels".
The benefits that can be derived from this heritage by reducing the risks of floods are many: starting from the positive effects of an organic and integrated management of the district's territory which is aimed to implementing actions that include "natural solutions", thus improving the resilience to disasters and playing an important role in the protection, reinforcement and preservation of our natural and cultural heritage. The creation of "green infrastructure" is a proven tool for achieving ecological, economic and social benefits through the use of "natural solutions". This helps us understand the importance of the benefits that nature brings to our society and that we should invest in actions to maintain and consolidate them. Green infrastructures form a network of natural and semi-natural areas strategically planned together with other environmental elements and designed to provide a wide range of ecosystem services in a rural and urban context. There are many benefits deriving from the creation of green

29. Santolini R., "Servizi Ecosistemici e Sostenibilità" ("Ecosystem Services and Sustainability"), ECOSCIENZA, n. 3, 2010.

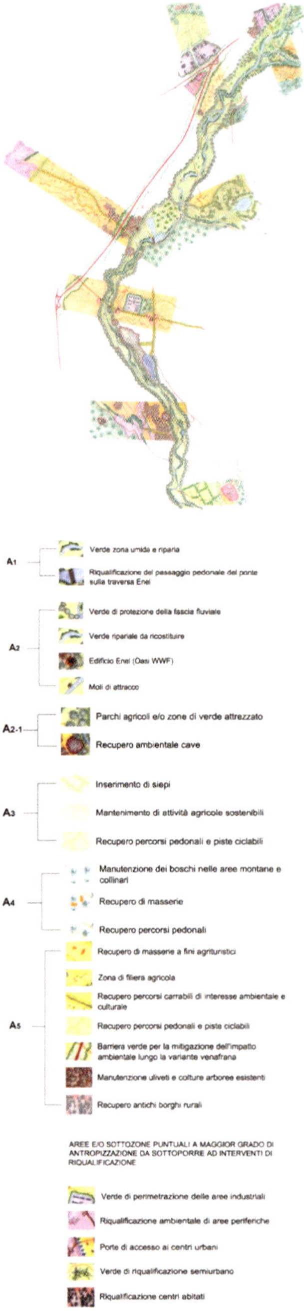

Preservation of wetlands: pilot area. Volturno's floodplain area: "Le Mortine" wetland.

infrastructures, such as: the improvement of the morphological quality of watercourses; water filtration; preservation of the water table; prevention of erosion; mitigation of the climate change effects; restoration of the watercourses' mobility and flooding areas and improvement of the quality of their waters and of the coastal habitats; flood lamination capacity-building; increase of the water retention capacity while respecting hydraulic invariance; protection of peri-fluvial areas; increase of river functionality; decongestion of the degraded, polluted and heavily exploited neighbouring areas and the protection of settlements and inhabitants; improvement of the aquatic ecosystems (including wetlands) and connected terrestrial ecosystems (including dune areas) conditions, optimising their capacity to withstand both catastrophic events and self-depuration; improvement of the soil's conditions (including plant covers), thus optimising its capacity to perform environmental and socio-economic functions; improvement of the forests' instability by optimising both its hydrological (reduction of run-off coefficient) and soil stabilisation functions; reduction of the ecosystem's fragmentation by enhancing the connectivity between the Natura 2000 sites.

Areas of National Interest: the Domizio Flegreo coastline and the Agro Aversano. District Management Plans

The Domizio Flegreo and Agro Aversano areas, located between the provinces of Naples and Caserta and object of the PRIN research, is a Site of National Interest including 77 municipalities and a population of over 1.3 million people, for an area of 157.025 hectares of land and 22.412 hectares of sea and coastal strip. This area falls largely within the Basin of the Volturno River and the rivers Agnena and Savone, in the province of Caserta. The Sites of National Interest (SIN) are areas defined by the Italian Legislative Decree 22/97 (Ronchi Decree) and Ministerial Decree 471/99; then re-defined by Decree 152/2006 on contaminated areas, in which the soil, subsoil, surface and underground water pollution is so severely widespread as to constitute a serious danger to public health and to the natural environment; these areas have been classified as the most dangerous by the government and their reclamation (soil, subsoil, surface and underground water) is necessary to avoid environmental and health damages. The SINs are identified by Decree of the Minister for the Environment and the Protection of Land and Sea (MATTM) in agreement with the regions concerned. They differ from other contaminated sites also because their reclamation procedure is assigned to the MATTM, which may also request the involvement of regional Councils, ARPAs (Regional Agency for the Protection of the Environment) and other subjects.

Law no. 426 of 9 December 1998, "new environmental interventions", provided for the launch of a national programme for the reclamation and environmental remediation of polluted sites by identifying interventions of national interest and priority interventions in industrial areas and sites at high environmental risk, as defined by the MATTM and the concerned regions. These sites are 57 (28 of which are located along the coastal strip) scattered throughout Italy. In 2013, they were reduced to 39 by Ministerial Decree of 1st January 2013. The remaining 18 were identified as Sites of Regional Interest (SIR) and the responsibility for their reclamatio was therefore assigned to the regions. The reclamation of these sites is, in part, underway but there is still much to be done about the soil, water and air pollutions.
Today, the Domizio Flegreo-Agro Aversano coastline is an Area of Regional Interest and, based on the updates on the reclamation actions performed in the scope of the management Plans, it appears to deserve a greater attention as very few interventions have been performed (including characterization, risk analysis and reclamations). The Domitian coastline, the Basso Volturno and the Basso Garigliano areas constitute Analysis Units for the application of management plan measures; they are pilot areas in which the methodological experiments carried out over the years can continue as an opportunity to improve territorial and social security from an environmental point of view. Moreover, they represent an economic opportunity to improve the production processes' efficiency, particularly the use of natural and cultural resources. After all, the different accomplished actions have already delivered positive experiences, such as the "Environmental laboratory of the middle course of the Garigliano river" (2007/2010) and the Providune project Life + Natura and biodiversity "Preservation and restoration of priority dune habitats in the Site of Community Importance Pineta Della Foce Garigliano" (2009/2015).
These experiences have proven that the actions envisaged, in particular by the local authorities, are oriented towards an integrated system of interventions, both immaterial and material, designed for an active environmental protection of the territory, the river waters and the sea.[30] Finally, in the scope of the two management plans, the existing emergency in the Campania region's plain (contaminated sites) and its related problems is monitored through the updating of data acquired by the responsible authorities, the quality analysis of surface and underground waters and the mapping of flood risk areas.

Internship at the Basin Authority

The internship programme for two students of the 'Urban Planning, Territory and the Environment' degree course from

30. Nappi R., Pennetta M., Solino G., 2010 - "Progetto Providune Life + Natura". (Providune Life + Nature Project). Quaderni Medio Garigliano. Environmental Laboratory Project - Liri-Garigliano and Volturno National Basin Authority, p. 260, 268 ISBN-889553404-02.

the "Università Federico II" of Naples was conceived among the University and the National Basin Authority and within the framework of the PRIN Re-cycle research. The aim was the correlation of objectives and the interrelation of "drosscapes" and recycling issues with those concerning the Basin and District planning.
The objectives set by the research thesis on "Recycling and regeneration of drosscapes and ecological remediation of the adjacent urban spaces for the creation of new territories" are to be considered an integral part of the authorities' mission, namely to "contribute to ensure, through adequate fact-finding and planning actions in their area of competence, such as catchment areas and/or river basin districts, the protection, preservation and management of water and related ecosystems; the protection, regeneration and valorization of land; the management of risks to reduce the negative consequences for human health, the territory and properties, the environment, the cultural heritage and the economic and social activities; the protection of the environmental aspects linked to the water and soil resources as well as to the hydrogeological risk; the environmental compatibility of the productive systems; the acquisition and the diffusion of data, including the information to the public."
The internship focused on the "drosscape" issue and their reclamation in the SIR (Site of Regional Interest) Domitio Flegreo-Agro Aversano pilot areas along the Volturno river, with particular regard tof its mouth and the coastal stretch overlooking it. The internship developed starting from the evaluation of the planning tools in force[31], then through a series of targeted on-site investigations; the identification of abandoned and/or degraded sites along the Volturno river; the territorial analysis of the issues described in the objectives of the Basin and District planning, as well as the regional and province's planning; and finally the interpretative summary: sensitive and areas with potential; formulation of possible actions which, within the District's planning framework, are aimed at the environmental requalification or urban regeneration; creation of green infrastructure; stimulation of biodiversity for flood risk prevention and improvement of water quality.
In this context, the pilot area's choice was determined by the specific problems and potentials of the Domitio-Flegreo coastal area, the Volturno river, and of the Municipality of Castel Volturno (Caserta, Campania region), in particular the area located on the right bank close to the river's mouth.
Coastlines are one of the most delicate environmental systems as it "represents an area of transition between the marine and the internal environments, between water and soil, in a continuous exchange of matter and energy; the dynamics regulating it depend on the interaction of the environmental processes

31. In particular, the examined plans are: 1. Coastal Erosion Plan; Environmental Protection Plan; Water Management Plan and Flood Risk Management Plan of the Southern Apennines District; 2. Campania Region Reclamation Plan; 3; Disused quarry recovery plan in the province of Caserta; 4. Territorial coordination plan for the province of Caserta;

with the ecological systems present. In addition, the dynamics induced by social and economic activities influencing the coastal strip must be considered as well". The coast is the balance between the demolishing action of the sea and the constructive action of the rivers and is, above all, sensitive to direct or indirect changes resulting from human activities. A sandy coastal system, such as the Domitian coastline, is therefore a very complex one, linked to the interaction of dynamic processes connected in a non-linear way, such as: the debris of fluvial or marine origin; the winds and vegetationìs activity; erosion processes; the raise of the sea level and human activities.[32]
Unfortunately, the coast is mainly perceived as a place of recreation and of economic, tourist and productive activities, while its high naturalistic value is not taken into account. This lack of consideration for the rarity (e.g. dune habitats) and importance of these areas is one of the most serious threats to their conservation, as the lack of knowledge can lead to harmful behaviours. In these transitional areas between land and water, nature is to be considered, more than elsewhere, an enormous value to be protected; History of mankind, including the pre-existing archaeological heritage[33], must be considered with extreme attention for its conservation but also to draw from the past and traditions effective ways of managing the territory. Therefore, the coastal system must be analysed in a simplified way and exploited without consequences; on the other side, for an effective management and protection of the coastal environment in a broader sense, it is necessary to consider all the processes, factors and phenomena within the system. The area of Castel Volturno is densely populated, with an evident phenomenon of generalized imbalance resulting in an increasing retreat of the coastline towards the inland areas. Besides causing the loss of beaches, dunes and of areas related to sustainable bathing activities, such issue has generated additional risks. In the context of a highly compromised environmental system, this area represents an important multidimensional resource in which densely urbanized areas, natural areas of great landscape interest and infrastructural networks are intertwined. In fact, this area stands out for its variety of ecosystem units - which define the historical landscape and the articulated environmental mosaic, representative of the territory's richness and complexity in terms of climate, hydrology, geology, morphology, soil and vegetation[34] - and for the wide range of cultural, historical, architectural and archaeological resources, as well as for its typical sustainable productive resources, which together constitute a heritage of great value and potential. The potential of this area is further enhanced by its geographical position as a crossroads between the rivers' mouth areas (Garigliano, Savone, Agnena, Volturno and Regi

32. Pennetta M., Corbelli V., Esposito P., Gattullo V., Nappi R., Environmental impact of coastal dunes in the area located to the left of the Garigliano River mouth (Campany, Italy). 3rd International Conference on the Management of Coastal Recreational Resources Beaches, Yacht Marinas, Coastal Ecotourism and Coastal Hazards, 27th - 30th Oct 2010, Grosseto (southern Tuscany), Italy.

33. Turri E., Il paesaggio degli uomini: la natura, la cultura, la storia, Marsilio, 2003.

34. Di Gennaro A., La terra lasciata, Naples, Clean, 2008.

Lagni) and the Domitian Coast (its natural completion), among hills, mountains and the volcanic area, and close to parks (Roccamonfina, Foce Garigliano, Monti Aurunci, Riviera di Ulisse), natural reserves (Falciano lake, Volturno-Costa di Licola river mouth, the Variconi oasis) and several archaeological sites in the Campania and Lazio regions.
"The Campania plain alongside the Domitian coastal urban belt presents areas in which some of the most extensive urban and illegal transformations are concentrated, with landfills located within peri-urban territories that are already strongly affected, in terms of soil and water, by the intensive usage of chemicals in agricultural practices". Its extension, however, infiltrates into the built environment and draws a punctiform but dense repertoire of opportunities for transformation, extending the critical points of the "brownfields" to a very extensive network of urban spaces matched by design responses that are, unfortunately, limited by a sectorial and fragmentary logic.
The internship experience, in terms of knowledge of basin/district planning, of the methodologies experimented in the pilot area's project and the on-site inspections, prove that knowledge - together with the environmental systems' exploration in a scope of complementarity which makes up the territory and the interrelation of resources - is an effective approach to fight the issues and and enhance the potentials of the territory and the environment, in order to define sustainable and far-reaching developments.
A number of highlights emerged from the examined plans: the "criticality and vulnerability" of the studied area in order to define its issues; "the territory's values", in relation to its valuable and protected areas, in order to define its potential; "the proactive ideas for environmental requalification" to identify strategic objectives and a mosaic of integrated actions aimed at preventing the risk of floods and coastal erosion, as well as promoting urban and environmental regeneration, which have been included among the criticalities and potentials of the area and adapted to this phase of climate change through proposals, guidelines and concrete actions.
The mosaic of possible actions range from: identification of abandoned, degraded and dangerous buildings; demolition of abandoned buildings in areas at risk (coastal and river); transformation of residential buildings' ground floors into open and green spaces; increase of naturalistic value along the river area; preservation, valorisation and management of residual areas of environmental value; creation of a recognisable gap between the built-up areas and the sea; creation of sustainable land-sea penetration axes; recovery and restoration of degraded dune habitats; regeneration of the dense fabric of seasonal houses, not only through demolitions but also through interventions of recovery, requalification and/or urban restructuring for the creation of green areas and public spaces.

Results of the exchange of information and other ideas

With regard to the purpose of triggering the dissemination of values and criticalities, the achievements can be expressed in terms of the information and contents made available to the PRIN Re-cycle research working group by the Authority and in the consequent follow-up debate and investigation of the risks emerged. The transmission of knowledge, in fact, was identified as the primary requirement for an effective circulation of information on procedures, decisions and the progress of the planning process. This knowledge is a fundamental basis for the drafting of the basin and district plans and a key element in the assessment of the effects on the territory and the environment. At the same time, this has stimulated a debate, fostered research, training and a greater attention for an approach towards awareness and responsibility in relation to the territory's vulnerabilities and risks.
Good planning is the first key action for risk prevention, for a better link between urban planning and security and for the adaptation to climate change. There is no doubt that climate change is a fact and that it has exacerbated the population's exposure to risk. The deep

changes in climatic conditions have brought exceptional factors closer to us today, impossible to delete forever; and the densely built territories also contribute to this situation.
Too often urban growth has not respected local territorial morphologies and has not considered the maintenance of water systems as a planning component. With regard to the goal of implementing the preliminary framework of measures/actions foreseen for the district's management plans, the research group's contribution and the investigations carried out during the internship certainly allowed further verification and research for new measures of prevention and protection. The following actions were added in the final framework of measures attached to the PGRA DAM:
Among the M2 Prevention Measures, M24 Other Categories for the safeguard of life, of the inhabited areas, of the economic activities, the environmental and cultural heritages, consisting of monitoring actions, studies, protection and safeguard. In particular, sub-measure M24.2 "Proactive management" refers to a process of cultural transformation in which the involved entities are involved in the process of changing the reality by promoting several shared actions which facilitate the implementation of the PGRA (non-structural actions, e.g. guidelines, manuals) and are designed to mitigate any existing conflicts that may arise in matter of competence, plans, projects, etc.
Among the M3 Protection Measures, M34 Surface Water Management relates to physical interventions aimed at reducing floods from surface waters generally, but not only, in urban environments. In particular, sub-measure M34.2 "Identification and implementation of urban regeneration interventions" in the river and coastal areas, aimed at providing ecosystem services. These interventions include: the creation of ecological networks and green infrastructure along river, coastal and wet areas that can reconcile urban and natural environments; the demolition of buildings located in areas at risk and delocalization in compatible areas; the recovery of resilient spaces, waste and marginal areas that are not yet under human control; and the ordinary maintenance of watercourses.
Finally, in the scope of establishing a constructive dialogue with the city and its institutions, the stakeholders, operators and investors (public and private) interested in the issues of research and, more generally, in the issues of recycling as declined in the territory's analysis, the prospect could be the establishment of an "environmental laboratory on the territory or/and river and coastal contracts". Tools for the shared implementation and promotion of participatory practices integrated in the planning processes, able to create a system of relations between institutions, local communities and citizens; to share knowledge and increase awareness among citizens and economic actors operating on the territory at risk on the issues of land care, preservation of natural resources, urban regeneration; to broaden the vision of individual municipalities that are meant to "rise in scale" and become confronted with the planning of large areas; to set in motion sustainable design capacities and generate integrated projects and implementation programs capable of interpreting the inside and outside, the vulnerabilities and values, limits and potentials, local and global, identity and development, rural ecology and urbanity, natural, cultural and economic resources, services and infrastructure, needs and requirements, demand and supply.
The real challenge is to understand, teach and share the knowledge that progress does not only mean producing more, selling more and consuming more. It is also about recognizing the signs of nature, healing ecosystems, recycling and recovering the "drosscapes" scattered throughout the territory, restoring urban spaces, pursuing a new and organic landscape, providing environmental and social services and sustainable economies.

The Domitian coast area: from waste to re-evaluation

Francesca Fasanino

Throughout urban history waste and cities have always coexisted.

However, we are assisting today to the invasion of the waste subject in social spaces and city planning. Waste that, likewise factories, become the mark of a deep change imposed by the first industrial Capitalism. Garbage, waste, and the recycling concept it involves become the key element of the productive process: this last aspect is ultimately meant to stress the difference with nowadays, where waste is inextricably bound to the driving force of mass consumption.

City planning in the first years of modernity expressed the idea of a project, it projected forward its present, in which waste was considered as a significant component, not yet concealed and removed. Contemporary cities instead push away and hide their waste to avoid seeing it. They throw it back instead of forward. This lack of planning for the future deeply characterizes the cities of late Capitalism.

In modern and contemporary cities, the waste issue continued to be a problem difficult to be solved. More and more areas started to be allocated to waste collection, with consequent issues that not always (if almost never) have been resolved. For decades, huge waste storage areas have been growing next to the cities, and often garbage was not collected separately, but indistinctly. The consequence of which has been the rising of real, often many hectares wide, dross cities, with no idea of their long-term destiny. For many years after the 50s nobody paid real attention to these areas, implicitly assuming that the two cities - the real and the waste one - could coexist ignoring each other. Finally, the need for a solution impetuously arose, but large portions of land were already compromised; whereas the cities' expansion to metropolitan areas had become unstoppable, frequently approaching the dross cities areas. The issue started to be perceived as

explosive also thanks to the diffusion of the environmental awareness.
The importance of abandoned areas in cities and the territory in general was first embraced by Kevin Lynch in the late 50s, fascinated by the opportunities of re-using these areas, such as rail yards, cattle pens or floodplains, to define new forms for the contemporary city.
In Wasting Away, published posthumously, he argues in fact that "useless wrecks mixed randomly can suggest new shapes while maintaining the pathos of their old meanings."
It results that the abandoned city, while considered as terrifying and degenerated in the collective imagination, acquires a pleasant look to those who engage in planning while using these materials to represent a wild reality: spontaneously savage, a mix of freedom and danger.
Bulwinkleland, an open-air museum in West Oakland, California, is an example of this land art, realised by Mark Bulwinkle, artist and owner of the place, recycling abandoned materials in the neighbourhood. As he stated himself in the New York Times[35], he arrived in West Oakland directly from England in 1972. The neighbourhood was degradated, mainly popular for prostitutes and criminals: Mr. Bulwinkle's house was paid 10.000USD only, and after attending a school for carving metals he started using dross material he found all around to make these pieces of art now considered unique and valuable. With the passing of time, West Oakland has been regenerated, and from abandoned it became a residential neighbourhood for rich people in which, however, 9000 squared meters are still occupied by the garden of iron wrecks.
"We know that the landscape is intimately linked to our own subjective and cultural interpretation, that the environment is an objective list of living components, and that the garden

35. Rafkin L., "Bulwinkleland", The New York Times, U.S., 3rd March 2012. Source: https://www.nytimes.com/2012/03/04/us/bulwinkleland.html

Bulwinkleland by Mark Bulwinkle, West Oakland, California, USA

is the place for dreams, a collection of the best and political project." With these words, Gilles Clément opens the chapter on recycling cities in the book Gardens, landscape and natural genius[36]. It is particularly important to underline that the author's reflection dwells on two founding aspects for the contemporary city's shape: the first is the absolute necessity of recycling used up territories; the second is the use of time in the construction site, its use, acquisition and abandonment.
We find ourselves to prove that a full-cycle ecology hands us back to matter, and to ask ourselves which could be the tools to be used to give back to the environment, therefore the landscape, what we subtracted. Recycling thus becomes an imperative for a consumed society: "the recycling city would then become the compromise between ephemeral architecture and permanent architecture transformed into a garden"[37].
In the meanwhile, the issues of waste and dross have determined new "landscape challenges" in areas previously not considered as such, like the Domitian coastline, where the system'scritical fabric, made of: quarries, landfills, industrial areas, canals, cultivated land, uncontrolled construction on the dunal and retro-dunal areas, the same system of dunes and area behind the dunes; all this has contributed to the creation of this framework of "new negative landscape territories" to which a solution must be found.
From here the need to offer a model of sustainable city able to optimize distribution and consumption through a reduction of the environmental impact is more and more gaining ground. Europe is among the most urbanized continents. Cities are not only economic engines but also play a unique role in providing the founding elements of our quality of life from all perspectives: environmental, cultural and social. However, all cities are called to keep the balance between economic activities and growth on the one side and the cultural, social and environmental aspects, on the other. Urban sprawl and the spread of settlements with low population density are the greatest threats to sustainable territorial development. Furthermore, in some regions, the incentives to re-use dismissed sites are insufficient, with a consequent increase of pressure on uncontaminated soils. Finally, there is a general lack of appreciation for soil and landscape's value, which is not yet considered as a limited, non-renewable resource.
In this respect, if we consider the European Union's strategy set for 2020 and 2030, the fight against climate change represents one of the five main pillars of the European's global strategy for smart and sustainable growth. In particular, this strategy seeks to ensure a 40% reduction of greenhouse gasses emissions by 2030 compared to the levels of 1990. Leaders of the EU countries have agreed to set the threshold of renewable energie to a minimum 27% of the total energy

36. Clément G., Giardini, paesaggio e genio naturale, Ed. Quod Libet, 2013.
37. Ibidem

Waste dumping in Giugliano (Naples)

Ex-Resit landfill and gypsy camp in Giugliano (Naples)

Regi Lagni Canals

consumption since these are essential for the transition to a sustainable, safe and competitive energy system. Also, given that energy efficiency is a key component of the strategy, an indicative 27% saving objective was approved for 2030[38]. Thanks to the pressure put by the EU, and carried out by all countries, according to the International Energy Agency (IEA) in 2040 the share of renewable energy will represent approximately 33% of the world's total mix.
A similar scenario, through an integrated and multisectorial plan able to combine in a sustainable way the Institutional dimension with the other fundamental ones concerning the environment, the economy and society, is more necessary than ever in the Domitian coastline. It should also consider an integrated vision of the territorial system and a strategic sustainability assessment, in a general framework in which all impacts are defined, from the economic to the social and environmental ones. Furthermore, this means finding, in a mostly urbanized, although not excluding natural and protected areas, territorial context a conceptual reference able to blend sustainable mobility, zero consumption and a new way of living.
In this context, the research project coordinated by Professors Giuseppe Scaglione and Mosè Ricci from the University of Trento called L.I.D.O. (Learning Island Design Opportunities) has the objective of building sustainable scenarios for Venice's Lido.
L.I.D.O. intends to support and restore the Lido's territorial general capacacity through the implementation of sustainable environmental scenarios, as defined by the European regulatory framework, in order to allow the island to be total Green by 2050. This means limiting, mitigating and compensating the current failings through a high standard territorial planning based on an integrated approach that involves all the competent public authorities, particularly the ones responsible for the management of the territory.
The L.I.D.O. Project proposes three sustainability scenarios, on which different transformation and change set-ups can be considered. These scenario do not exclude each other, they trigger complementary actions instead. In this sense, they define qualitative targets which, depending on the contingencies and opportunities, can be paramount or not: "Nuovi isolani" ("**new islanders**", meaning the Lido as an island for "smart communities"); "Mobilità dolce" ("**smooth mobility**", the Lido as a "slow island"); "Consumo zero" ("**Zero consumption**", a "carbon-free Lido").
The transformation and change set-ups for the Lido process sustainable scenarios that combine compatibility and integrated actions with new uses and operations serving the community, with a specific transition approach used as best

38. EU policies on Climate Action, VV.AA., ISBN: 978-92-79-41350-6

practice for participation, management and surveillance of performances through time[39]. A better standard in the quality of life perceived by old and new islanders, by inhabitants as well as tourists, is therefore assured in time by the spatial reconfiguration at the urban and landscape levels through actions that, starting from waste and residues, and through physical and intangible networks, achieve a "smart" and "green" model.
In conclusion, the L.I.D.O. research project stands as a symbol of a cultural operation where commitment is considered as ethical need, willingness for reinterpretation and innovation in planning. The proposed example considers the sensitive components in matter, state, location and community, and acts through clear spatial classifications to transform environmental passivity in a resource for all. If we saw the Domitian coastline not like a landfill but as a value, we would shift from extreme situations to the critical recognition of how the qualities of an abandoned and discarted area can make it possible to save the landscape itself.

Fuel pallets ("ecoballe") in the landfill "Taverna del Re" in Giugliano (Naples). The pictures in these pages and the previous one are by Davide Di Martino.

39. Guidelines proposal for the Lido, edited by Prof. Arch. Consuelo Nava, Università di Reggio Calabria.

Land remediation network: ecological and social networks for new dimensions of the open space

Antonella Senatore

The methodological contribution indicates the criteria and approaches needed to bring out a new design vision for contemporary territories, "rethinking" and "re-reading" the dynamics and themes that characterize them, and highlighting new ways of using and perceiving the open space (public and for public use).

Construction of knowledge

The complex environments and metabolic dynamics underlying contemporary cities and territories require a change in the method of investigation, as well as innovative representations capable of returning a level of shared and transferable information. Building a framework of knowledge which allows the re-reading of the city in its spatial forms, in its functioning and in its complex processes which determine its infinite trajectories of growth and contraction, is a necessary condition for the support of innovative projects.

Sometimes it is a question of identifying the territory's implicit structures of narration. Although they are not immediately available and cannot easily be deciphered, they can be recognized as clues and traces that must inevitably be brought to light in order to reconstruct a latent story. Recognizing this implicit writing means re-reading the places, starting from the actual ways of using the city, from the actors who actively intervene in it and, above all, from the requests for change that need to be intercepted and interpreted in order to create new and representative images of a project for change.

Possible trajectories

The objective of the proposed contribution[40] becomes that of understanding how a thorough knowledge of the places and questions is able to influence a proactive design dimension

40. The work presented was developed by Antonella Senatore within the degree thesis in Urbanism (CdL PTUPA, DiARC - University of Neaples "Federico II"). Supervisor: prof. arch.Carlo Gasparrini. Co-supervisor: arch. Anna Terracciano.

which is more aware of contemporary territories, by designing "possible trajectories".
The area covered by this dissertation is the Domitio-Flegreo coastline (north of Naples, reaching the border with Lazio region), where a quadrant measuring 25x25km has been identified, with a total surface area of 802,19 square kilometers in which there are 36 municipalities with a total of 823.012 inhabitants. The coast is part of one of the 6 former SIN (Siti di Interesse Nazionale/National Interest Sites) of the Campania region.
Work was done on the construction of drawings capable of relating the character and the many configurations that the open space assumes in the territories of urban sprawl, where the explosion of the city has given rise to a rampant consumption of land with a consequent loss of identity value and the formation of generic landscapes. These landscapes, detached from the context, have contributed to weaken, disperse and therefore compromise the environmental and landscape systems with which they interact in a less righteous way, transformed by processes that need to be observed as a whole in order to grasp the complexity of the mutual relations between the natural and the constructed environment.
The proposed readings include the "drosscape", that is, the contemporary city's archipelago of waste areas, sometimes wedged in compact and consolidated structures to form an unusual porosity in the settlement fabric, sometimes in the fragmented form of urban sprawls.
A systemic reading of this landscape of waste returns a reticular conformation similar to an urban framework that is parallel and inverse to the city we are used to considering, which includes not only drosscapes but also the connections and infrastructures through which the diffusion of the sometimes negative effects that the latter generate are conveyed.

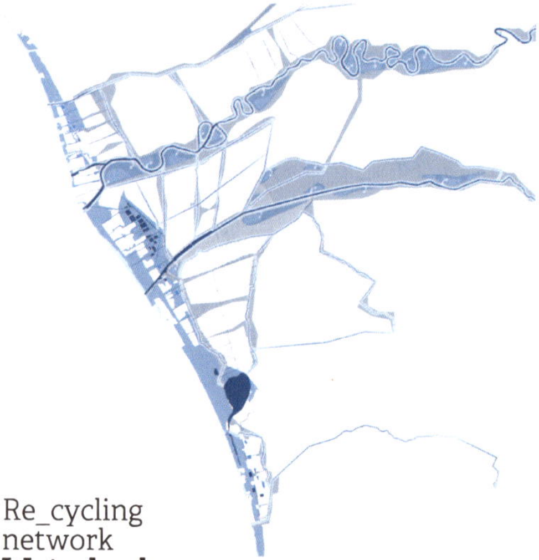

Re_cycling network **Waterlandscape**

Re-generating/Re-programming
The recycling of water landscapes consists of the re-naturalization of the river basin to increase its self-purifying capacity; of the agronomic and industrial reuse of the purified water that will contribute to the recharge of the aquifers and to the mitigation of the phenomenon of ingression of salt wedges found in the coastal strip, and of the re-appropriation of the blue landscapes.

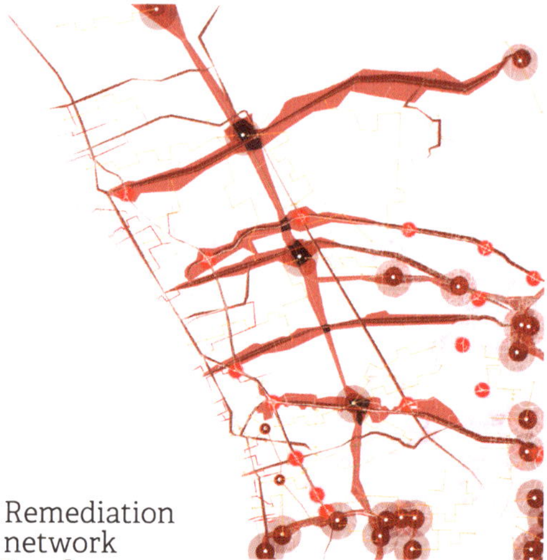

Remediation network **Landscape**

Re-generating/re-programming
Infrastructural interventions aimed at strengthening the network of road and rail transport, promoting the primary network, favouring the production systems, and the secondary network - the connections between railway stations and urban centers. Re-qualify the landscapes that the infrastructure has configured as marginal areas of infrastructural compression through operations of re-stitching.

Attractions **Infrascape**

Renew / Re-connect
To create a territorial infrastructure able to re-connect the coast and the hinterland through a multiplicity of actions on a large and small scale, and at the same time re-stitch the meshes of the plain, re-using and re-destining those unproductive, contaminated, abandoned areas that have reached the end of their life cycle (quarries, landfills, compromised land, and so on) by assigning them new functions and filling those large voids that today encumber the territory in question.

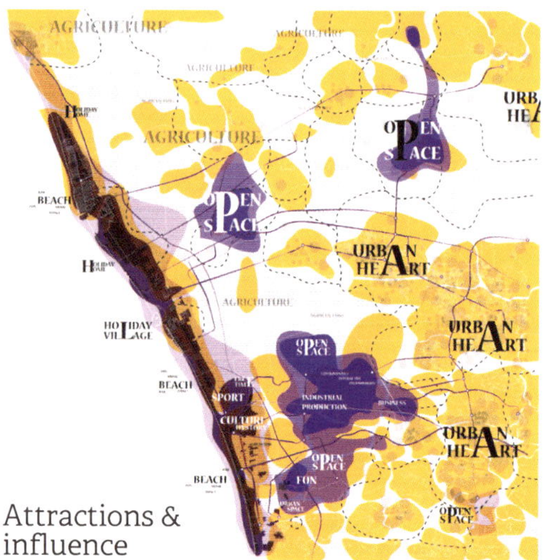

Attractions & influence **Communityscape**

Integration/Re-destination/Formalization
To give shape to the latent demands of the territory, through the formalization of practices already present in the territory. To create a structure not only physical, but also social and cultural, that supports the development of new cities and their union with the coastal city and with those already existing, by promoting a coordinated and homogeneous system of actions throughout the territory.

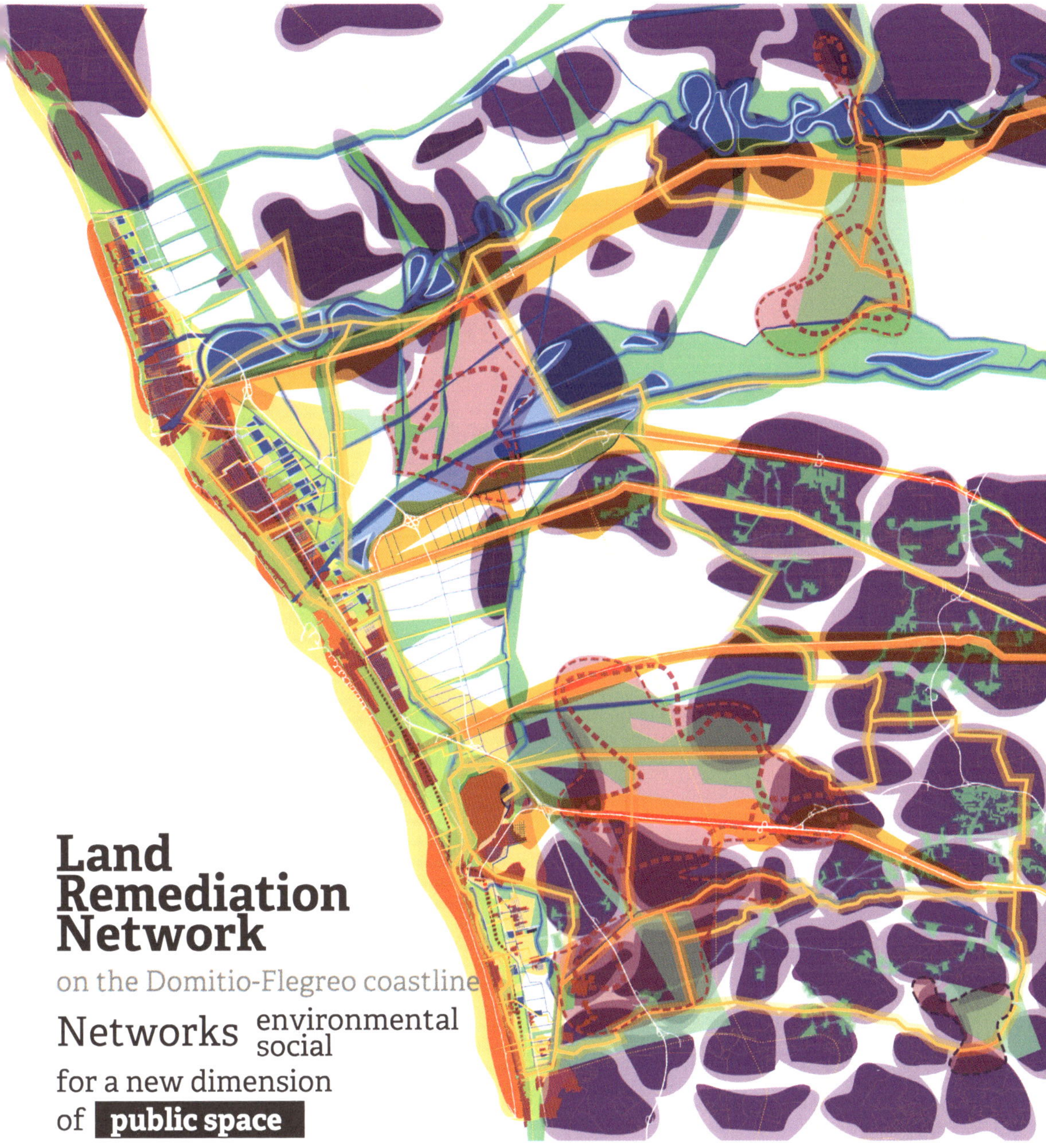

Land Remediation Network

on the Domitio-Flegreo coastline

Networks environmental social

for a new dimension

of **public space**

ECONOMIC and SOCIAL MIXITÉ

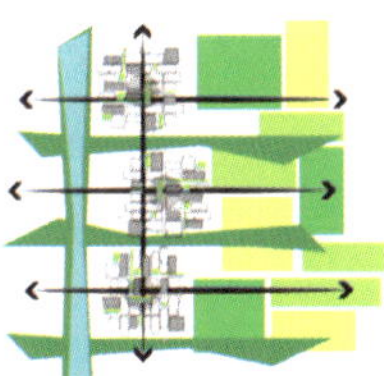

INTEGRATION with NATURE

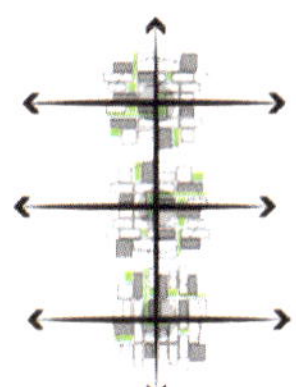

MULTIPLE CONNECTIONS

URBAN REGENERATION AND REQUALIFICATION

PRODUCTIVE AGRICOLTURE/ ECONOMY

Network Composition

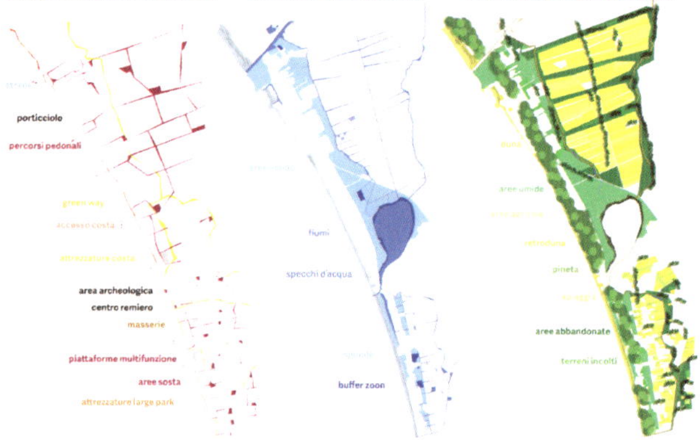

Land Use Distribution of Project

Concept Project

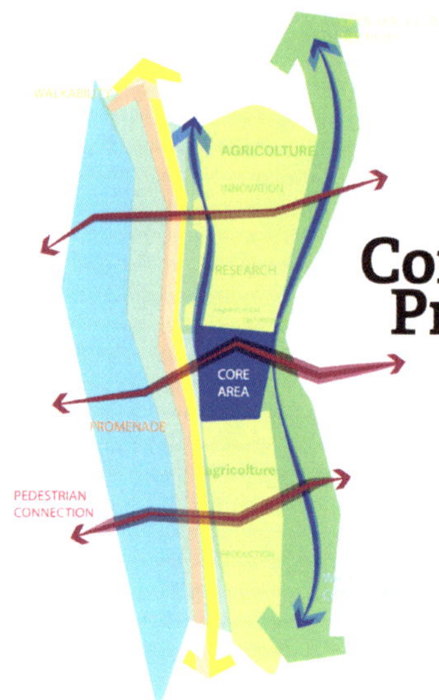

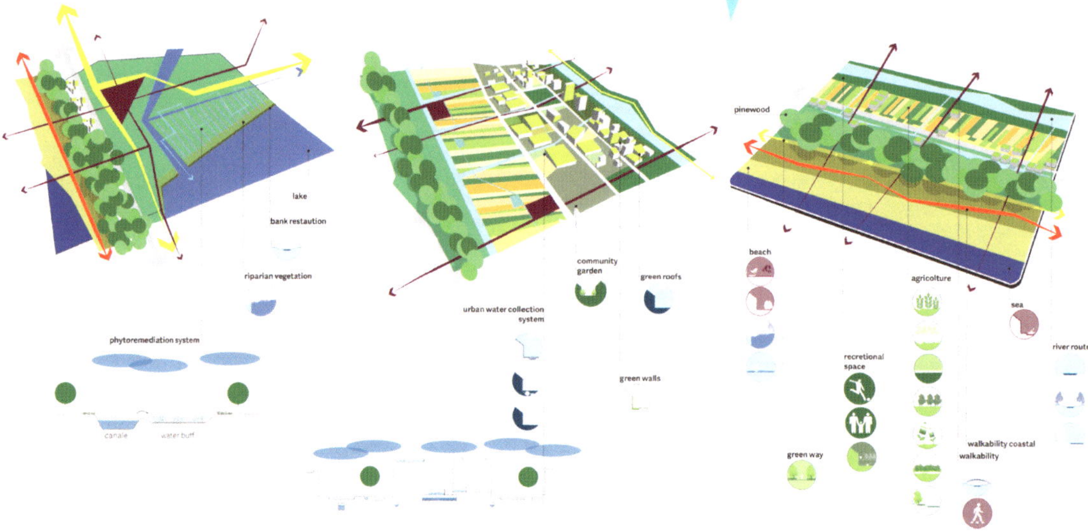

Composition of Landscapes

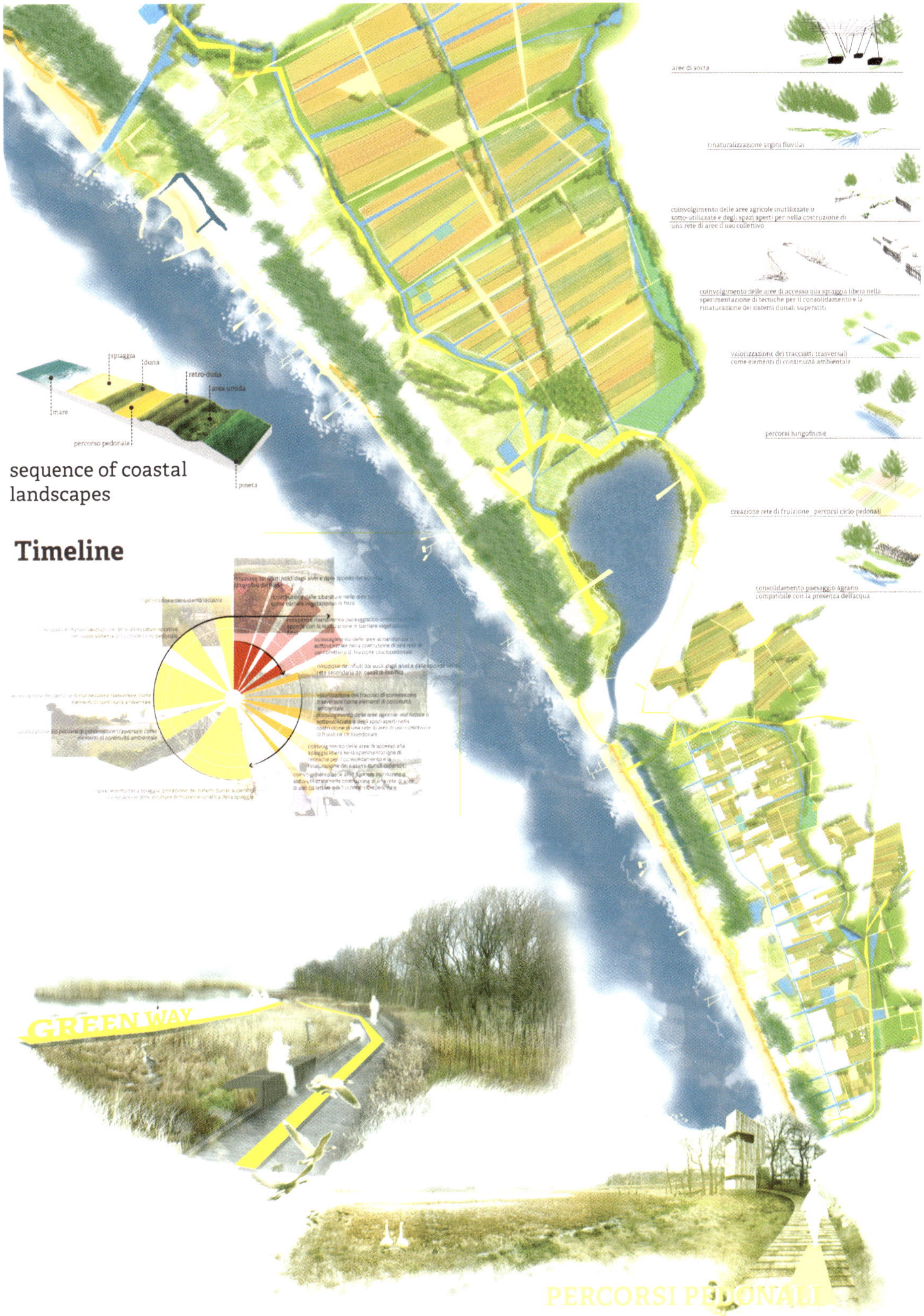
aree di sosta
rinaturalizzazione argini fluviali
coinvolgimento delle aree agricole inutilizzate o sotto-utilizzate e degli spazi aperti per nella costruzione di una rete di aree d'uso collettivo
coinvolgimento delle aree di accesso alla spiaggia libera nella sperimentazione di tecniche per il consolidamento e la rinaturazione dei sistemi dunali superstiti
valorizzazione dei tracciati trasversali come elementi di continuità ambientale
percorsi lungofiume
creazione rete di fruizione _ percorsi ciclo-pedonali
consolidamento paesaggio agrario compatibile con la presenza dell'acqua
spiaggia
duna
retro-duna
area umida
mare
percorso pedonale
pineta
sequence of coastal landscapes
Timeline
GREEN WAY
PERCORSI PEDONALI

These are therefore urban machines that are functional to the city's metabolism and that the city itself tends to expel, together with the system of networks to which they are connected.
In this research, we attempted to give an interpretation of this phenomenon in the Piana Campana, trying to denounce its intensity by means of the design of its inverse dimension.
Through the re-reading of the specific geographical contexts, a new structure made up of nodes, areas of influence and networks of relationships has been identified.
The drosscape areas, starting from this reticular vision, are therefore the new "fertile spaces for the project", because they play a central role both in spatial forms and in urban processes, generating connections and relations to macro and micro scale and constituting a new territorial network that can potentially trigger new processes of urban re-generation in a proactive perspective.
They can act as waiting spaces, conveying a demand for transformation that must be re-read beginning with the identification of the latent design implicit in the structure of such places - the only element capable of identifying the questions and requests that they pose.
A circumstantial analysis of urban signs and the way of living in such places is fundamental for the reconstruction of that pattern of landscapes and environments that depicts new urban realities as relational structures, both existing and potential. Beyond the city's physical space, there is the network's virtual space in which informal actors can express their commitment. All this contributes to the understanding of the different use of the territory and how it defines its space.
The (re)readings have highlighted their social, political, spatial and environmental character.
These elements are becoming increasingly important in the construction of new design paradigms, to the point of becoming the backdrop for every possible reflection. These highlights have been used to construct some selective images, used to identify the priority themes of intervention, while trying to draw out possible trajectories:
-Waterlandscape
The redesigning of the coastal hydrological system sees, through a series of micro-interventions on marginal, residual and abandoned areas, the conformation of a widespread system of purification of polluted water, introducing new devices for the virtuous treatment of water.
- Landscape
The construction of a complex, continuous and usable landscape, safeguarding and enhancing its productivity in economic, environmental and social terms.
-Infrastructure Attractions
The re-organization of the infrastructure system through the strengthening of the road and rail transport network.
-Integration_re-destination_formalization
The construction of a complex and articulated collective infrastructure that exploits the urban fabric and pays attention to the size of the area; the creation of living spaces and the concrete possibilities of access to services. The territory in its overall articulation should be a place of reflection on collective welfare, understanding the collective living space as a new and contemporary great public work.

Urban and environmental regeneration strategies for the right side of the Volturno river

Mara Longobardi

Claudia Pengue

The area covered by this study is located on the Domitian Coast, in the province of Caserta and specifically in the area of Lake Piatto, on the right orographic side of the Volturno river. The landscape scenario is marked by the contrast between the large open agricultural territory and the compact outline of the coastal urbanisation. The study was born in the context of an internship at the National Basin Authority of Liri Garigliano and Volturno, the body responsible for drawing up the Basin Plan for its organisational and functional re-organisation aimed at soil protection.

The project research focuses on three aspects that are typical of the area: site degradation and the causes behind it, as well as the associated impacts; hydrogeological issues, in particular flood risk and coastal erosion; the ecological dimension, exemplified by the coastal dunes and the "river habitat".

The project is aimed at identifying strategies for the solution of environmental and settlement issues by providing social services and eco-systemic re-qualifying measures.

The territorial analyses refer to territorial plans regarding the greater area whose physical scope is the river basin, capable of overcoming the fragmentation produced by the adoption of reference areas with exclusively administrative boundaries. It is a landscape whose environmental elements, such as beaches and recent dunes, a dense network of water bodies, as well as various natural and protected reserves, are, on the one hand, subject to a phenomenon of diffuse pollution of the soil and contamination of lake water sediments. Meanwhile, on the other hand, they are subject to a strong increase in the consumption of ground.

In addition, the road infrastructure that runs through the territory has a significant impact, marking a highly fragmented landscape, as it draws net margins and generates resulting spaces whenever it come into contact with the surrounding rural environment. In particular, the Via Domitiana marks two different areas of landscape, agricultural and urban. The latter, with the exception of a small part of historical heritage, is the result of the phenomenon of soil occupation for construction purposes which took place as a result of a lack of appropriate urban planning tools. The building fabric thickens along the coast and is concentrated in plates, shut off from the outside by the large infrastructures.

Today, there are evident traces of underuse in the settlements, as well as their dismissal and housing "filtering": many second homes have, under some circumstances, become shelters for migrants and agricultural labour, while in other cases they were abandoned, and so gradually deteriorated.
As far as open spaces are concerned, the interweaving of the environmental damages already produced by pollution and those due to the excessive sealing of the soil create a landscape of degradation that develops close to the inhabited centres. The quality of the environment depends on its degree of "resilience" which can be increased by experimenting with new strategies for the development of abandoned sites. These strategies would be linked to their ecological conversion and would be able to "circularize the metabolism" of production processes, also through reuse practices. All transformations of the landscape can be understood as moments of a continuous and unstoppable process of anthropization, bound to form a new habitat that we must deal with in order to reharmonize both the natural and the built environment, to restore a hydrogeological balance, to redesign the landscape, to create new urban quality and to develop alternative economies.
In this sense, the various abandoned, degraded, disused and abusive sites become new central aspects in the field of land reclamation and are also, both in economic and landscape terms, an opportunity for wealth for the territory, as they are able to represent a "new place" for the design of new natural spaces. Moreover, through the insertion of functions and the introduction of a "green tertiary sector" into a system, they dwell within a condition of possibility that transforms them from a "threshold" to a "place of value".
The project[41] works on two scales. The larger scale contains the river belt, the green network and the paths that serve it. These are made up of green areas that will preserve the current vegetation and cultural characteristics through experimentation with new "green economies" able to regenerate and strengthen the flora and fauna systems, making it also enjoyable for naturalistic tourist excursions. The other scale, which includes the new flood strip, and is close to the equipment, services and the built-up area, will configure a park for public use and leisure activities in which to locate the main gathering hubs.

41. The project has been developed by Mara Longobardi and Claudia Pengue as part of a dissertation in Urban Planning (CdL UPTA, DiARC - University of Naples "Federico II"); supervised by prof. arch. Michelangelo Russo and co-supervised by arch. Enrico Formato.

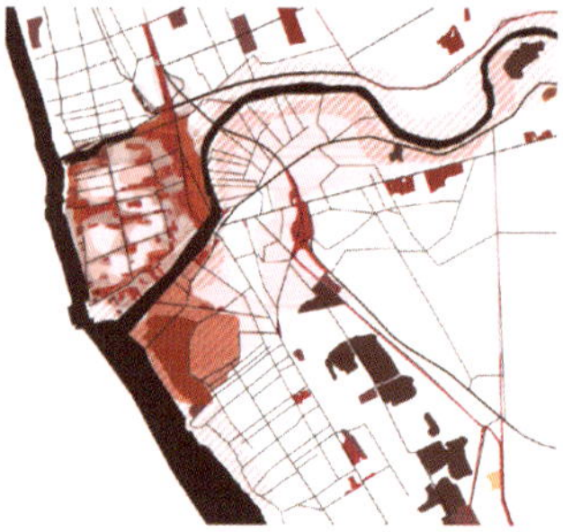

The Volturno river's mouth and drosscapes

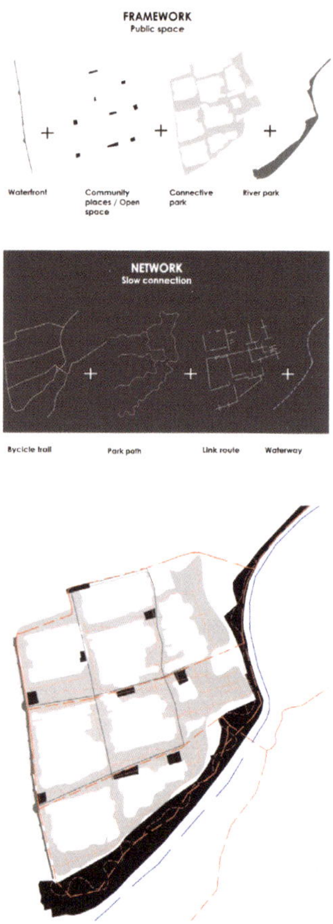

Fram_net: network of connections among different landscapes

To the east of such an area there are some buildings, many of which are degraded, whose proximity to the edge of the Volturno river is a strong point for developing a further relationship with the proposed river facilities. Currently, in addition to factories and warehouses, there are farms and rural businesses which are intertwined with the current structure of agricultural plots and thus are linked to the agricultural-environmental sector. This link, although not part of the specific site, is of strategic interest for the purpose of a virtuous connection of mutual support and development between the river park and the territorial redevelopment. The project outlines the first strategies for the environmental requalification of the territory.
To a first phase of the regeneration of ecological systems correspond demolition actions, aimed at the recovery of soils and spaces for new urban performance within the site: one of the ideas is to partially shrink the building barriers that were abandoned or placed in critical areas, so as to redefine the space and give the neighbourhood a new design. The main strategies are aimed at the creation of the "dynamic" area of flooding, the regeneration of the "dune habitat" and the re-permeabilization of the site through the spread of greenery.
The public space system is arranged along a green crown that acts as the limit of the built environment. The distribution of driveways has been designed at the edge of the greenery so as not to interfere with public spaces. In correspondence with the new urban centres, parking facilities are planned, from which paths leading to other areas shall depart. The potential value of open spaces has emerged from a concept of 'negative' urban space that can unveil new dimensions, resulting in relationships, exchanges and functional possibilities.
The idea of intervention aims to create a link between the agricultural and urban landscapes, inserting new possibilities for active enjoyment of the landscape. Through improvements that aim at a transversal connection with the embankment and the placement of a series of equipment and collective spaces, the goal is to relate the urban environment to the river. On a project scale, a continuity of greenery has been hypothesized, the characteristic element of which is represented by the topography of the areas; through the use of a differentiated system of land levels, the aim is the creation of open or introverted spaces while the possibility of wide views of the landscape is still available. This green system becomes the container of the new fabric of collective spaces, establishing a morphological and spatial continuity that guarantees the fluidity of relationships. Cultural centres, places for meeting and collective discussion as well as neighbourhood workshops are the territory of social exchange, able to nurture the neighbourhood's identity.

Focusing attention on the two main projects, the seafront and the river park, we propose, on the one hand, to restore the "dune habitat" through a new use of the coast through the pier and the walkway. On the other hand, the purpose is to transform the space between the urban and river into a real park, projected on the Volturno river and aimed at reducing marginality

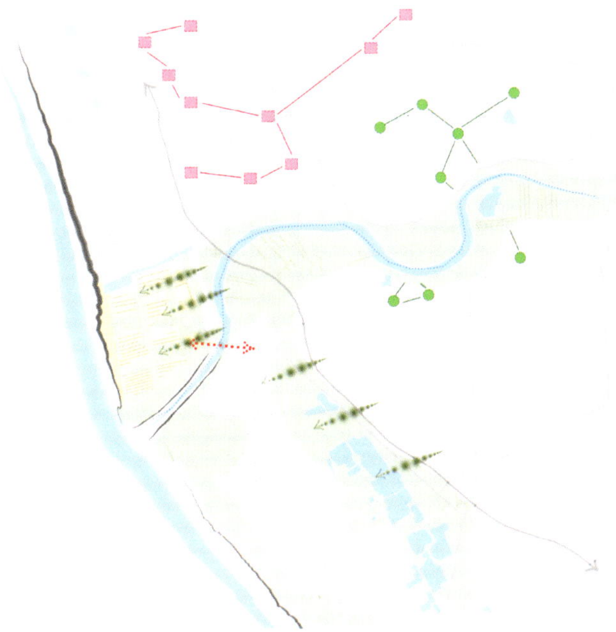

Concept: Network of interventions and interlinking of open spaces

River park

Green tertiary

Recovery of old warehouses and quarries

Enhancement of agricultural land

Pedestrian connection between the old town and the Volturno's right bank

Increased permeability/ accessibility

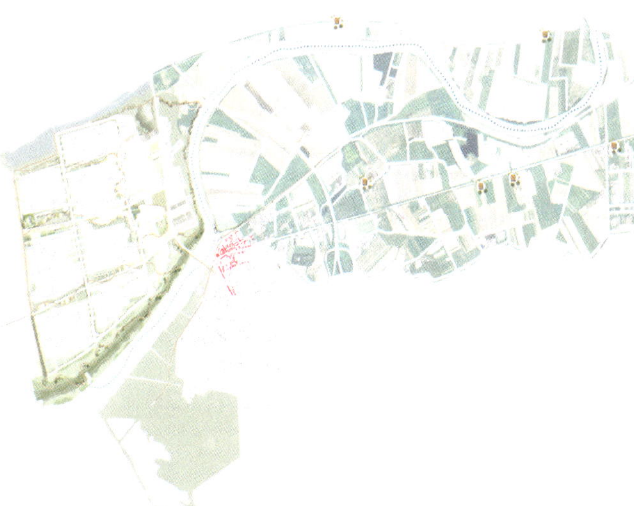

Masterplan and planning interventions

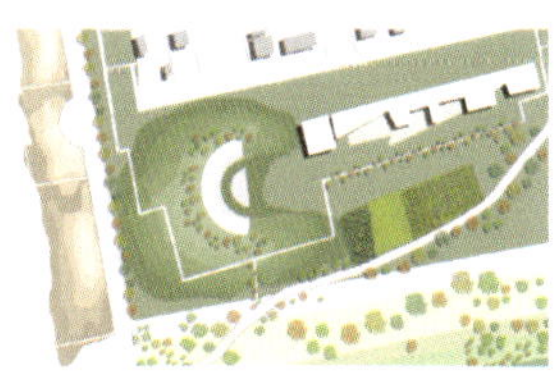

Three visions and a project for Caserta's landscape

Gabriella De Luzio

The peculiarity of the territory of Caserta lies in the simultaneous presence, within its palimpsest, of monumental complexes and large abandoned areas. For this reason, prior to interpreting the territory, the first operation to be carried out is a turn around: for too long the "grandeur" of the Royal Palace of Caserta and San Leucio has represented the city's privileged point of view. On the contrary, contemporaneity requires a broader horizon of meaning, which includes and gives dignity to those places that are ignored, abandoned, contaminated or residual, whose recycling can give life to new cores and to a widespread well-being.

Three dominant themes have been extrapolated from the information collected: the Royal Palace as an autonomous system, infastructure fractures and waste areas. These, in addition to identifying precise ranges of spaces where certain physical, historical and morphological characteristics prevail, are also the result of planning and land management methods that have failed.

First of all, the Vanvitellian complex of the Royal Palace represents a monumental fenced-in asset with no relation to its urban context.

The perimeter of the Ministerial Quarters is as rigid and impermeable as the edge of the Park: it virtually precludes any relationship with the urban context, thus minimizing the desired reverberation in economic terms and in terms of the use of the surrounding spaces. In addition, the legislation protecting the Landscape Plan translates into a mere preservation of the "status quo", even when this is represented by degraded spaces such as those along Viale Carlo III.

The second theme, that of infrastructural fractures, is the product of the sectorial logic with which the territory has been built in recent decades and is represented, in addition to the railway and the long-standing issue of moving it underground, by the presence of a fast flowing ring road that connects Maddaloni with Capua and that cuts through the so-called "cupe", the historic roads that ascend towards the Tifatini hills. By virtue of their presence alone, the infrastructures create residual, uncultivated and unused areas.

The final theme is that of waste. This concerns all those places that have not been considered worth recovering and have lost their original value by being deliberately fenced off and

forgotten by the city. Among these, there are the mountain quarries used for the extraction of limestone, which are of great landscape value as well as an ecological-environmental asset. The former railway station and the Saint Gobain area, the latter of which has been partially reclaimed in some parts over the years, although never through a single project, are among the places of industrial retraction. The barracks are another typical element of Caserta. Sometimes they consist of large expanses for military training, such as the former Macrico. There are also prompt elements of historical importance, such as the mill of San Benedetto. In order to address these problems, some figures of transformation were used, i.e. interpretative urban images on which it is possible to structure strategies, projects and policies.
The monumental element consists of the Palace's Vanvitellian complex and is based on the image of an axis. The proposal[42] is to recompose the latter by burying the railway, thus restoring the dignity of the landscape through the transformation of Viale Carlo III into a linear park. At the surface, the idea is to work on the permeability of the park border, opening all existing accesses and creating a further four access points at significant locations. The second element is that of the variant, which is complex because it intercepts a series of very different places (quarries, the cemetery, agricultural fields) but without relating to them. The idea is to transform it into a landscape infrastructure that holds an agricultural "green-belt" together. The functions, sequential and linked to land-use development, are concentrated in three significant parts: the first includes the area of the Policlinico and Saint Gobain, to which are assigned research and experimental activities in the food industry within a perspective of reusing empty industrial spaces; the second is located across the street and is intended to display quality products, allowing the rethinking of the vast exhibition area that is currently an expanse of concrete. Finally, the third area is located at the junction of San Leucio and, given its proximity to the Park and the Belvedere, has a tourist-recreational vocation. Finally, the civic element is based on the future dismantling of the Caserta-Benevento railway branch: the route will become a backbone for light mobility and will link together a series of disused spaces, which, once

42. The work hereby presented has been developed by Gabriella De Luzio, in the context of a degree thesis in Architecture (CdL Architettura 5UE, DiARC - University of Naples "Federico II"), supervised by prof. arch. Michelangelo Russo. Co-supervisors: architects Enrico Formato and Raffaele Cutillo.

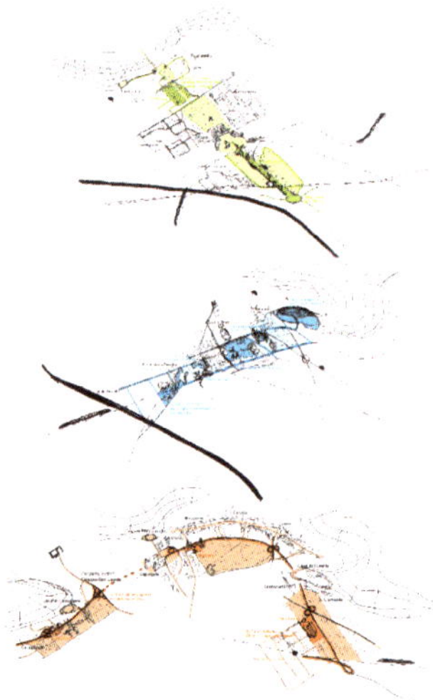

Interpretative scheme for the three dimensions of transformation

One of the ecological actions: the driving area for hydraulic resilience and protection of the residential area

One of the urban actions: the new permeability along the river and continuity of green areas

A possible repurposing of the Casagiove quarry

redeveloped, will assume a public function at the disposal of the citizens and students of nearby universities. The proposed use of the three areas is: student residences and equipment in the former railway station; a park for sports and leisure time within the Macrico area; an energy and educational park in the former quarry of Luserta, which serves as a backdrop to the landscape and regains its productive vocation. In terms of design, it seemed interesting to explore the potential of the intersection of two of these systems, represented by the area between San Leucio and Casagiove to the west of the Royal Palace, since it contains many significant categories of places: the monumental Belvedere and the Royal Park, the natural slopes of the Tifatini, the "dross" with the two abandoned quarries and the agricultural-residual category with its pattern of fields and the cut of the variant. The proposal translates into a series of ecological and "urban" actions that are explained in the form of sections in which the activities that could be related to them are considered. As far as the ecological ones are concerned, we have imagined the creation of areas for hydraulic resilience which leave room for recreational activities in periods of shallow water, while allowing the natural development of biodiversity in periods of high water. On the other hand, in areas where there is a risk of landslides, reforestation is planned, followed by the creation of sports routes that lead up the hillsides. Finally, the recycling of the quarries will imply the implementation of safety procedures and subsequent reuse projects, with the inclusion of stables, equipped spaces for climbing and hiking and, in the case of the S. Lucia quarry, an auditorium. The urban actions include, in the first place, the pedestrianisation of the historical villages and the opening of the courtyards to the public, encouraging the widespread hotel concept. The opening of the entrances to the Park was also considered, allowing continuity in the use of green spaces thanks to the boundary's new permeability.

Also planned are systems that will overcome the variant with the aim of reshaping the uneven agricultural fabric, creating a network in which farms, agricultural patterns of foothills and plains, together with small educational pavilions, constitute a new path for "leisure" and consumption at zero km. The master plan is based on the design of these actions on the territorial palimpsest and on the identification of a band of intervention within the chosen project areas. The shared space is therefore halfway between urban and ecological materials, in which both visitors and citizens fit. It has been represented through a red band that expands and narrows according to the elements it meets, creating a crown that extends the tourist-recreational use of the Park and the Belvedere to the surrounding area.

Driving areas for hydraulic resilience and barriers protecting the residential areas

Reforestation and development of sports tours

Securing and creative re-use of the "Casagiove" quarry

Securing and insertion of an auditorium in the "Santa Lucia" quarry

Pedestrianisation and fruition of spaces in the historical boroughs as a function of tourist-receptive activities

New entrances to the "Parco della Reggia": continuity of green spaces on both sides of the border

New entrances to the "Parco della Reggia": continuity of green spaces on both sides of the border

Km 0 consumption network: restoration of Masserie (farmhouses), agricultural production and pavilions for didactic purposes

Restoration of the Italian gardens and enhancement of the "Borgo San Leucio" as a centre of excellence for silk production

Perimeter of the shared area and the continuous park

NE/ NAPOLI EST

"Brin69: apre Eccellenze Campane, il modello Eataly a Napoli"
Nell'area post industriale una grande fabbrica della produzione metallica manifatturiera si trasforma in una fabbrica della nuova produzione delle idee e della creatività."

fanpage

TRASFORMAZIONI URBANE

RIGENERAZIONE TESSUTO INDUSTRIALE

DIFESA E PROMOZIONE DIRITTI CIVICI

PROMOZIONE CULTURALE E CREATIVA

- Variante generale al PRG

2004

- Iniziative in corso nella città di Napoli

Dossier Comune di Napoli

Riqualificazione delle aree ex-industriali e infrastrutture per lo sviluppo

Riqualificazione degli interventi di edilizia residenziale pubblica

Infrastrutture della conoscenza

Interventi in corso e pianificati

Miglioramento offerta dei servizi e spazi pubblici e qualità residenziale

Progetto sistema delle infrastrutture e dei trasporti

Ambiente e difesa del suolo

Progetti di trasfromazione urbana

Dossier ACEN

East Naples, a hybrid machine

Enrico Formato

Context

East Naples is a settlement consisting of large mono-functional enclosures with high levels of underutilization and decommissioning, caught between a recent industrial past and a residential, service-based future. The site, flat and wedge-shaped, is located between the western slopes of Somma Vesuvio and the Poggioreale ridge and is crossed by the Sebeto river - a watercourse reduced to a "canal" during the viceroyal reclamations (16th-17th centuries), and today largely intubated[43]. From the mid-19th century the valley, transformed after the viceroyal reclamations into the "Orti di Napoli" (Vegetable fields of Naples), became the object of a rapid and massive industrialization process, based on the design of the "Recovery and expansion plan" following the unification of Italy. In this way, East Naples became an important energy-chemical centre (filled with deposits, refineries, processing factories), directly connected to the merchant port by means of various types of network infrastructure (roads, railways, oil pipelines). Industrial urbanization thus overlapped with the pre-existing infrastructures of modern reclamation: first it laid an abstract road network, with a north-south position, on the impluvia rotated for hydraulic reasons; then, over the decades, this resulted in a widespread artificialization of the soil. As in a sort of "delirious reclamation" - in the sense that Rem Koolhaas attributes to the New York skyscraper[44] - the viceroyal reclamation, brought to the extreme in the context of the processes of industrialization, achieved its perfect expression but, at the same time, marked the beginning of decadence. The Sebeto Valley was replaced by East Naples: the "river" became a "tube", an element of an invisible network of relationships, a machine that acts in the thickness of the mineralized platforms, an integral part of the production cycles: under the arches of the Ponte della Maddalena now runs a tube bundle that carries - in pipes of different colors - both mineral oils (from the dock to the refineries) and water (from the refineries to the sea). At the end of the 20th century the industrial phase-out went hand in hand with City's attempt to regain possession of East-Naples (the 2001 P.R.G. proposed for the "disused areas " includes new integrated settlements and parks). Twenty years later, while waiting for the implementation of some urban plans of considerable extension, the Sebeto Valley presents itself as an area of

43. See A. Migliaccio, "Il sistema idrografico della piana orientale di Napoli quale componente strutturale delle trasformazioni ambientali", in Leone Ugo (The hydrographic system of the eastern plain of Naples as a structural component of environmental transformations", in Leone Ugo, edited by, "The eastern area of Naples. Contributions for a project", AMRA, Naples, 2005)

44. R. Koolhaas, "Delirious New York. A retroactive Manifesto for Manhattan", Monacelli Press, Inc., New York, 1978

extraordinary interest, rich in territories in transition and spaces waiting for a project - "drosscapes"[45] or "dead-spaces" that have completed their life cycle, but that have a strong regenerative potential. There are plenty of polluted soils and water, compromised ecosystems, special or industrial buildings that have been abandoned or are in the process of being abandoned, infrastructures (some of which have been abandoned), interstitial areas to which various types of production enclosures have been added; technological systems and new settlements, which are neither controlled nor planned. Their presence has also led to an almost total erosion of agricultural soils, limited to some areas on the territory and strongly hybridised with depots, storage areas, car parks and logistical areas.

East Naples, for an artificial nature project

In East Naples, the life cycles of settlement and environmental systems - the former swamp system, the subsequent hydrographic network and the most recent industrial production plant - have long been exhausted or lie in an advanced state of crisis. Agriculture and spontaneous residential dwellings have also been affected. The area is almost completely artificial; only locally do fragments of "natura naturans" emerge from the disused enclosures, an extension of the 'Orti di Napoli' wedges along the motorway strip from the Piana dei Regi Lagni to the east of the large mineralised plates. Even where nature seems to "return", colonizing disused urban areas, we have to deal with the heavy legacy of the energy-industrial past. This legacy marked East Naples as a polluted site of regional and national interest. To understand this process, it is interesting to try to bring back the biological clock of the different systems to which each waste refers, investigating the moment of crisis of each of the main life cycles: the swamp (1400); the completely reclaimed countryside (1890); the industrial city (1966); the waiting city (2014). The procedure aims to analyze the "time of crisis" of each particular system, a moment before the end of a life cycle. This is a "retroactive conceptualization", based on the deconstruction of the palimpsest detected in a reasonable set of "plates"[46]. A "hyperrealistic"[47] condition, where the gaze captures permanence and traces latent overlaps, a potential for the image of new projects. In this way, time becomes the material of the project. It helps to use several references at the same time

On the following page, specialised precincts in the "Ambito 13", East Naples

45. A. Berger, "Drosscape, Wasting land in urban America", Princeton Architectural Press, New York, 2006

46. P. Viganò, "I territori dell'urbanistica", Officina Edizioni, Rome, 2011

47. R. Barthes, "Diderot, Brecht, Ejzenštejn", in "L'ovvio e l'ottuso", Critical Essays III, Einaudi, Turin, 2001; Orig. ed. Paris, 1982.

in order to avoid the simplifications and antinomies that would be generated as a consequence of simple mapping of the data and their static interpretation. The fourth dimension takes on the hybrid "naturartificial" character as its essence, and thus unfolds the cycles of life in a procedural perspective. In this process East Naples recycles in the present (yet with an eye to future) parts of its past lives: industry and the swamp co-exist once again, symbolizing a living ecosystem and identity; the service-based / residential future is reconciled with urban agriculture. We think on different levels - topologically continuous, conceptually opposite, temporally connected. If, on the one hand, we work on the residual rural void - the "wedge" of the "ruralscape" - on the other hand we do not neglect the need to deal with the residues of the industrial past and the city waste. It is necessary to contribute to the long-term objective by generating new hybrid "metabolisms" with different aspects (socio-technological-spatial) inserted into structured, yet flexible, metabolic chains. Looking at the theme from the vegetable gardens perspective, because it is a marginal area, there is the possibility of immediately using residues, waste, forgotten micro-areas that constitute a potential, extraordinary, re-connective fabric; creating temporary public uses, renaturalising canals and building roads, supporting the reconquest of the "third landscape"[48]; and finally creating a "natural hydraulic machine", a counter-reclamation operation such as the "wetland machine" of Berger's Agro Pontino. However, when 'your feet are on the tarmac' and an acrid stench comes out from the tanks, we can not underestimate the contribution provided by a strongly artificial purification machine (purifiers, water pumps, canalizations, artificial soils and impluvia) capable of contributing to the formation of new metabolisms, especially during the phases of the reclamation process. The combination of nature and artifice thus becomes the theme of the project, to technically deepen the solutions and describe possible combinations, times and rhythms. The starting point of this process, while waiting for the great post-industrial transformations to take into account the inertia that is due to the complexity and relevance of the interventions, inevitably concentrates the "middle city"[49] system on the margins of the great transport infrastructures, the "terrain vague" of the public residential plates, the fragmented and conspicuous set of ex-rural areas subject to more or less significant contemporary transformations (greenhouses, fences, soil transformations, etc.). Starting from these, waste has some advantages: it can give energy (through a catabolic metabolism: that is, the aggregation of elementary parts in a systemic logic) to transform the most difficult urban parts[50]. Since they are marginal areas neglected by real estate developers, this approach also makes it possible to immediately use the residues, the waste, the forgotten micro-areas that constitute a potential, extraordinary, re-conceptual fabric. At the same time, acting like a scissor,

Overlayer Mapping in East Naples

On the following pages, synoptis of the pollution of surface and underground waters in East Naples, with the dam's reclamation diagrams.

The images in this contribution are by Davide Di Martino

48. C. Gilles, Manifeste du Tiers Paysage (orig. title), "Manifesto of the Third Landscape", Quodlibet, 2003

49. M. Russo, (2012), "L'esperienza come progetto: conoscere l'area est di Napoli", in Lucci Rejana and Russo Michelangelo, (edited by), "Napoli verso oriente", Clean, Napoli, pp. 144-163, 2012; M. Russo, "Terre di mezzo", in Ricci Mosè and Gausa Manuel, "Med.net.rep.01", ListLab, Trento, pp. 208-2013, 2012.

50. E. Formato, 'Recombinant' hybrid ecologies and landscapes: Piana Campana, South Italy", in Lieto Laura, Beauregard Robert A., "Planning for a Material World", Routledge, London-New York, 2016

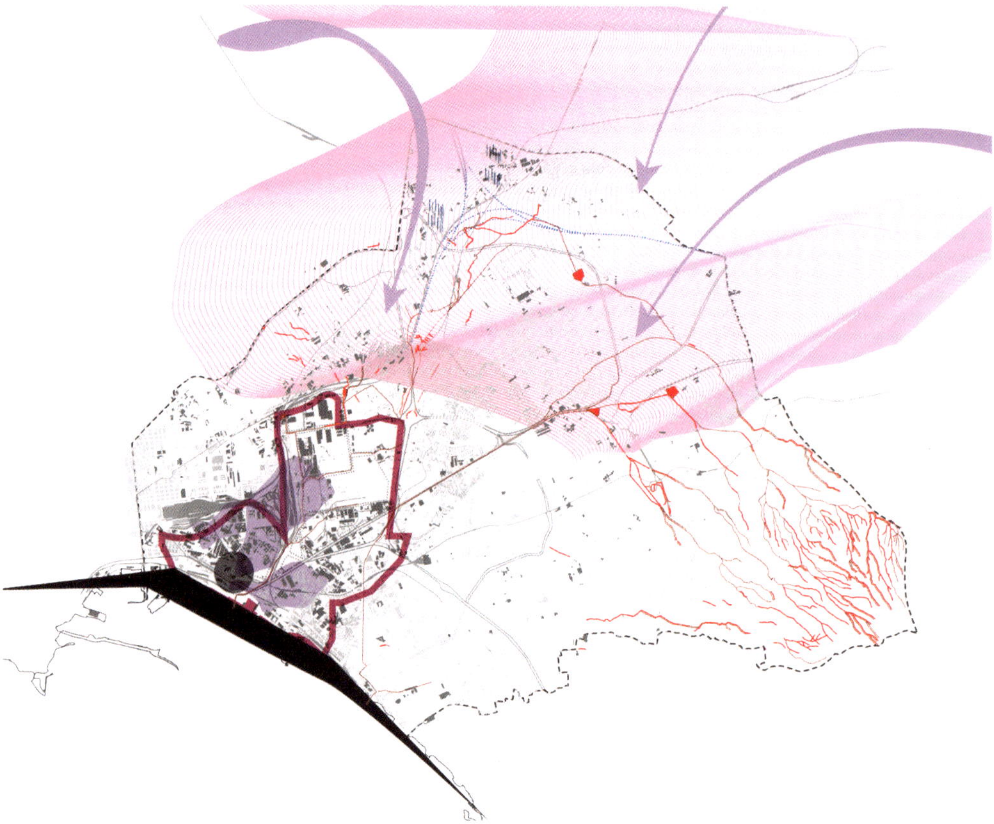

pollution of the groundwater: tetrachloroethylene

pollution of the groundwater: nitrate

diagram: hydraulic barrier and wastewater treatment plant

water system

outline of the S.I.N. East Naples

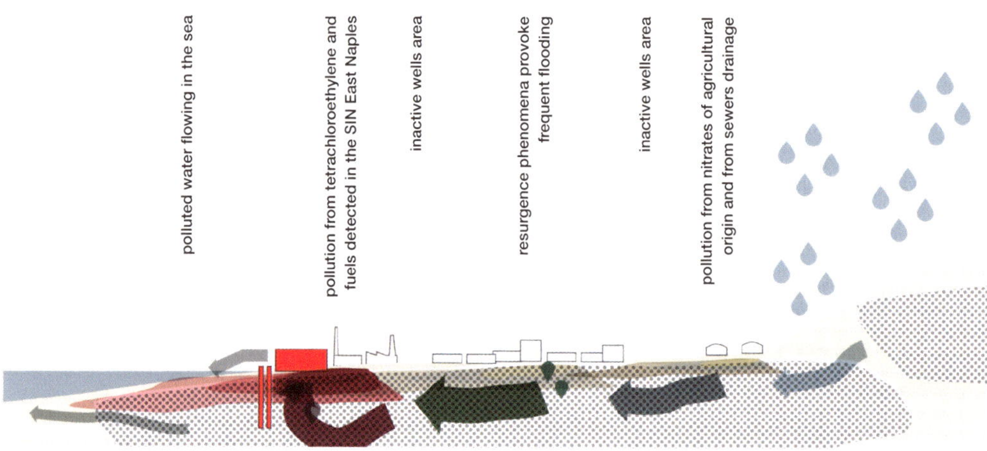

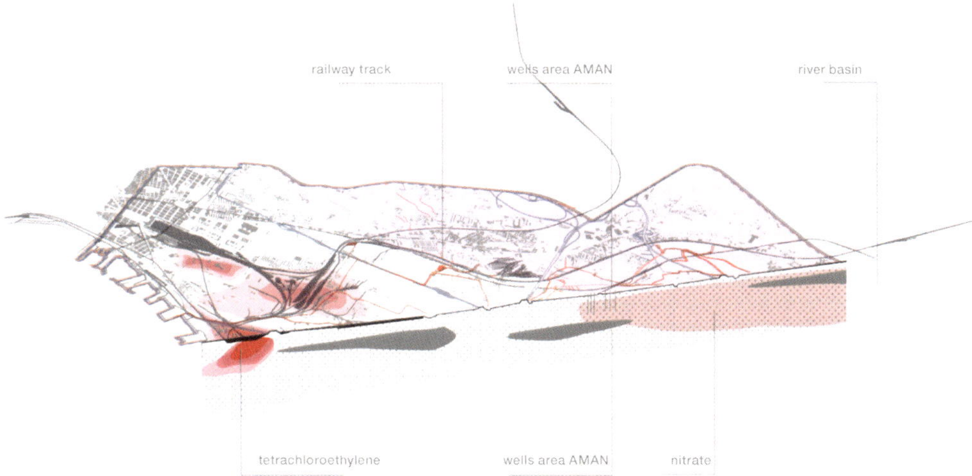

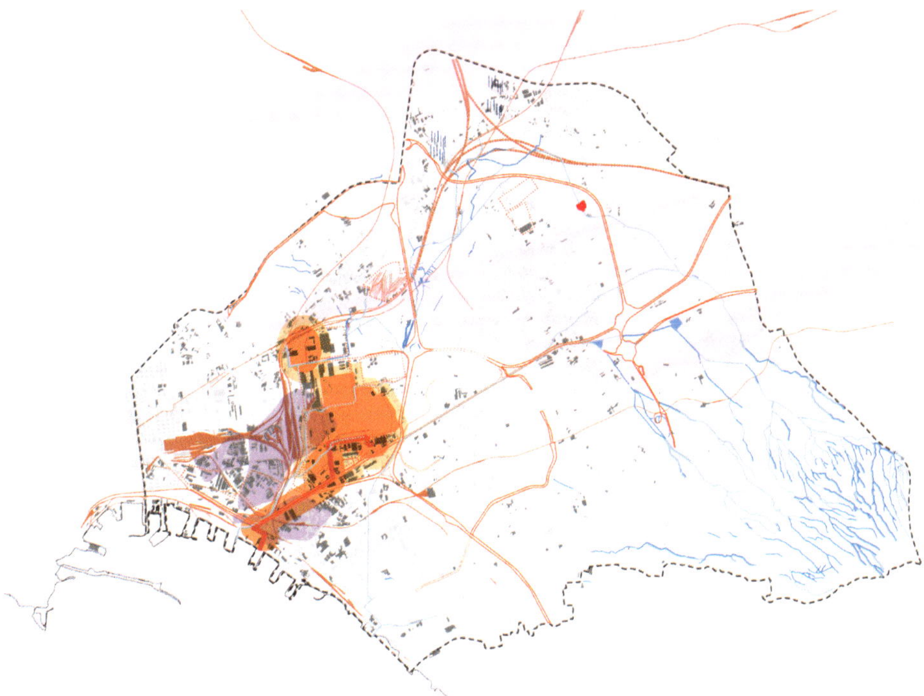

pollution of the groundwater: tetrachloroethylene

pollution of the groundwater: nitrate

diagram: hydraulic barrier and wastewater treatment plant

water system

high risk of accident

pipeline

one has to force the large industrial plates and the storage fences, trying to open up the gates of urbanity within the grey-pattern of the refineries, hangars and disused offices: promoting connections between city areas, the reclaimed agricultural wedge and the emergence of temporary uses, even spontaneous ones, capable of triggering the production of new imagery[51].

East Naples: a project for the present

The rationality of the recycling project for East Naples, in search of systemic efficiencies and rationalization of blue and green nets, establishes a critical relationship with history, extroverting itself at the imminence of the present time. The resulting design hypothesis, for example, given the heavy pollution of the former industrial soils, envisages a new canal which - by collecting the white and grey waters of the urban and rural systems (going north as far as Regi Lagni) - is capable of considerably reducing the loads of the purification plant. The plant is thus left with the task of "cleaning" the groundwater infiltrated by the former industrial soils, in addition to the black water of urban origin[52]. This channel becomes a structuring element of a new peri-urban public space, of connection between the settlement and environmental systems of the interior and the coast of East Naples. In particular, we would like to underline that - in relation to some theoretical positions taken in the PRIN research - the topic is dealt with by avoiding any allusion to the "conservation" or "restoration" of pre-modern, pre-existing structures. The concept of context, although central to the setting of the recycling project[53], may shift the focus from the echoing of the past and the obsession with memory, to the present and the immanence of the urban. In this way, any contemplative, romantic or formalist temptation is avoided, and the practice of re-cycling is immunized from the Ruskynian sense of ruin that every object - if extrapolated from its context of belonging - risks taking on. In this hypothesis, it is the sense of the canals and the hydraulic machine that comes as a legacy of history, not the archaeological research on the traces of a dissolved time. A particularly evocative image can be taken from the film "Fahrenheit 451"[54], where the winning strategy against the destruction of books is their incorporation into the memory of "book-men", able to pass on, word for word, every page. Not the object or its history, but only the stories extracted philologically from the texts, deprived of their own material character and elevated to collective memory. A sort of "preventive dissolution", accompanied by a solid strategy oriented towards the future, for a conscious transformation: the presumably safeguarding character of the archive and of scientific knowledge can thus merge with a powerful and simultaneous production of concepts, capable even of destabilizing the current state of things[55].

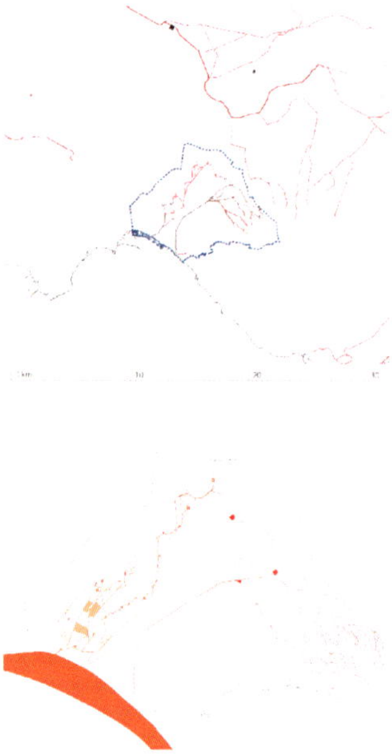

River basin framework in the East Naples plain and revision of the pre-unitary hydraulic system's reclamation.

On the opposite page, recycling of territorial frameworks and design of the "hybrid machine" in East Naples. The figures are by Massimo Lanzi.

51. S. Zizek, "Che cos'e' l'immaginario", Il Saggiatore, Milan, 2016

52. See the following contribution in "Acque, suoli, spazi vegetali ed energia per il riciclo dell'area orientale" by Carlo Gasparrini.

53. See my other contribution in the book

54. Fahrenheit 451 is a 1966 film directed by François Truffaut, based on Ray Bradbury's 1953 dystopian novel bearing the same name.

55. For further reflections on the subject see E. Formato, L. Amenta, S. Castiello, C. Di Marco, "Naturartificiale, verso nuovi metabolismi ibridi", in R. Pavia Rosario, R. ecchi Roberto, C. Gasparrini, (ed.), "Il territorio degli scarti e dei rifiuti," Aracne, Rome, pp. 163-170, 2014; E. Formato, M. Russo, "Re-Use/ Re-Cycle Territories. A retroactive conceptualisation for Est Naples", in "TEMA", Special Issue Input 2014, pp. 431-440, 2014; C. Gasparrini, M. Russo, "Campania. Napoli Est", in "ECO WEB TOWN", 5, pp. 1-30, 2014

de-ingegnerizzazione della rete dei canali e delle vasche di raccolta

nuovi bacini di assorbimento

nuovi bacini di depurazione

EST

nuovi paesaggi dell'energia

nuovi paesaggi dell'acqua

nuovi paesaggi delle infrastrutture

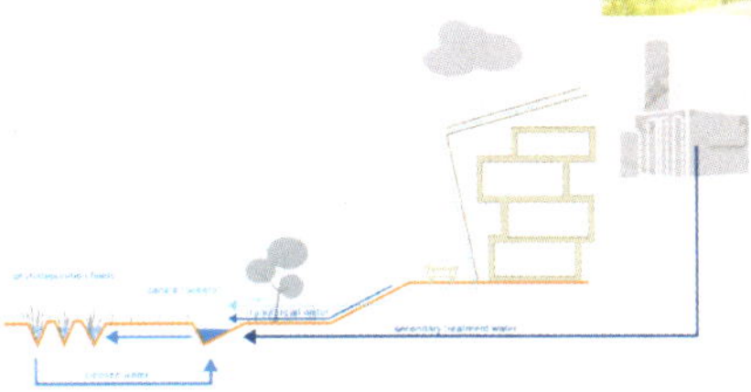

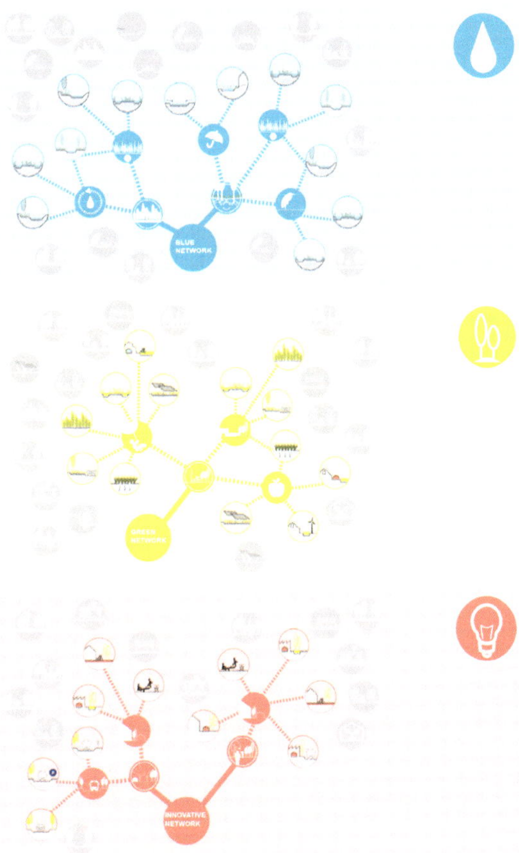

The recycling of Gianturco's critical fabric

Paola Galante

In the eastern area of Naples, considered a structuring hiatus, "impluvium-cerniera" of the territorial systems that converge therein, the district of Gianturco stands out in its specificity of being a "left-landscape".
The expression means a limited landscape, made up of urban fabrics that have registered (and in some cases continue to register) changes in production activity, through the accumulation of artefacts that have arisen even in different eras: in these landscapes, different types of spaces dedicated to production coexist. They can easily be attributed to the variations in the market that have imposed, with the evolution of time, functional and thus also spacial models that are different: Gianturco offers a vast array of workspaces, which - through accumulation over time - have saturated the area.
The set of production complexes, consisting of open and closed spaces, constitutes a "critical fabric": many of the spaces have arisen from needs that no longer exist, and there are several cases of abandoned buildings or entire disused areas.
A careful analysis of the current use of space shows that the neighbourhood is a "left-landscape" undergoing reorganization, due to the recent establishment of production activities related to the high-tech sector.
The way in which spaces are adapted to new uses, and the location of specific activities linked to the service sector and open to the public, show an attractive scenario in which the critical aspects of the fabric become the protagonists of its regeneration and activate the desired "civic network".
Within the case study of Gianturco, the third framework, dedicated to the re-appropriation of the dimension of collective use of the area, thus assumes a character closely linked to the physicality of the places and yet capable of structuring innovative relationships and projecting itself outside, beyond the precise boundaries of Gianturco, to contribute to the

re-texture of the eastern area in the very delicate point where it meets the historic city.

The area is characterized by the presence of fences that have settled down over time. Its characteristics, formed and transformed by work, are well described in a short film by Francesco Jodice. The photographer chooses to describe the area through a series of interviews filmed in some indoors. The camera is stationary, and placed at human height: the area is therefore described by a finished series of photographs animated by the stories of these people, which are - more specifically - stories of work. Interviews are edited into pieces and put together with direct transition, from one indoor space to another.

Between one piece of the interview and the next, there are visions of the general frame of the area, always shot with a static camera, but this time from above or from an unspecified place. The footage is edited with accelerated time, thus showing fragments of entire days (from dawn to dusk), where the increase and decrease of vehicular traffic on the roads are evident, Above all, the striking presence of natural elements such as Vesuvius, or artificial elements such as infrastructures, act as geographical references useful for orientation.

Francesco Jodice's East Naples appears as a piece of the city made up of non-communicating interiors. The difference in the type of building and activities within this set of workplaces also tells the story of the evolution of working methods. A finished series of blocks that do not interface with what is contiguous, but directly with the largest scale, the geographical one to which Mount Vesuvius belongs, as well as the hills and the area beyond the walls, the fences and barriers, and the sea.

In the image proposed by Jodice, the absence of anything that could act as a connective element is evident. We have no news of the quality of the space between one interior element

ROYAL CANIN
GRUPPO
APRILE
MATERIALE ELETTRICO
ILLUMINOTECNICA

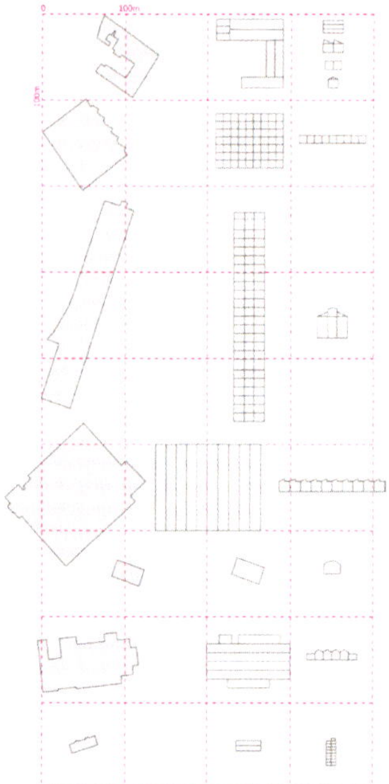

Abacus project

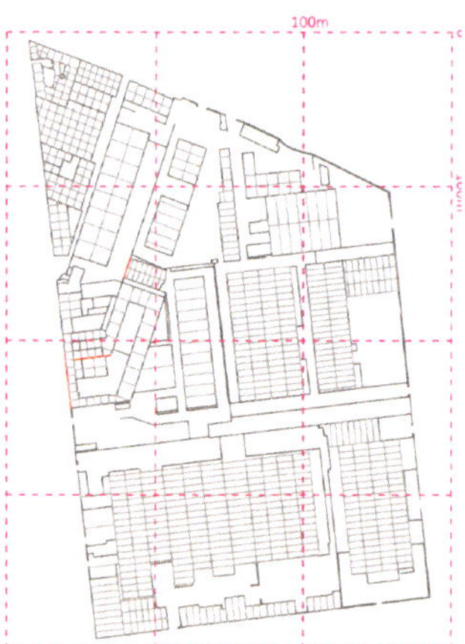

Project of the abacus structure

In the previous pages, view of the port of Naples from the Brin parking

and another. We do not know the section that characterizes the streets, the predominant materials, we have no news of the possible presence of sidewalks, shops on the street, meeting places, lighting: what is absent - as it is not at all represented - is the public space. Moreover, it is not known how the elements that govern the relationship between indoors and outdoors are made, or if there are filters that mediate this relationship. The direct transition technique used in the video recalls sequences of opacity; blind walls. Roads, therefore, seen as corridors between walls in which one can only pass through, but without stopping - functional but not relational vectors.
The cartographies, as well as the zenithal images, show a saturated sector, documented by the lack of significant transformations of the public space from 1957 to today.
This figure, in spite of the common thought, is not matched by a 'freezing' of the area, which is prevented by variations in the use of the artefacts and adjacent open spaces, and by minimal variation in density within the fences.
In other words, the saturation of the area does not correspond to an absence of physical mutations, even if they are inscribed in the blocks. Currently, in the enclosures of the apparently immobile area, in addition to a portion of "unused space" that is not stationary, but moves alternately from area to area, there are different uses related to different categories: traditional production, production with high added value, storage areas, wholesale trade, equipment, accommodation areas, parking lots, residences.
The different activities are distributed in an isotropic way inside the - previously mono-purpose - clods, to the point that the map of the land use is variegated. The same variety of colours also appears in the backgrounds of buildings, that are very different from each other. In other words, the same uses can be found in different categories. The urban "mixitè" is a predominant feature in the area, as is the permanence of productive spaces that arose in different eras.
Therefore, the observation of the map of use shows a regeneration of the sector, which was first mono-functional. It is worth noting that there are no large projects or plans at its base, but rather small transformations that involve buildings, or portions of buildings.
In the case of Naples, this is due, on the one hand, to the economic situation (the impossibility of major investments) and, on the other, to the implementation methods provided for in the Variant to the General Regulatory Plan, which makes it particularly challenging and burdensome to undertake building or urban restructuring works, since it requires the drawing up of implementation plans that must involve entire urban sectors - areas or sub-areas - which are often fragmented in terms of properties.

The changes, therefore implemented as part of the direct interventions allowed by the regulatory plan, are silent. They are not advertised, not published, not recorded in the list of ongoing interventions (continuously updated by the City of Naples through the Department of Urban Planning), and fall mostly in the category of extraordinary maintenance.
The notation is rather important, as it explains the nature of the transformations in the area of Gianturco by identifying potentialities and limits. It explains the persistence of buildings that have changed their intended use, but not their form; it explains why the urban structure, and in particular the relationship between public and private spaces, has remained unaltered; it identifies the load-bearing structure of buildings as a limit and resistance to transformations that must necessarily be investigated, and adapted to the regulations and to the new loads.
Since the beginning of the 1990s, many companies linked to the "high-tech" circuit have been set up in the Gianturco area. The need to adapt spaces to new uses has triggered changes in use. An examination of the companies linked to the "high-tech" circuit shows that the presence of the actual productive activity is confirmed, whether it is the production of tangible goods (goods to be placed directly in "output", rather than the production of prototypes) or whether it is intangible goods (research laboratories, recording studios, etc.).
There is, above all, an increase in the variety of activities. The traditional production spaces (warehouses) and management spaces (buildings) are added to or replaced by laboratories, studios for professionals (mostly engineers) but also classrooms, conference rooms, meeting rooms; that is, there are more space dedicated to meetings or reception, composed of open and covered spaces that develop from the concierge. These activities have to satisfy, first of all, the need for hospitality, and encourage exchange. Thus, they have common characteristics: they are located in the most valuable places of the complexes (where they get light); they are positioned in identifiable and recognizable places.
The analysis shows the percentage growth of tertiary functions within the production enclosures. These construct urban sequences that, as a whole, act as an element of connection between the inside and the outside, configuring a unique type of collective spaces: the interzones. These are sequences of spaces that regulate and manage the relations between the different areas of the companies and between these and the city: the relaxation area, the canteen, the "showrooms" can be read individually, but also as a concatenation of places of interaction of different intensity.
The transformation - born and developed within companies to adapt to the market - therefore concerns the entire

Intersection

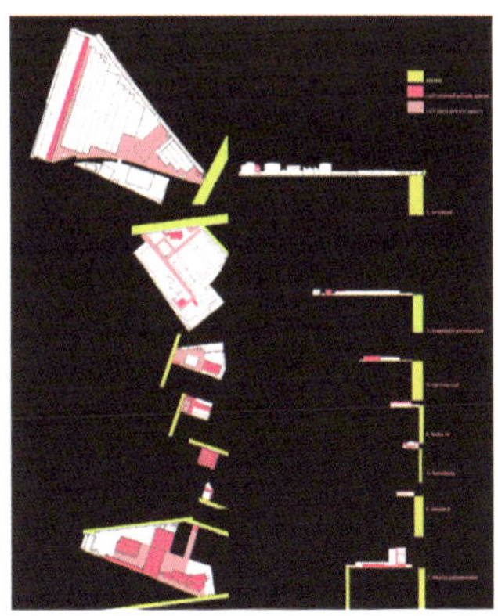
Urban sequences

Land Use in the area of Gianturco

area. It certainly requires support, but directly or indirectly offers numerous resources and opportunities to this area. It is possible to select the public and semi-public spaces of the Gianturco area. It is easy to see how the selected spaces build an urban sequence that leads from the public road to the innermost parts of the fences: we are in the presence of semi-public (or semi-private) spaces that build the link between the inside and the outside. The selected system puts into operation the noisy activities on the road, the concierges, the internal routes, the closed outdoor spaces, the spaces dedicated to accommodation and reception, the elevate spaces (stairs), the accessible points of view from which it is possible to observe a strongly "geo-referenced" panorama (terrace) and the explicit presence, position and size of the intersections that act as a filter between the public spaces of the city and the private spaces of the companies.
The selected spaces are not all available at the same time. It is for this reason that the size of the "layer", which includes the set of public spaces and "open indoors" may vary throughout the day, and also depending on the days.
The interzone expresses a potential not used at full capacity, and at the same time it is an indication of development that shows scenarios in which different uses - but also different users - alternate over 24 hours. The conference rooms could host evening shows open to a wide public, libraries could open their doors to school groups, courtyards could be used for moving vehicles and loading and unloading goods to leave room for markets, terraces could lend the view to cocktails bars, houses could host hotel rooms.
The transformations implemented, those underway, and those planned will change the face of a part of the city according to a metabolic mechanism that recalls the constant, simultaneous, minimal mutations, dictated by adaptation needs, recently observed in Tokyo.[56]
The updating of production sites and the above-mentioned silent changes can ultimately have "positive effects on the urban sector, an added value that manifests itself in a gradual recovery of the building stock and, in some cases, in the appearance of new centralities that act as catalysts for the public ".[57]
The case study shows existing opportunities, and poses the problem of their unveiling.
The steps to be taken are: knowing what is already there, putting resources in the system; identifying places and buildings that can be easily transformed; building a map to identify areas of (much needed) interventions.
From this point of view, the theme of the recovery of artefacts acquires a meaning that goes beyond the scale of the building and is linked to the "re-cycle" of an entire sector, whose life cycle seems to have been interrupted.

56. K. Kitayama, Y. Tsukamoto, R. Nischizawa, "Tokyo Metabolizing", TOTO publishing, Japan 2010, p. 11

57. M. Ragonese, "Spazi condivisi, luoghi ritrovati", in A. Bertagna F.Gastaldi, S. Marini (edited by), "Luoghi di lavoro, nuovi compiti e nuovi luoghi del progetto", Quodlibet, Macerata, 2012

East Naples: for the reuse of the “middle city”

Fabio di Iorio

Marco Norcaro

This thesis[58] dissertation aims to regenerate the urban fabric of the eastern area of the city of Naples. This objective shall be met by mending and restoring fragments, spaces, buildings and soil which have already consumed their life-cycle in this part of the city's metabolism and that now lay unused waiting for a new function.

All these materials, that will be called the "middle city", are rejected by the city of Naples and represent the outcome of the loss of its "industrial identity".

In this dissertation, the waste or drosscape (adopting Alan Berger's[59] definition) are a possible category of materials that could be useful for the reconstruction of a complex "network" of open spaces in which water returns to be the key element.

The starting point was the rereading of this part of the city's history and ongoing development. Through the centuries, the eastern part of Naples has changed its identity many times: maps of the time helped identifying, from 1775 to the present day, how this area went from being a swamp to an agricultural land thanks to the reclamations and finally an industrial area in the 20th century.

To this day, different local systems co-exist in Eastern Naples, each of which presenting well-structured features shaping this part of the city and divided in 4 broad categories: infrastructural, agricultural, of the waters and of the industries, which struggle for their existence and sometimes conflict between them, creating an hybrid territory, full of ambiguous and contradictory elements.

The water system is the best example to explain this conflict: while the historical maps show the river Sebeto to be a still visible element, today it has completely disappeared instead, apart from a few underground streams. Through a process of historical re-enactment it is now possible to affirm that water was the key element in Eastern Naples. In fact, the "Cosenza

58. The work presented in this article is by Fabio Di Iorio and Marco Norcaro in the scope of the thesis degree in Urbanism (CdL UPTA, DiARC – Università degli Studi di Napoli "Federico II"); thesis advisor was Prof. Arch. Michelangelo Russo and co-advisor Arch. Enrico Formato.

59. A. Berger, "Drosscape, Wasting land in urban America", Princeton Architectural Press, New York, 2006

plan" included the implementation of a navigable channel over the Sebeto river. The current crisis of the water system is also the good example of the "status of imbalance" that the other systems mentioned before are experiencing too; water, in particular, went from being a central to a progressively contained and finally invisible underground element.
This area is also strongly characterized by infrastructures giving shape to the surrounding area, with a complex system of residual spaces, often with no identity. Agriculture, central in Eastern Naples until the 50s, has suffered significant changes in the following decades, producing a fragmented and discontinuous mosaic of isolated areas in the urban environment. The current unprecedented agricultural system is made up of a dense agglomeration of fixed greenhouses which have completely modified the landscape and hardened the soil.
Another undeniable indicator of change is the growing tendency to dismiss industrial premises, in particular the ones that are at the end of a life-cycle in which the industrial system has been the product of overlapping metal plates and fences which have created closed off spaces, devoid of dialogue with the surrounding territory, in an alienating context. This study examines thoroughly this phenomenon of dismissal through a map of the industrial life-cycles, in which the factories have been analysed following the perspective of time, from the heavy industry's cycle, the Ford model and the post-industrial cycle, which have created huge complexes for the production of goods and services.
All these systems described create spaces with no real function, often difficult to recognize and interpret. Eastern Naples is therefore full of drosscapes and it's exactly from these that it will be possible to start new projects of urban regeneration aimed at transforming the interstitial spaces, frequently left on

Infrastuctural buffer areas

Greenhouses

Industrial precincts

Uncultivated land

Logistics warehouses

Drosscapes in East Naples

the sidelines of the great areas of transformation (P.U.A.). These spaces in fact can act as catalysts for other spaces waiting to be transformed because "the dross is considered as a natural component of each city developing dynamically. It is an indicator of the city's health and development"[60].

From Berger's remark, these "critical elements" become the starting point of the research project. Following the drosscapes' mapping, an abacus of the interventions to apply to these elements was developed, in order to have a structure for this concept. The project's aim is to create a "linear public park" which crosses from north to south the Eastern plain of the city; this park would reach the sea in the area of San Giovanni a Teduccio, starting from the agricultural area next to Casoria and crossing all those dense peripheral neighbourhoods of Eastern Naples. The park would simultaneously retrace the Sebeto river course and the underground canals, of which the re-emersion is foreseen, therefore becoming the connecting element among the different "networks". The new canal will have the dual objective of collecting the grey water coming from the neighbourhoods and the fields and of purifying them through damp tanks using natural phytoremediation systems. In this way, water would flow into the sea with no polluting substances.

Through the "land's design" the project aims at creating/ building a "linear urban park" supporting the new water system and working as a big green facility serving the surrounding neighbourhoods and all the Eastern area of Naples. The park would spread within the dense urban fabric, providing the citizens with a multitude of different scenarios: in the north, the agricultural area, which represents a means through which the territory shall be protected and safeguarded; the urban park, which alternates spaces with different dimensions and functions, such as the sport and leisure activities, community gardens, bike lanes and buildings conceived for events and cultural activities. The razionalization of agriculture and the use of "skyfarms" close to the urban fabric have the objective of optimising the production and restore the soil, previously hardened by the complex system of fixed-structure greenhouses. Finally, the re-use of specific elements, such as abandoned industries, aims at transforming these warehouses in meeting places and forums, collective spaces hosting different activities that promote the social and territorial regeneration. Int this way, these industries become places and workshops for activities linked to the environment, nutrition, art, handicraft, besides being meeting places for students and young families.

Following the analysis of the territory's specificities and the adaptation of the conceptual elements to the existing physical spaces, with all its rules and functions, the masterplan

60. A. Berger, "Drosscape, Wasting land in urban America", Princeton Architectural Press, New York, 2006

Industrial precincts
Logistics warehouses
Greenhouses
Infrastuctural buffer areas
Abandoned areas

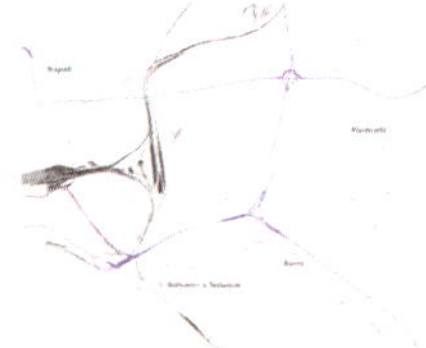

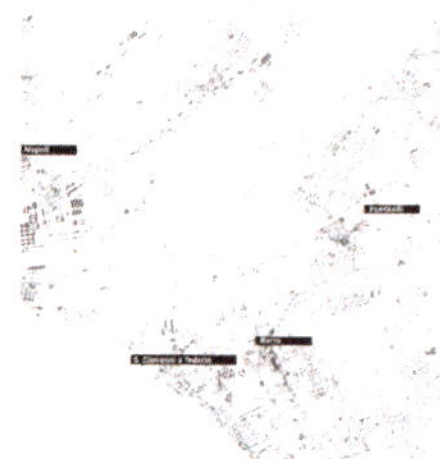

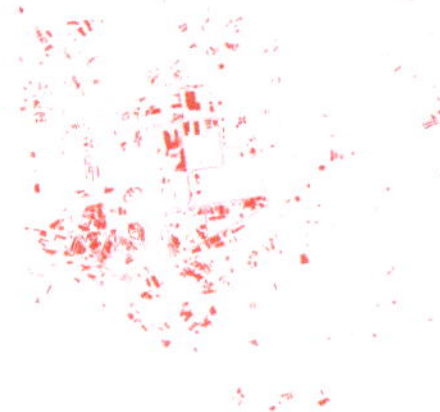

sistema agricolo

1775

carta del Duca di Noja

1840

Carta dei Dintorni di Napoli
Real Officio Topografico

1907

Istituto Geografico Militare

1957

finally gave birth to this project. This complex design grafts on other large scale elements and projects like the "Ambito 13" and the one in San Giovanni a Teduccio, eventually becoming the key element to mend this part of territory, isolated and withdrawn so far.

On the opposite page, Masterplan, territorial systems and restoration of the hydraulic network

Below, Energy-rich landscapes, San Marco dei Cavoni (Benevento)

From the "labyrinth" to the "porous city". New living spaces for the eastern area of Naples

Francesco Stefano Sammarco

Ciro Sepe

Danilo Vinaccia

The objective of the Laboratory is to develop the interpretative and design tools for the definition of an urban project in the eastern area of Naples, in implementation of the P.R.G. (General Regulation Plan) in force since 2004. The project area is part of the so-called "Area 13" (also known as the " Ex Raffineria", a former refinery) - a large area to the east of the railway tracks that includes oil depots, disused areas, industrial activities and residential areas, for which an extensive and complex process of urban redevelopment is planned.
The work undertaken - called upon to deal with the dimension of waste and exclusion typical of large urban areas as a result of industrial decommissioning - proposes a clear and effective way of constructing an urban project, capable of proposing a new idea of a city. While producing newly habitable spaces, it also gives a new meaning to these places with which one can collectively identify. The enormous complexity of the eastern area of Naples is determined by its huge amount of those values and traces, which define a meaningful space. This complexity requires a dedicated investigation of the area in the scope of its dimensions, which leads to the recognition of an urban structure that has been built over time as a dominant characteristic.
This stratification has given life to specialized islands that overlap each other, creating undefined spaces, caesuras (cut spaces), and dead-end alleys, configuring a situation that can be interpreted through the sedimented "figure" of the "labyrinth".
Here, nevertheless, there are still points of strong and clear symbolic content, the "monuments", which commemorate history - a territorial memory narrating the identity of a certain place. At the same time, there is also a thin territory consisting of fences and paths, difficult to read, with different characteristics and resources. This complex overlapping of traces and memories represents a value to be preserved and appraised. These places are not erased by modernity, but are placed in the background. They are like indicators of the flow and survival of time. They last over time, as do the words that express them, and will express them again[61].

61. M. Augé, "Non-places", Eleuthera, 1993

This focus on grasping some evidence of a latent demand for the use of public space indicates the presence of a script, one that is not immediately perceptible, but that allows us to glimpse some possibilities for transformation. The idea is to bring out the image of a "porous city" by exploiting the symbolic elements of the monuments and the walls - the distinction between what is inside and what is outside - used to interpret the same space in the opposite way. The design of the ground and the organization of the open spaces derive from the expansion of the new road system. It identifies two types: "Weft" and 'Warp'; the first is the set of multifunctional and multi-social living spaces that interact with the environment in daring combinations. The second is a system of areas with a strong structure of public open spaces. From a spatial point of view, therefore, the subversion of the consolidated image into a new condition of porosity focuses on the system of open spaces, by erecting them as a new supporting infrastructure that serves the purpose of the project.
For Walter Benjamin, porosity is a feature of the city's architecture, reflecting its "communal rhythm" and transferring into its material structure. "What is final, formed, is avoided. No situation appears as it is, thought forever, no form is declared to be such and not otherwise"[62].
Re-reading and re-interpreting the territory gives us the opportunity of including this condition of porosity within the urban project[63].
As mentioned above, this project begins with the transformation of the road networks surrounding the polder into load-bearing infrastructure frames. The system of north-south roads (which quickly connect the various parts of this new organ of the city) and those of the neighborhood that form a cross (at the service of the internal parts of the polder spaces, and promoting slow mobility) creates a matrix of organization of the space, constituting complex sections capable of accommodating functions and equipment, new places and new ways of life. In order to respond to the laboratory's objectives, great attention must be paid to existing relationships, by trying to

62. W. Benjamin, "Neapel", Frankfurter Zeitung, 19 August 1925 / Italian edition: W. Benjamin, "Immagini di città", Giulio Einaudi Publisher, 2007

63. The work presented was developed within the Laboratory 2 'Piano Urbanistico Attuativo' ("Urban Implementation Plan") (Cdl PTUPA, DiARC - University of Naples "Federico II" a.y. 2013/14); Teachings: urban planning prof. arch.C. Gasparrini, architectural planning prof. arch. R. Amirante, landscape design prof. arch. V. Cappiello; integrative didactics: arch. D. Buonanno and arch. A. Terracciano

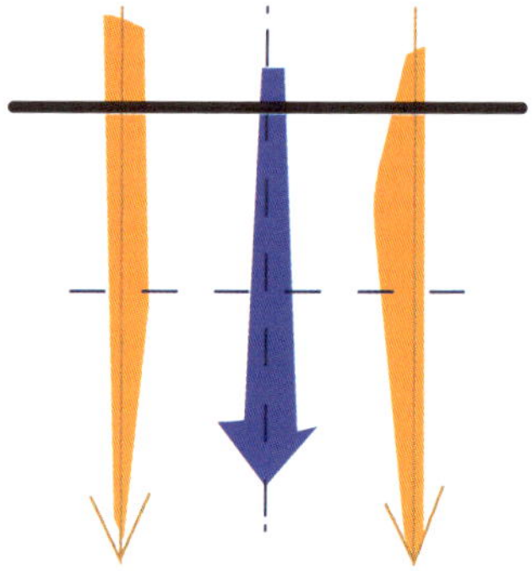

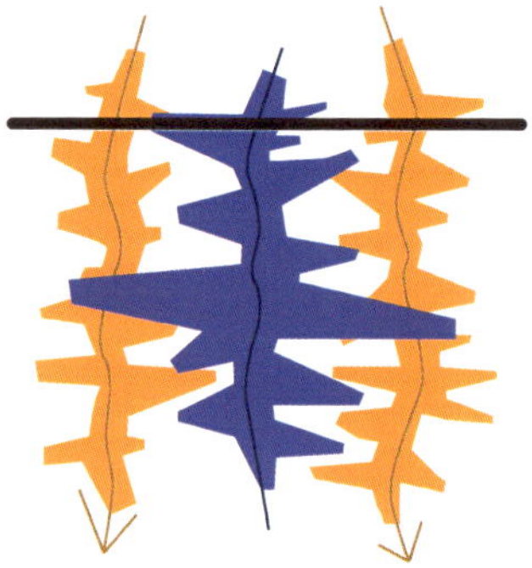

Warp and weft of main roads

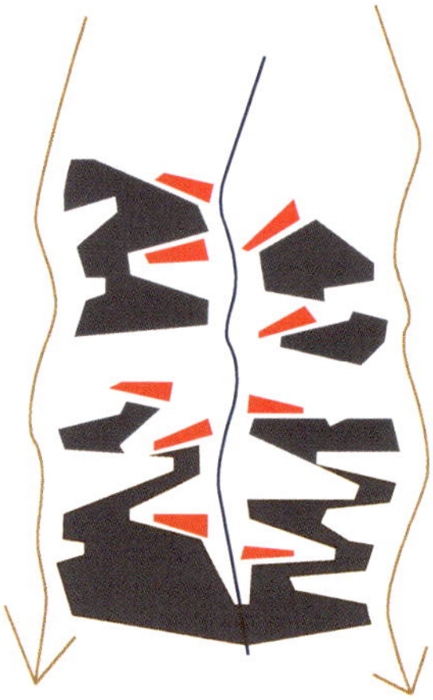

Settlement splinters

Concept

64. D. Dottorini, "Rileggere Benjamin: la forma della città, la doppia immagine della modernità ", Cahiers d'études romanes, 19, 2008, pp. 153-166

propose a model that favours a broader rethinking that is not only capable of relating to nearby realities, but is also able to plan and make connections outside them.
The weft and the warp define, therefore, the strong structure around which further spatial levels are articulated, building new hierarchies and new possibilities of relationships, capable of moving at different levels. These range from the large spaces of the park to the urban courtyards, up to the level of the micro-courtyards and of the car parks embedded in the thickness of the polders.
For the design of the polder area, we start from the definition of the block which is framed by the road and from which the retreats and variable sections of the new public places and the "slow" cycle-pedestrian mobility derive. The matrix of the sections draws the "expansion of open spaces" which flow along the main streets of the polder, through spaces at different altitudes, both public and private, to be directly connected with the large public park south of the area. This design of open spaces includes the "schegge insediative" (literally: splinters of settlements), a new building system which includes residential functions as well as facilities and trade services.
Furthermore, the project aims, through the creation of attractive, recreational and resting places, at the integration and re-functionalization of the degraded spaces below the SS 162 dir viaduct. This is to favour not only the continuity of the north-south crossing of the area but, and above all, to give new dignity and life to those marginal, interstitial and abandoned areas generated by infrastructural management.
The construction of a public park available to the population is part of the context and consistent with the overall design of the project, making this dialectic of compression and expansion of space a metaphor for a much wider need and latent demand for the re-appropriation of this part of the city. This time, the expansion in variable sections acts along the beds of watercourses emerging from the aquifer, creating new usable areas. Investigating and recognizing all the sensitive material and social aspects of the contemporary city and territory is fundamental for urban planning, for triggering design processes that are fully assimilated by the various parts of the city and shared by the people who live there.
However, it is not a question of elaborating an abstract, scientific and rigorous method of decoding the multiple signs of contemporaneity, nor is it a matter of analyzing them according to pre-established categories, keeping them, so to speak, at a distance. It is about making these signs, their interpretation and their new life interact through a single language[64].

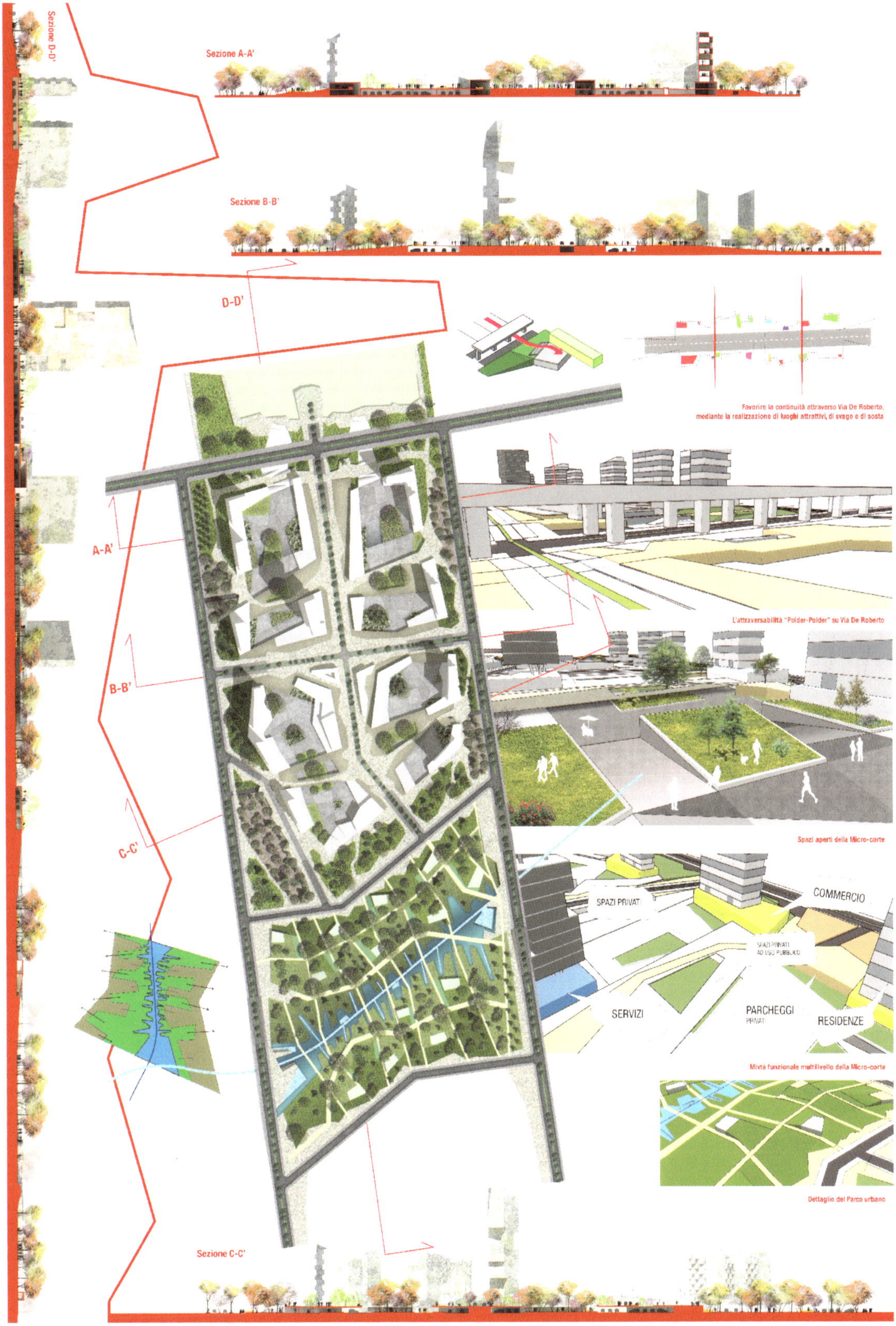
Sezione D-D'
Sezione A-A'
Sezione B-B'
D-D'
A-A'
B-B'
C-C'
Favorire la continuità attraverso Via De Roberto,
mediante la realizzazione di luoghi attrattivi, di svago e di sosta
L'attraversabilità "Polder-Polder" su Via De Roberto
Spazi aperti della Micro-corte
SPAZI PRIVATI
COMMERCIO
SPAZI PRIVATI
AD USO PUBBLICO
SERVIZI
PARCHEGGI
PRIVATI
RESIDENZE
Mixtè funzionale multilivello della Micro-corte
Dettaglio del Parco urbano
Sezione C-C'

Zero impact

Andrea
Felo

Project area classification

The project area is located in the eastern area of Naples, more precisely, with reference to the tables of the zoning plan, it corresponds to sheets n.16 and n.9. These areas of the city are currently industrial areas, and in particular they are characterized by the presence of hydrocarbon deposits of the companies Q8, Esso, Eni, Shell and Total. The company with the largest storage area is Q8, which since the variant of P.R.G. (Piano Regolatore Generale/Urban Development Plan) 2004 is considered an area of great urban interest.

Analysis of the variant at P.R.G. 2004 (Area 13)

The scope covers an area of 400 hectares, bounded to the west and north by the railway embankment, to the east by the agricultural area and to the south by the motorway link road. A large part of the territory, 120 hectares, hosts oil plants, while the remaining part consists of active and disused industrial plants. The environmental upgrading of the production area involves the relocation of oil plants and the conversion of disused industrial plants. The project involves the construction of infrastructure, residences, and a park on an urban and territorial scale, enhancing the 'hinge-like' geographical position that the area has with respect to the immediate surroundings. (a)

Analysis of the preliminary P.U.A. 2009

The preliminary version of P.U.A. (Piano Urbanistico Attuativo/ Urban Implementation Plan) has been drawn up to provide design guidelines to the P.U.A. that will be implemented at a later date, as - given the extension of the area - action will be taken in parts. This strategic move will make it possible to obtain a even project over the entire area, an important aspect that, if it is left out, could generate problems of great importance in the future. From the project plan, it is possible to detect the strong will to make green areas, infrastructures, equipment, new generation industrial plants and residences coexist. Particular attention is paid to the reuse of industrial components, such as the "refining tower" and the "park of natural deposits". (b)

Analysis of the final P.U.A.

The final P.U.A. version in analysis is the one concerning the Q8 areas, i.e. the largest plant on which to focus the intervention and from which we would draw more benefits in terms of development of the entire area of interest. The reuse envisaged for the structures located in the area is based on different methods: maintaining structural integrity, eliminating structures or maintaining traces of their persistence.
In the project in question, a choice was made to safeguard the integrity of these structures. In particular, the focus was on the "tanks/silos" of hydrocarbons, which are considered the key element of the terminal part of the park, becoming the central theme of the latter. In accordance with the guidelines that have been drawn up previously, the issue of the reuse of tanks will be connected to the natural systems and equipment aimed at improving the livability of the park. (c)

Current area status

Currently the area is divided into operational and non-operational areas. A specific plan has been drawn up for the second area to reclaim the soil and aquifer. The presence of tanks is much lower due to the dismantling of the non-operational area, where the former refinery plant was located. Considering the new layout of the park area, it is still possible to see the persistence of the tanks through the traces on the territory. (a)

Area Extrapolation

By zooming in on the area of the deposits[65] it is evident that the strong presence of the tanks is an integral part of the landscape. In particular, this is useful to underline the fact that they may become part of the future reality as a historical memory, as they are evidence of a system dating back to the Fascist era. Until the 1990s, the plant remained fully operational. After that, the refining process was suspended, generating the largest hydrocarbon deposit in Italy. The large number of tanks, now a symbol of the area, has become an integral part of a green area called, for this reason, "area of natural deposits". (b)

65. Work by Andrea Felo within a degree thesis in Architecture (CdL MAPA, DiARC - University of Naples "Federico II"). Supervisor: prof. arch. Alfonso Morone. Co-supervisor: prof. arch. Carlo Gasparrini.

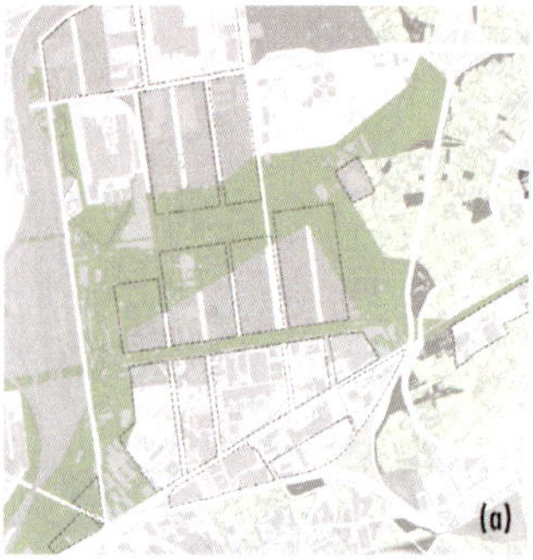

Green park

Agricultural areas

Areas of interest

Preliminary PUA in "Ambito 13" (2009)

The tanks in the park as elements of a molecule

The park will be composed of a series of ex-tanks which, given their flexible structure, can be used as large containers for different uses. The park is then seen as a well-defined structure, within which each tank has its decisive function in order to achieve the final result, like a molecule, in which the union of elements with different characteristics gives life to the structure. The urban layout of the park can be compared to a molecular structure, as the planimetric shape of the tanks reproduces the "shape" of it. (c)

Q8 system cycle - Pre intervention

The Q8 system operates in a linear manner: the starting point is the oil extraction phase, which - since oil platforms are used - causes serious damage to marine fauna and compromises its biodiversity; the next phase is the refining phase, which generates a strong impact on the atmosphere by increasing greenhouse gases; after refining, the substances are loaded onto the tankers and then arrive at the port where the discharge phase takes place; the hydrocarbons, after having travelled the oil pipelines for 4 km, reach the tanks, which - by means of a distribution system - fill the tank trucks, which in turn will provide for the supply of the filling stations; finally, the distribution reaches the means of transport by road, that are the major producers of CO_2. This process generates a very high environmental impact (soil, air, water), causing further damage to human health. (a)

System cycle after intervention

The proposed intervention in the area consists in converting the tanks into hangars whose new function will be to become components of a composting plant. The choice of the destination of use was made in line with the strategies envisaged in the reclamation plan. Such plan foresees the recovery of the land, which now has no organic matter and requires large quantities of "soil improver", which is the main product of a composting centre. The other advantage is to encourage agriculture without chemical fertilizers, since the surrounding areas are characterized by the strong presence of crops. The composting process thus gives life to a virtuous cycle that involves inhabited areas and surrounding agricultural land, to the extent that the citizens are the producers of the organic fraction and the parks and farms are the producers of the so-called "wood cellulosic" material. The collection is carried out within a radius of Km n that allow to consider the process at Km0 (zero miles); subsequently the "compost" produced can be re-used in the sectors of origin, increasing employment, reducing the environmental impact, boosting the economy and greatly reducing the impact on the health of the inhabitants. (b)

Social function

The social function associated with the composting centre is to enable students and groups of people to use it in order to promote the importance of waste treatment and the related concept of "recycling". A problem that has been rooted for years, waste disposal can be made easier by the availability of urban treatment plants, as has been done in many countries of the European Community. In particular, with the concept of "smart cities" and the strong desire to reduce consumption by looking for Km0 systems, it is possible to adopt strategies in which both the manufacturer and the citizen, through the use of innovative communication methods, can play an active role in the process. In this regard, the system adopted is inspired by the installation of the Russian Pavilion at the Venice Biennale 2012/13 by Sergey Tchoban, which, considering the theme of "Common Ground", has created an attractive communication system to stimulate the knowledge and use of QR-CODE (quick rapid code) readable via tablet or smartphone. This system would allow the use of a workspace without the aid of a guide. It would only require a reading tool and a path along which it is possible to see and know the working processes in a totally autonomous way. This system can be used as a pilot project. (c)

Project data

Number of inhabitants: 20.000 - Specific production R.S.U (Rifiuto Solido Urbano=Municipal solid waste or MSW): 1,5 Kg/ab/d - Organic fraction present in the waste: 35% - Efficiency FORSU (Frazione Organica del Rifiuto Solido Urbano or Organic Fraction in MSW): 45%
From these data we obtain the "FORSU mass flow rate" to be started at the plant: 20,000 ab x 1.5 Kg/ab/d x 0.35 x 0.45 = 4,725 Kg/d -> 4.72 t/d

The material to be treated, of course, is only partly made up of FORSU. A part of the material to be treated, in fact, will be made up of "cellulosic wood" material, used as a structuring material. It was decided to add 0.5 kilograms of structuring material to 1.0 kilograms of organic material from separate collection. Given the mass flow rate of FORSU, it is therefore necessary to use 2.360 Kg/d of "cellulosic wood" material; the plant has thus an overall potential equal to: 4.72 t/d + 2.36 t/d = 7.08 t/d

Delivery and storage

For the "cellulosic wood" material to be stored on a paved area, the weight of the unit of volume is assumed to be 0.15 t/mc, a residence time on the storage area of 2 weeks and a height per heap of 3 metres.

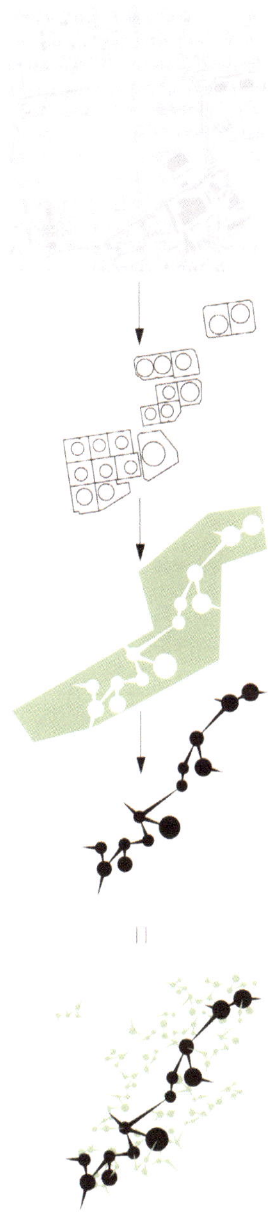

Structural composition of the park's project

Project data

FRAZIONE	PORTATA MASSICA(t/mc)	PORTATA IN INGRESSO(t/dlav)	GIORNI DI STOCCAGGIO(d)	Vtot(mc)	ALTEZZA(m)	AREA(mq)
Forsu	0,7	5,56	2	15,90	3	5,30
Strutturante	0,45	2,78	2	12,35	3	4,10
Area miscelazione						145
Area pretrattamenti						270
Somma						424 + 30%
Totale						550

Pre-treatments

PARAMETRI DI FUNZIONAMENTO	VALORE
Turni giornalieri	1
Ore di Funzionamento al giorno	6
Giorni lavorativi settimana	6
Giorni lavorativi annui	310
Potenzialità a regime giornaliero(t/d)	7,08 t/d
Potenzialità a regime annuo (t/a)	7,08 x 365,4 = 2.587,03
Potenzialità a regime lavorativo - giornaliero (t/dlavorativo)	2.587,03 / 310 = 8,34

Structural composition

Old tanks

Green park

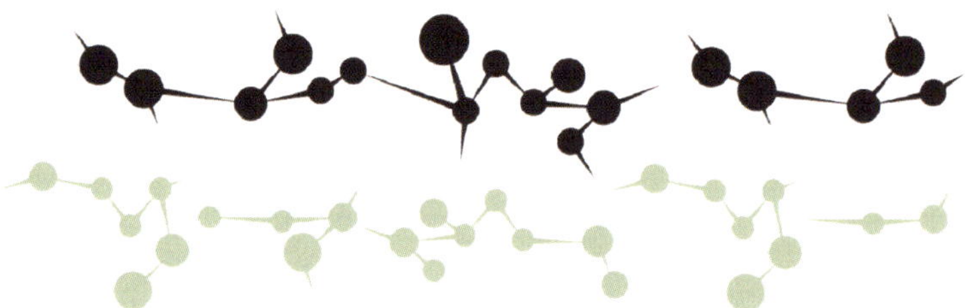

Agricultural areas

Homes - services

Functioning scheme

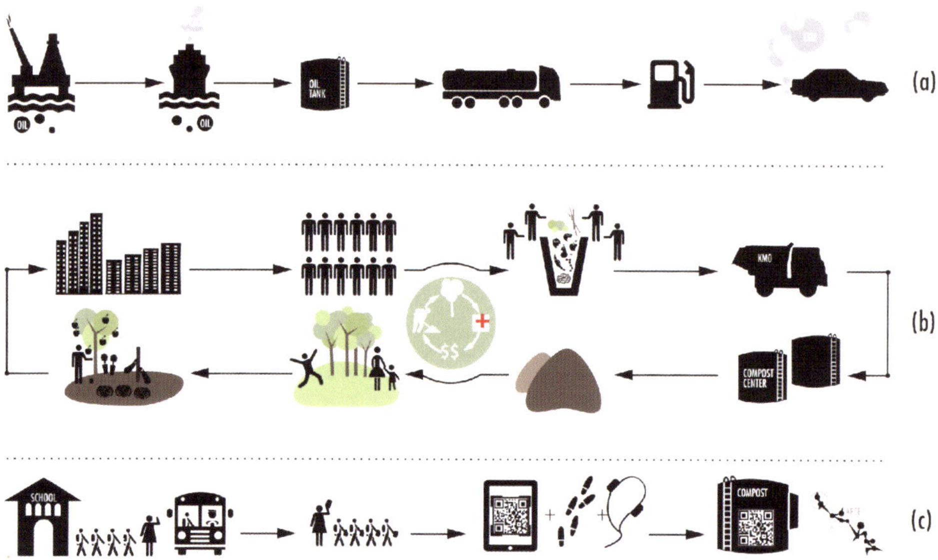

The "total area" (Sl) required is therefore equal to: Sl = {[(8,34 t/dlav):3] x 12 dlav/2set x 1/0,15 t/mc}: 3m = 74 sqm

The structuring material must be shredded and then sent to storage. The weight of the unit of volume after the shredding is assumed to be 0,45 t/mc and the permanence time is 2 months and, therefore, 52 working days. The "storage area" (Slt), assuming that nothing is lost during the shredding operation, is equal to: Slt = {[(8,34 t/dlav):3] x 52 working days/2months x 1/0,45 t/mc}: 3m = 107 sqm

Adding 80 square meters as the space necessary for the shredding operations, to store the incoming structuring material, for the shredding and for the storage of the shredded structuring material, the following "total surface area" is required:Ssl = Sl + Slt + 80 sqm = 74 + 107 + 80 = 261 sqm + 20% (manoeuvring spaces) = 313,2 sqm

Pre-treatment

The operations dedicated to the pre-treatment of the mixture to be composted take place in a closed structure. Part of the available area is dedicated to the temporary storage of lignocellulosic matrix and organic fraction, for a period of about 2 days. All the machinery necessary for the preparation of the mixture to be treated (trito-mixer, bob-cat with wheel loader) is expected to be available.

Bioxidation phase

The pre-treatments produce a quantity of mixture to be subjected to Accelerated Bioxidation or ACT (active composting time), equal to 80% of that at the entrance to the plant. The bioxidation system of choice is the mobile reactor system: a modular system consisting of n reactors to which a biofilter is added; each container has a capacity of 25 cubic meters (6.2 m x 2.3 m x 1.75 m). The number of biocontainers required is calculated in relation to the material to be treated.

Qi,ACT = Qi,imp x 0,8 = 8,34 x 0,8 = 6,67 t/working days -> (considering that the weight of the volume unit is 0,59 t/d) -> Vi,ACT = 6,67/0,59 = 11,30 mc/working days
Value that - in relation to the 25 mc of the "biocontainers" -with a use of 3 weeks / 4, leads to: 11,30 x 18/25 = 8 Biocontainers + 1 Biofilter

Maturation and refining

The area dedicated to maturation is measured here below, considering that the biomass that reaches maturation is equal to about 60% of that at the entrance to the plant with a stationing of the mass of 76 days in the form of heaps 10 m wide at the base and maximum 2.5 m high, with a section of 20 square meters, after which they are refined through screening systems (sieve analysis/gradation test).

Maturation: Ptot = 0,6 x 8,34 = 5 t/working days -> Vm = 5/0,6 x (76 -76/7) = 540 mc -> Ltot = 540/20 (heap section) = 27m -> 27 x 10 = 270mq

Refining: Machinery area = 200 sqm - Packaging area = 100 sqm - Sum = (200 + 100) + 30% (manoeuvring space) = 330 sqm

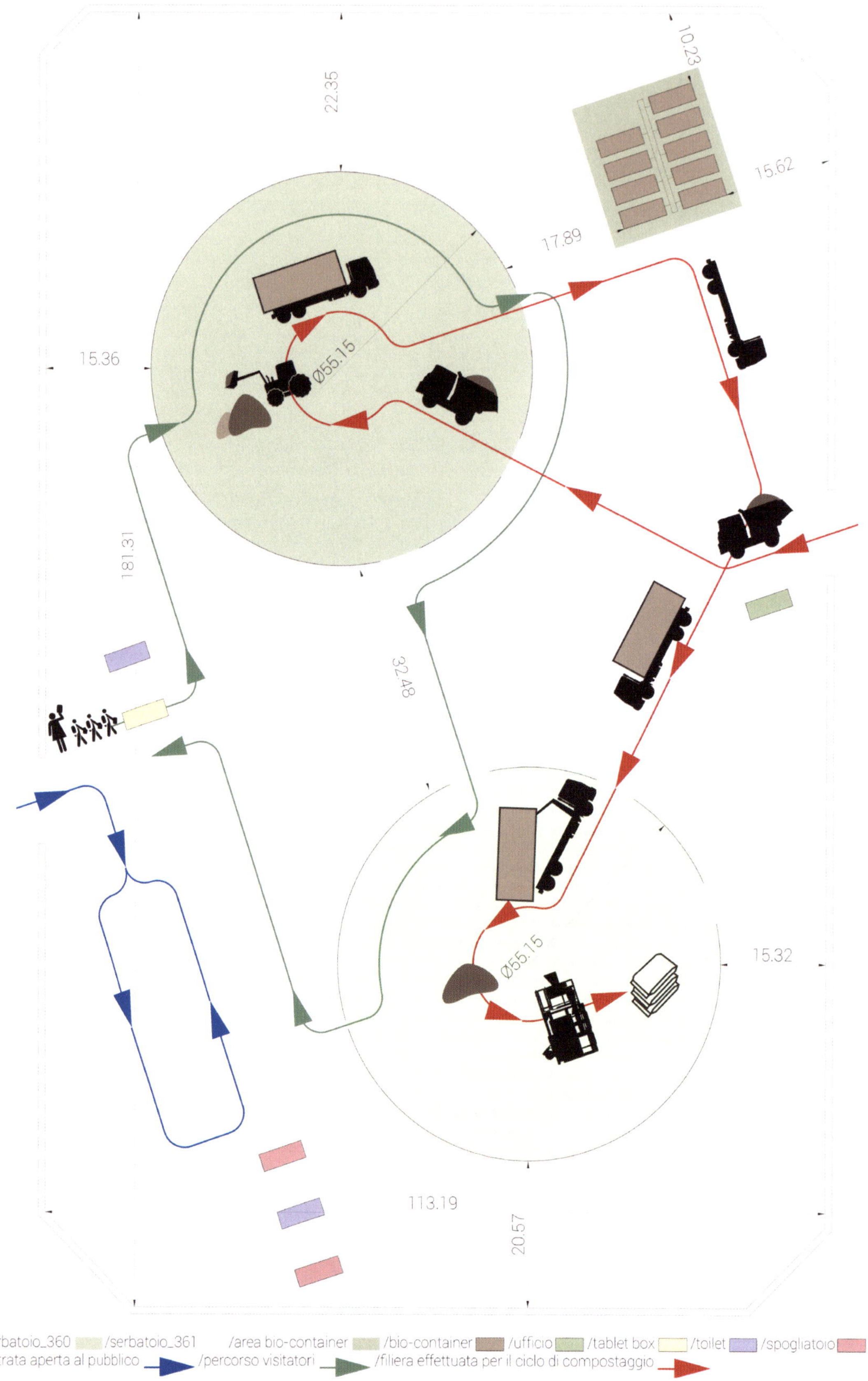

22.35
10.23
15.62
17.89
15.36
Ø55.15
181.31
32.48
15.32
Ø55.15
113.19
20.57
serbatoio_360 /serbatoio_361 /area bio-container /bio-container /ufficio /tablet box /toilet /spogliatoio
entrata aperta al pubblico /percorso visitatori /filiera effettuata per il ciclo di compostaggio

360
IMPATT
ZER
BIO-CONTAINER

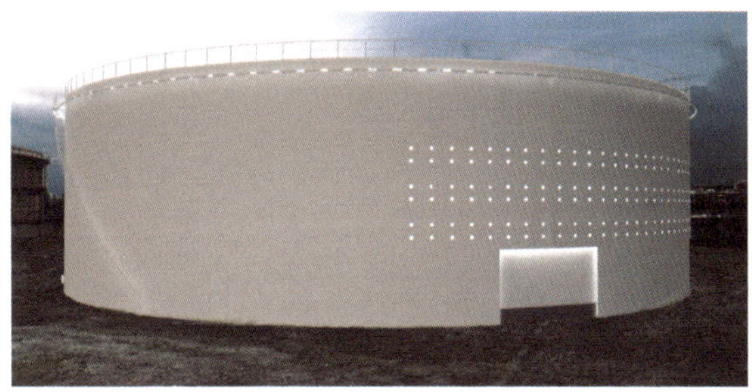

TABLET BOX

TABLET BOX

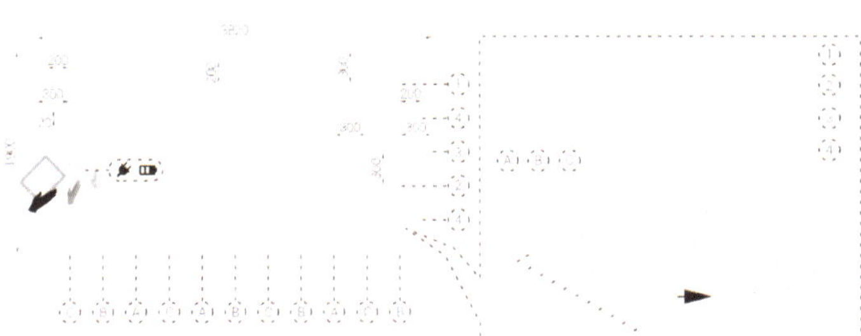

TABLET BOX

PIANA del fiume SARNO

Lavori fermi in molti cantieri e il grande progetto è nel caos. Ma dov'è la bonifica?

paralleloquarantuno.com

Operatori di settore

- REGIONE CAMPANIA
- EMERGENZA FIUME SARNO E ARCADIS
- ENTE PARCO REGIONALE DEL FIUME SARNO
- AUTORITÀ DI BACINO DEL FIUME SARNO
- CONSORZIO BONIFICA FIUME SARNO
- FERROVIE DELLO STATO
- PROTEZIONE CIVILE
- COLDIRETTI SALERNO
- CONFEDERAZIONE ITALIANA AGRICOLTORI
- PATTO TERRITORIALE DELL'AGRO S.P.A
- LEGAMBIENTE
- CIRF CAMPANIA

SEL: "Canale Conte di Sarno? Una bomba da disinnescare"
"..4 incontri per denunciare lo stato di abbandono e di degrado in cui versa il canale Conte di Sarno e la bomba ecologica che è già scoppiata e che rischia di compromettere un territorio oramai già compromesso."

StabiaChannel.it 29/01/2014

- Dischiarazione di "Altorischio di crisi ambientale" — 25.08.1992
- Dichiarazione dello stato di emergenza socioeconomica-ambientale — 14.04.1995
- Alluvione di Sarno e Quindici — 04.05.1998

"Sarno tour, un viaggio seguendo il corso del fiume più inquinato d'Europa dalla foce alla sorgente"

Senato della Repubblica, XVII Leg.

"Allagmaneti e veleni nelle vasche, l'ira di Poggiomarino, Terzigno e Striano: Fate presto!"

Il fatto vesuviano

Associazione regionale amici del Sarno, Gruppo archeologico terramare 3000, Movimento 5 stelle "Sarno tour" e "difendiamo il fiume Sarno, Gente del Sarno, No alla seconda foce di Rovigliano, Comitato per la salute, No vasche Si alla messa in sicurezza, associazione cittadinanza attiva, Onlus fucina

Comitati

Rischio di natura sanitaria e idro-geologica per le vasche di Poggiomarino

21.12.2011

Emergenze

Maltempo, il Sarno rientra negli argini: resta lo stato di allerta

Il fatto vesuviano
22.01.2014

The Sarnese City

Anna Terracciano

Emanuela De Marco

The historical area, its transformations through time and the water system's role

The Sarno Plain is a morphologically well-defined area of ancient colonization, favoured by the richness of the waters, the territory's fertility and its strategic position. Between 2000 and 2004, a proto-historic reservoir was discovered in the area of Longola-Poggiomarino[66], confirming the stable anthropic presence between the Middle Bronze Age and the Final Bronze Age and bringing light to an area of extraordinary historical and archaeological interest. The first settlement probably was a river port dating back to the second millennium BC consisting of a village built on stilt houses. The canals delimiting the islets, built as part of real regimentation and reclamation interventions, prove that at that time the population already possessed a high knowledge of hydraulic engineering and made a careful selection of materials for the construction of housing structures, as never documented before in this part of Italy.

In classical times the Sarno plain, thanks to the intersection of three important consular roads - the Popilia (also called Aquilia), the Puteolis-Nuceria (or Domitiana) and the Stabiana - assumed an important role as transit area for Rome's trades with the South. The Sarno river is described to us by Strabone[67] as a busy waterway, whereas the plain was dotted with rustic villas which controlled vast farms for the milling of cereals and legumes. The intense area's activity was, however, completely nullified by the eruption of Vesuvius in 79 A.D., which covered the plain with a thick layer of pyroclastic deposits. It took five centuries for the area to be populated again, also due to the Sarno river's re-emergence.

The first historiographical mention, by the milling industry, was however only after year 1.000.[68] In the absence of records, during the following five centuries it is only possible to assume the existence of several small mills along the river, such as not to alter its outflow and built to meet the daily needs of the local communities.

66. D'Angelo S., "La Sarno protourbana e perifluviale dei Sarrasti: l'insediamento protostorico finora sconosciuto in località Affrontata dello Specchio, sul principale porto fluviale a monte del Sarno", Edisud, Salerno, 2004.

67. See Biraschi A.M. (ed.)," Libro V and VI", in "Geografia. L'Italia", B.U.R. Biblioteca Universale Rizzoli, Milan, 1998.

68. See "Regestum Sancti Angeli ad Formas".

It is starting from the end of the 16th century that the spreading of mills and paper factories brings an increase of not always legitimate interventions aimed at changing the course of the river, not only for agricultural reasons but also to use it for industrial purposes. In this way, the future urbanization of this area - until then sparcely inhabited because the water runoff from the Somma-Vesuvio and the periodical floods of the river Sarno made it unhealthy - was affected in a decisive way.
The first industrial-scale exploitation plan in this area saw the construction of the artificial Count of Sarno Canal[69] for the collection of water from Santa Maria della Foce to Torre Annunziata, thus allowing the mills to work. The difficult and expensive excavations began in 1592 and incidentally marked the discovery of the Roman city of Pompeii, whose site had been lost in memory. The construction of a tunnel, expensive and complex due to area's low slope inclination, produced a profound morphological change as reclamations were being done throughout the area making a large quantity of land fertile and ready to be farmed. From this moment on, a slow, progressive but unstoppable process of colonisation of the plain began. The constant increase in population, however, was not met by a corresponding building expansion based on regular urban planning.
A second important intervention on the river's hydrographic network was the construction of the so-called "Traversa di Scafati". It still works today and consists of a barrier with mobile bulkheads that narrows and diverts the course of the river towards the artificial bed of the Bottaro Canal and the Scafati Canal, a natural branch of the Sarno river, thus allowing the mills of Torre Annunziata to work.
The "Traversa di Scafati" radically modified the hydraulic functioning of the middle and lower course of the Sarno, draining the main course of the river and affecting its navigation. The consequent increase in the water level and

69. The Count of Sarno, Muzio Tuttavilla, commissioned the Conte di Sarno canal, associated with the origin of the first settlement of Poggiomarino due to the arrival of a large number of workers. The Count appointed the pontifical architect Domenico Fontana, at the time involved in the project of the Regi Lagni, who finished the work in 1605. Currently, the Conte di Sarno canal is dry because of the collection of water from S. Maria della Foce to supply the Sarno aqueduct, and is sadly relegated to the role of sewer for the daily discharge of thousands of cubic meters of wastewater from the homes of the municipalities north of Boscoreale. A recovery project in the early 80's included the hydraulic arrangement of the canal with the construction of a C-shaped box in the lower part, from a piling and a covering slab, which was to collect the white water and sewage and convey them from the town of Sarno to Torre Annunziata. The works, which began in 1981, were suspended in 1995 due to the impossibility of completing it without causing damage to the archaeological area. To date, the box-shaped structure works like a draining trench collecting all the water coming out of the absorption tanks and, since the canal is interrupted in some points, and in any case has no outlet to the sea, it is subject to frequent unhealthy flooding. See Franco A., Squitieri M., Miranda A., Mazza G., Martorelli M., "Studi Storici Sarnesi", Franco A. (ed. by), Il Chiostro, Benevento, 2012.

→ Direttiva quadro sulle acque 200/60/CE

› Prevenire il deterioramento qualitativo e quantitativo

› Migliorare lo stato delle acque

› Assicurare un utilizzo sostenibile

"cit."

→ I Bacini Idrografici in Italia

› 39 Bacini idrografici

› 8 Distretti idrografici

› Individuazione delle aree degradate del territorio

› Fissazione vincoli per la conservazione del suole e la tutela dell'ambiente

→ Il Bacino Idrografico del Fiume Sarno

Il Bacino comprende 38 Comuni appartenenti a 3 Province divese, su un'area di 483km², equivalente al 5% della superficie della Regione Campania.

Il reticolo idrografico è arricchito da un gran numero di affluenti secondari e di fossi, per uno sviluppo lineare complessivo di circa 1.630 km.

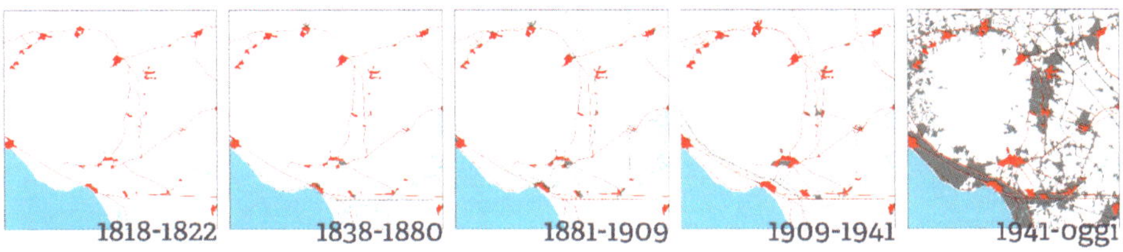

"cit."

*"Né lo storico né il cartografo sono in grado di **riprodurre la realtà** che cercano di comunicare."*

› Urbanizzato

› Territorio naturale

↓ Boschi

↓ Pascoli

↓ Arbusteti

↓ Spiagge

› Territorio agricolo

↓ Seminativi

↓ Colture

↓ Frutteti

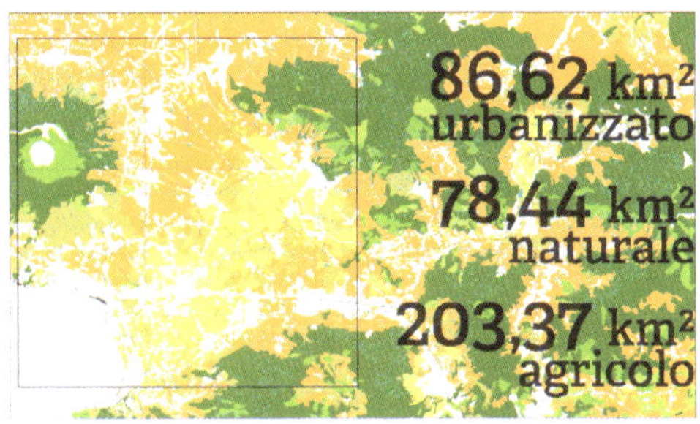

↓ R1 Rischio moderato

↓ R2 Rischio medio

↓ R3 Rischio elevato

↓ R4 Rischio molto elevato

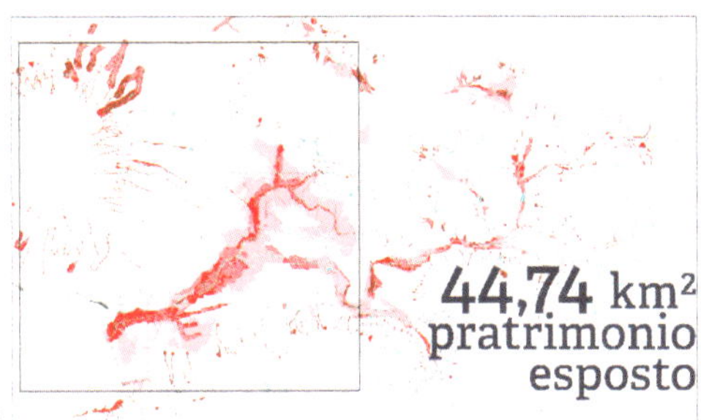

↓ R1 Rischio moderato

↓ R2 Rischio medio

↓ R3 Rischio elevato

↓ R4 Rischio molto elevato

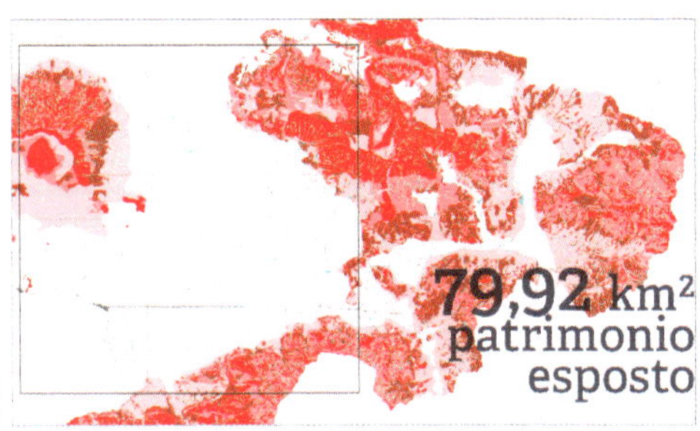

the landfills upstream soon caused a crisis in the existing drainage system, provoking swamping and the formation of unhealthy areas. It was not until the 19th century, in the scope of the major works carried out by the Bourbons following the establishment in 1855 of the General Administration for Land Reclamations, that the Plain's reclamation and sanitisation issues were finally resolved.

The proposed solution involved the construction of channels on both sides of the river in order to intercept the waters coming from the plain by conveying them, with a single delivery, downstream of the Traversa. Meanwhile, in the upper part of the river, a system of artificial channels was built in 1803 by the Superintendence for Bridges and Roads, aimed at conveying the water from the Cavaiola and Solofrana in order to solve the flooding issue of the Nocera's area farmhouses. This intervention produced a first substantial alteration of the river's upper part, artificially expanding its flow and basin, and became permanent in 1857 with the construction of the artificial canal named "Alveo Comune Nocerino".

Still today, the "Traversa di Scafati" puts the lower section of the Sarno in crisis because of its low efficiency. For this reason, in March 2012 a memorandum of understanding was signed between the Region, Arcadis[70] and the Sarno Basin Authority for the Great Project to complete the redevelopment and recovery of the Sarno river (GPS)[71].

On the Somma-Vesuvio side, on the other hand, the soil's great permeability, which has caused the deep absorption and distribution of rainwater, has not allowed the formation of a stable superficial hydrography and has determined the conditions for the creation of a widespread hydraulic risk along its slopes, particularly in the contiguous foothills. Historically, the abundant rainfalls, flooding the riverbeds of the Somma-Vesuvio, ended up transporting enormous quantities of water downstream, often with solid loads, such as to seriously compromise the safety of the inhabited areas located downstream.

The widespread "hydraulic disorder", particularly affecting the top sector where the torrential water lacked a final delivery, favoured the dispersion of water in the plain, making it become marshy. For this reason, hydraulic safeguard measures were periodically adopted, through the construction of canalization works, absorption tanks, artificial collectors, etc.[72]

The urban fabric has thus conformed over time to the water network's morphology and role[73], arranging itself linearly along the road axes as it typically occurs in agricultural areas with a strong demographic development. This settlement principle is still visible today as its characteristics have been unaltered since the 1950s, notwithstanding the construction of the Circumvesuviana in the early '900s. With its ground-

70. The ARCADIS (Agenzia Regionale CAmpana per la DIfesa del Suolo / Campania Regional Agency for Soil Protection) was established by Regional Law 12/11/2004, n.8, supplemented by Article 33 of Regional Law 30/01/2008, n.1 and governed by Regional Law 8/08/2014 n.19. It is an instrumental body of the Campania Region with legal personality under public law. The body has organizational, administrative, patrimonial, accounting, technical and managerial autonomy and is in charge of implementing interventions in the Campania Region in the field of soil protection, reclamations and water protection.

71. This project foresees, among other things, the completion of the lower Sarno's hydraulic and environmental upgrading through the construction of a second mouth.

72. These interventions were the result of a complex hydraulic project started in 1855 by the "Corpo degli Ingegneri del Regno di Napoli" (Engineers of the Kingdom of Naples) with the dual purpose of limiting the descent of volcanic material and making the lower riverbeds capable of draining flood water. See Simonetti R., "La Bonifica e la sistemazione idraulica dei torrenti di Somma e Vesuvio", lithographic plant of the Civil Engineers, Rome, 1912, available at: http://www.consorziobonificanapoli.it/wp-content/uploads/2012/01/Bonifica-e-Sistemazione- Somma-e-Vesuvio_COMPLETO.pdf; Ruffo F., "La Campania antica. Appunti di storia e topografia", first part: "dal Massico-Roccamonfina al Somma-Vesuvio", Denaro Libri ed., Naples, 2010.

73. For further information see: Sarno Basin Authority, "Quaderno AdB ", Year 1 n.1/2003 Periodical report; and " Quaderno AdB " 2/2004, "Il rilievo dell'asta fluviale del fiume Sarno e dei suoi affluenti", Series of studies, documentation and research.

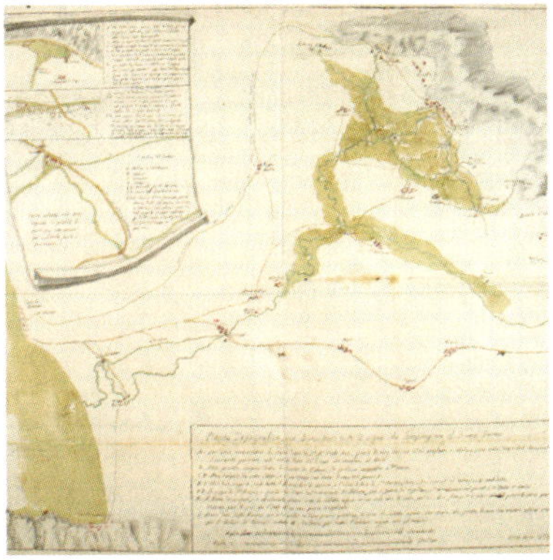

Pianta Topografica delle acque che compongono il Fiume Sarno
Nicola Anito

Carta del Litorale di Napoli
Antonio Rizzi Zannoni, Scala originaria 1: 114.000

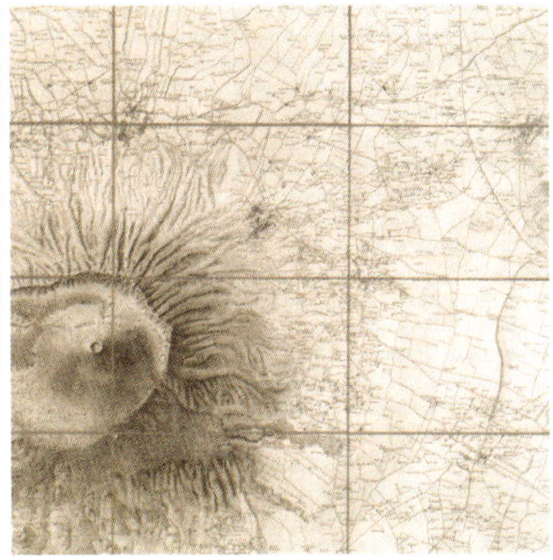

Carta Topografica ed Idrografica dei contorni di Napoli
IGM, Scala originaria 1:25.000

Carte preparatorie da Portici a Torre Annunziata

Carta Topografica ed Idrografica dei contorni di Napoli

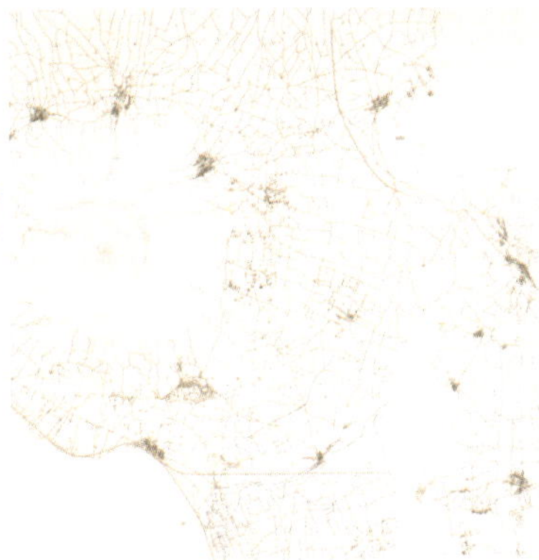

Vesuvio - Foglio LXII
IGM, Scala originaria 1:20.000

Carta delle province continentali dell'ex Regno di Napoli
, Scala originaria 1: 103.680

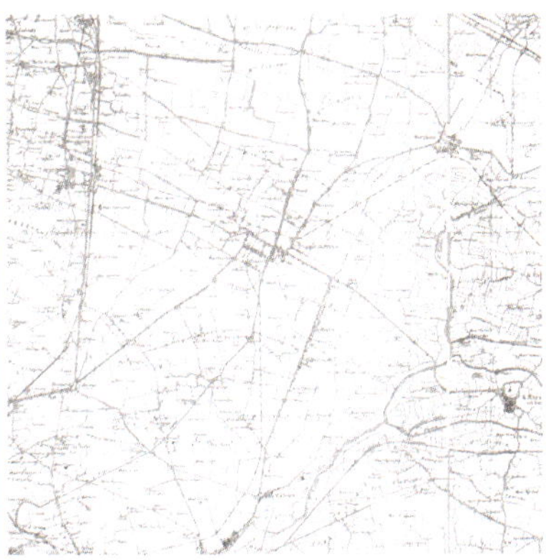

Carta Topografica Istituto Geografico Militare
IGM, Scala originaria 1:25.000

Carta Topografica Istituto Geografico Militare
IGM, Scala originaria 1:25.000

On the opposite page: reenactment of the human settlement's evolution compared to the one related to the water system. Illustrations by Emanuela de Marco

In the following pages: on the left, industrial activities at the port of Torre Annunziata (Naples); on the right, water exploitation, water needs and production of wastewater in the Sarno river's plain. Sources: AdB Campania Centrale, ISTAT, Consorzio di Bonifica del fiume Sarno; pictures by Giovanna Ferramosca.

level crossings, the Circumvesuviana constitutes one of the main caesura elements in urban areas. This structure was permanently compromised by intense building in the last fifty years. This has led to the systematic saturation of large agricultural areas of the interior through the replacement of fringe constructions with modest single or two-family units, frequently illegally built, completely devoid of any relationship with the pre-existing agricultural vocation of the territory and not connected to any urban planning.

Hydraulic network and water cycle. Contemporary disruptions

Today, in the Sarno river's plain, the effects of hydrogeological risk sum up to the volcanic ones[74] but also to those resulting from the degradation of environmental resources[75], due to chemical contamination of agriculture, anthropic pressure, high soil consumption and the abandonment of long-established protection practices. Furthermore, the failure to complete the sewerage system at municipal level (as in the case of Poggiomarino) and the collection system at district level lead to a general and continuous discharge of urban and industrial wastewater into the river Sarno and throughout the superficial hydrographic network. Finally, all this is combined with the waste areas, buildings and infrastructures of an increasingly incomprehensible urban metabolism.

The main vulnerabilities in settlements at high soil consumption, such as those of the Sarnese City characterized by urban sprawl, include the soil's waterproofing, which causes problems linked to the reduction of available surface area for the performance of soil functions, such as the absorption of rainwater and urban "heat island" phenomena, which generate environmental discomfort.

The traces of the above-mentioned ancient hydraulic works are still visible in the unified and rational vision of the water system, at the base of which laid a tendency to "take care" of the territory, as well as an ambitious industrialisation project envisioning canals capable of ruling the dynamics of the surface and deep waters and mitigating the hydraulic risk, through the construction of devices and areas dedicated to the expansion and collection of natural and artificial water. Today, such interventions have lost their rationality due to poor maintenance and improper use. The stories regarding the Piana, in particular of Poggiomarino[76] which is its centre not only in geographical terms, is therefore inseparable from the metamorphosis suffered by its water network. As it often happens, in fact, the water network has played a decisive role not only in the design of the landscape but also at a social, economic and productive level.

74. The Civil Protection Department and the Campania Region have begun the review of the Emergency Plan, with an extension of the Red Zone compared to that identified in the 2001 Plan. It includes the territories of 24 municipalities and three districts of the City of Naples. In addition to the 18 municipalities already included in the Red Zone, the new Plan are includes the districts of Barra, Ponticelli and San Giovanni a Teduccio in the Municipality of Naples, as well as the municipalities of Nola, Palma Campania, Poggiomarino, San Gennaro Vesuviano, Scafati and the enclave of Pomigliano d'Arco in the Municipality of Sant'Anastasia.

75. The Site of Regional Interest (S.I.R.)River Sarno Hydrographic Basin is part of a broader territory belonging to the authority of the River Sarno Basin (now AdB Central Campania) which covers a total area of about 42,000 hectares, including the areas crossed by the river Sarno from the mouth to the springs. With the approval and publication in the Official Gazette of 12/03/2013, the Ministry of the Environment downgraded the site of the River Sarno's Hydrographic Basin from S.I.N. to S.I.R.

76. See next contribution in this chapter.

Pre-Roman time/Iron Age

- Antico corso del fiume Sarno e relative sorgenti
- Nuclei urbani
- Insediamento su palafitte di Longola, molto probabilmente un porto fluviale, risalente al II millennio a.C
- Necropoli dell Età del Ferro
- Insediamenti minori
- Porto di Pompei

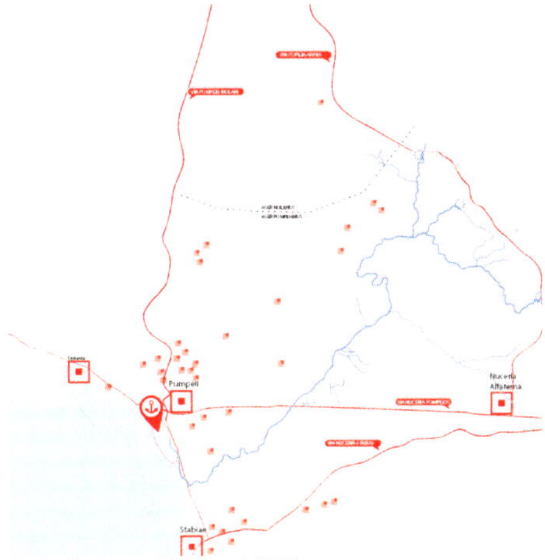

Period between 79 AD and 1592

- Orientamenti delle maglie centuriali
- Principali tracciati di collegamento dell epoca romana
- Principali centri urbani
- Ville romane
- Porto di Pompei

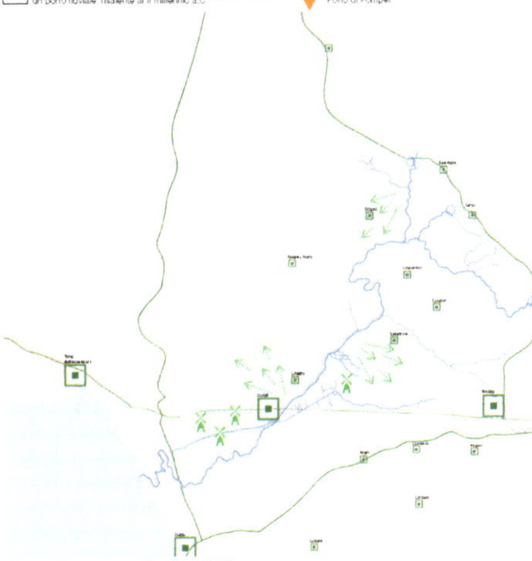

Period between 1871 and 1956

- Centri urbani
- Costruzione del canale Bottaro che alimentava i mulini
- Aree interessate da fenomeni di impaludamento
- Avanzamento della linea di costa

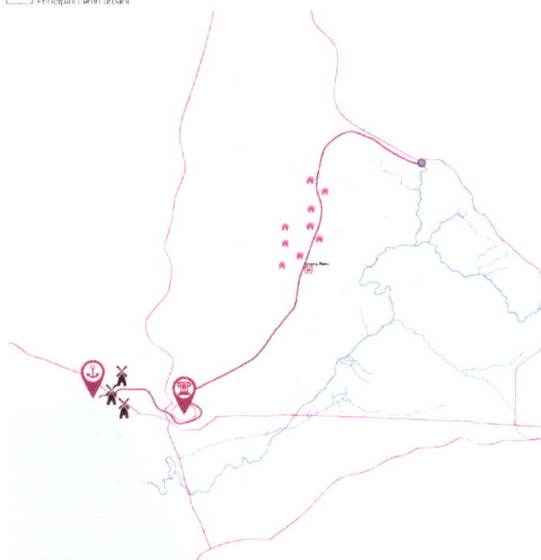

Classical era until 79 AD/Roman time

- Opere di costruzione del Canale del Conte di Sarno [scavo 1592-1605]
- Insediamenti industriali per la molitura dei cereali
- Primo insediamento di Poggiomarino in località Taverna Penta
- Rinvenimento dell'antica Pompei
- Porto di Torre Annunziata

Period between 1592 and 1631

- Opere di canalizzazione, vasche di assorbimento e di decantazione, collettori artificiali sul versante del Somma-Vesuvio
- Tombamento del Canale del Conte di Sarno
- Linea ferroviaria Torre Annunziata-Cancello
- Circumvesuviana

Contemporary period, from 1956 to this day

- Raccolta delle acque di versante e degli scarichi fognari abusivi
- Adeguamento della sezione del canale del Conte tramite la realizzazione di uno scatolare a sezione a "C"
- Interruzione del canale in prossimità dell'area archeologica di Pompei
- Costruzione dei degli impianti di depurazione
- Rinvenimento del villaggio protostorico di Longola
- Dismissione della ferrovia Torre Annunziata Cancello
- Linea ferroviaria AV
- Viabilità di grande comunicazione

Approvvigionamento acqua dai pozzi

comune	n. pozzi	totale prelievi (mc/anno)
Sarno	7	3.084.480
Nocera Inferiore	7	1.522.800
Pagani	1	231.840
totale	15	4.839.120
totale pozzi consorzio di bonifica	58	14.694.480

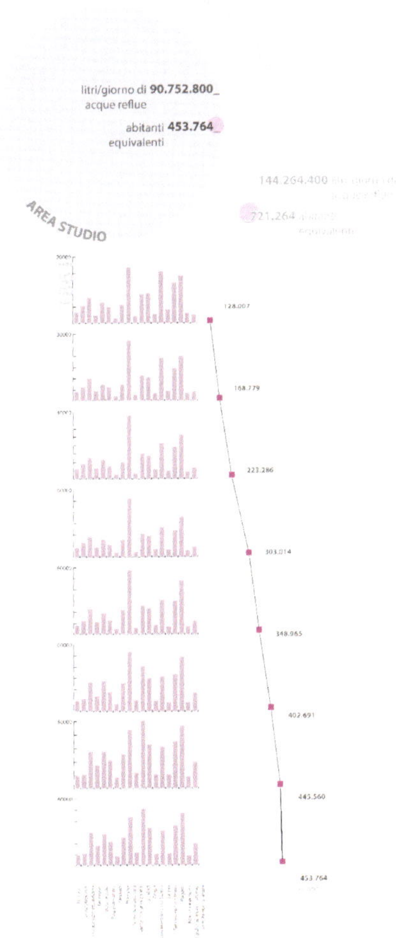

Exploitation and disposal of water resources

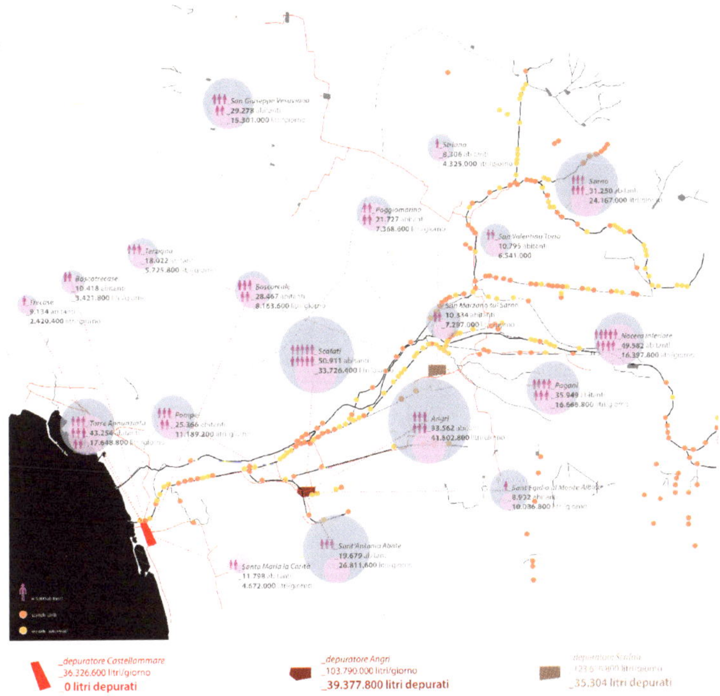

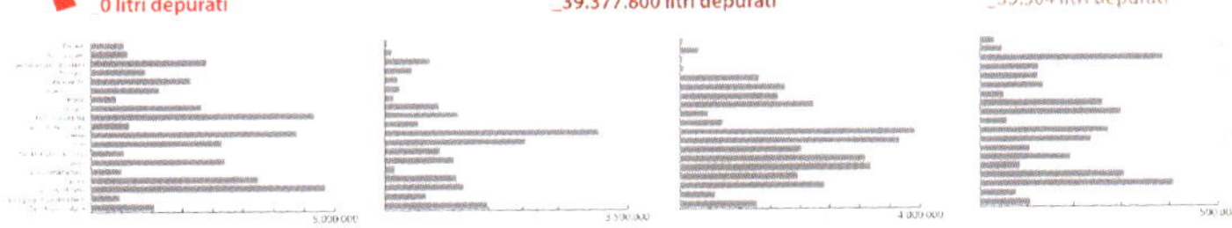

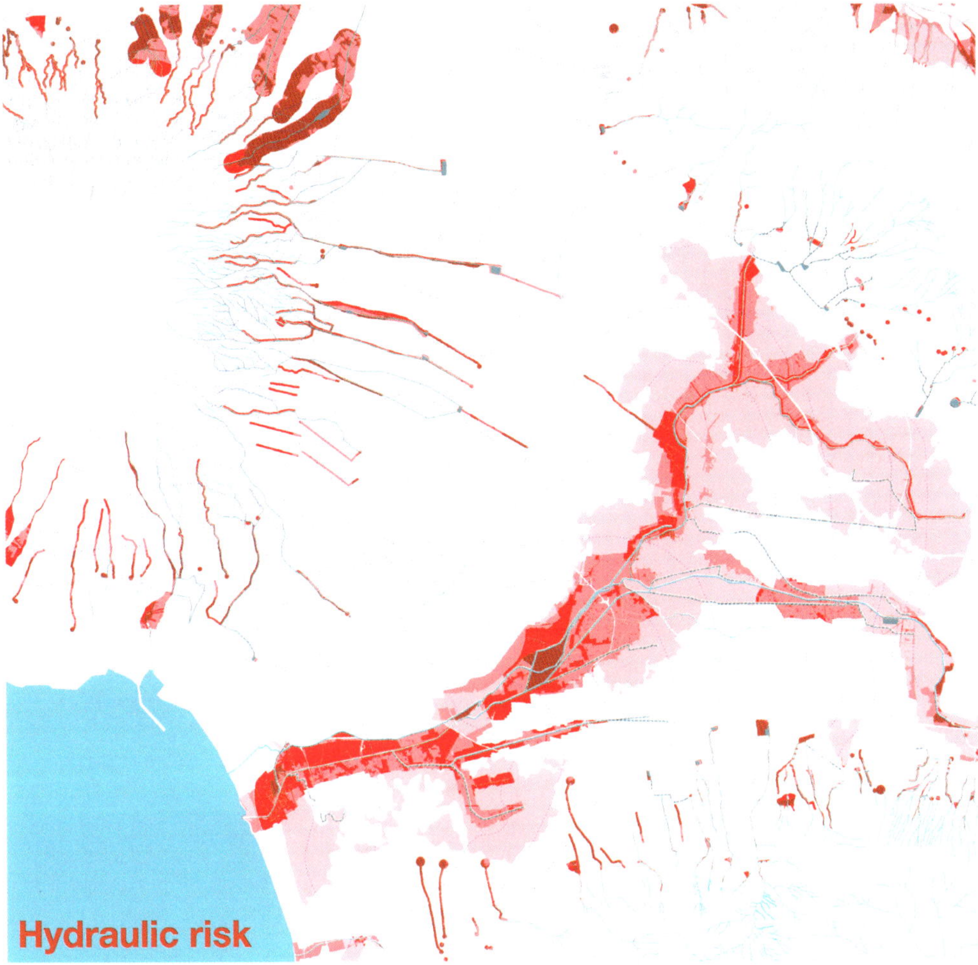

ciclo delle acque e uso del suolo

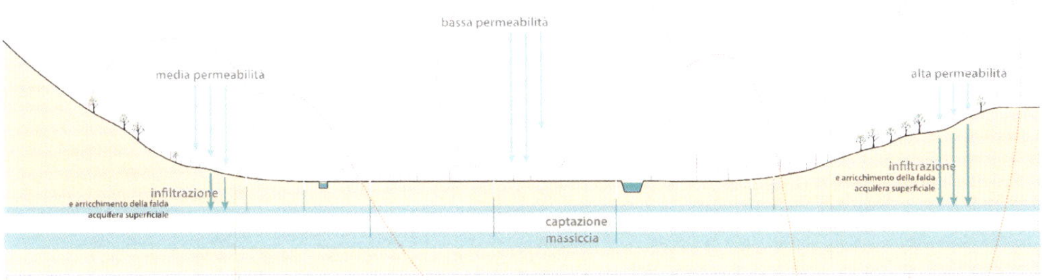

Water cycle

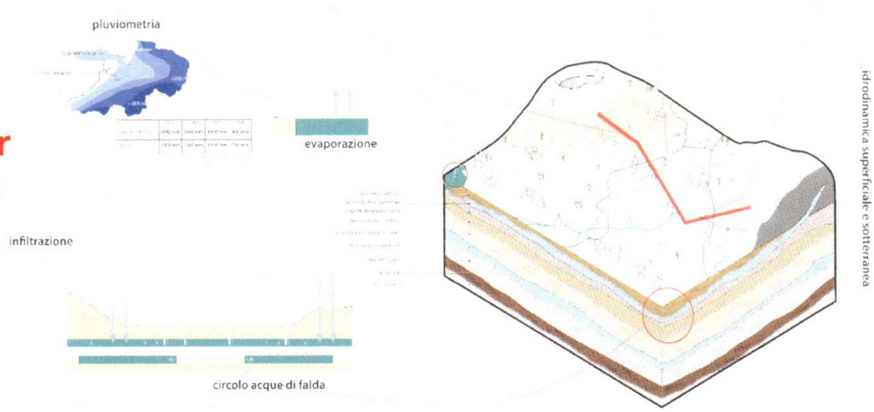

In the previous pages: on the left, Pianillo tank, San Giuseppe Vesuviano (Naples); on the right, above, Hydraulic Risk in the Sarno river Plain. Source: Piano Stralcio di Bacino per l'Assetto Idrogeologico (PSAI) AdB Campania Centrale, adopted by the Institutional Committee with Resolution n.1 of 23/02/2015 (B.U.R.C.n.20 of 23/03/2015); below, functioning of the water cycle and structures of the soil.

In the opposite page: Drosscape geography in the Sarno plain. The images are from the degree thesis by Francesco Sammarco, Ciro Sepe and Danilo Vinaccia (see their contribution in this chapter).

Massive industrialization and anthropization of the territory has, particularly in recent decades, progressively changed the virtuous relationship between man and water and led to serious imbalances. The availability of water resources, first used for agriculture through a rigorous design of the irrigation channels network and then acting as a driving force for the first factories, with the construction of powerful hydraulic devices from the '600s, has been permanently depleted by soil consumption which, starting from the second half of the last century to the present day, has finally created a fragile and damaged territory.

In the same way, the extensive hydraulic works begun by Ferdinand II of Bourbon to regulate the Vesuvian mountainside's runoff, have been distorted, as from the end of theSecond World War to date, by a generalised neglect that has affected their original function. The chaotic planning; illegal use of land; proliferation of crops in the riverbeds and tanks; illegal spillage of all kinds of sewage; and the Lagni (man-made channels) riverbeds used as roadsides or garbage deposits have altered the drainage system's functioning, making it completely ineffective. Moreover, in case of bad weather, yet of modest entity, the collection tanks overflow invading and polluting agricultural and urbanized areas. This is what happens, for example, in Poggiomarino because of the presence, at the municipal borders, of the Fornillo (Terzigno) and Pianillo (S. Giuseppe Vesuviano) tanks.

The main production centres are the tanning centre in the Alto Sarno area and the canning, paper and printing centres in the Agro Nocerino Sarnese area. They constitute the main source of pollution, as their production cycles generate hazardous waste, and the main condition of discontinuity and imbalance existing between highly industrialised and infrastructural areas with fringe and underdeveloped areas.

In addition to the environmental damage caused by the highly polluting production activities[77] affecting large portions of the territory, there is a significant number of critical situations linked to the presence of disused factories and abandoned mines, whose sites have been used as landfills for the authorized or illegal waste storage (of high chemical contamination). These situations have produced repercussions not only on the hydrographic system[78] but also on urban and agricultural soils.

Thus today, the historical continuity of the ecological networks is opposed by their fragmentation; the principle of minimum distance between the human activities and the waters is contrasted by a condition of overlapping; and the recognition of the river Sarno's founding and identity values is contrasted by disorientation and exploitation.

77. Industrial Archaeological Association: "Manifatture in Campania. Dalla produzione artigiana alla grande industria", La spirale n. 57, Guida editori, 1983.

78. Campania bonifiche s.r.l. (l.l.c), "Il bacino idrografico del fiume Sarno e il territorio sarnese - nocerino". Source: http://www.campaniabonifiche.org/archiviostorico/sarno/ricercasarno.pdf

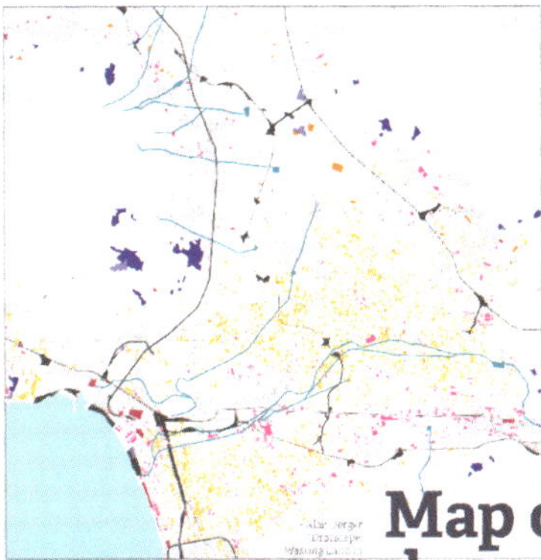

disegni di cose

"Restituiscono una rappresentazione fedele all'oggetto dell'oggetto in questione: ***disegni che assomigliano alla realtà che rappresentano.****"*

"cit."

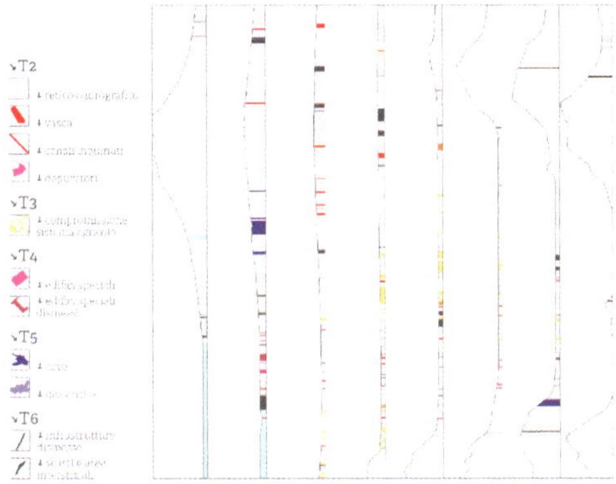

"Non si può creare ***una mappa perfetta.*** *Mai."*

"cit."

"cit."

Map of the drosscapes *

La parola ***drosscape**** (scarti) ha relazione con i termini: **vast** (vuoto) e **waste** (improduttivo). Il dross è la combinazione di processi **naturali** e **antropici**, diventando quindi un prodotto generato da pratiche esterne ad esso. Ampliando il ragionamento allo spazio urbano, Berger afferma che lo ***scarto*** è considarato come un componente naturale di ogni città che si sviluppa dinamicamente. Si può quindi assumere il dross come un indicatore dello sviluppo della salute urbana.

T1

suoli inquinati

T2

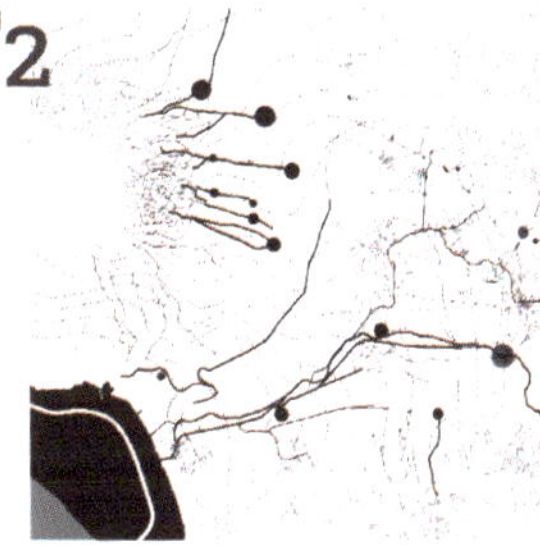

acque inquinate e dispositivi idraulici

T3

ecosistemi compromessi

T4

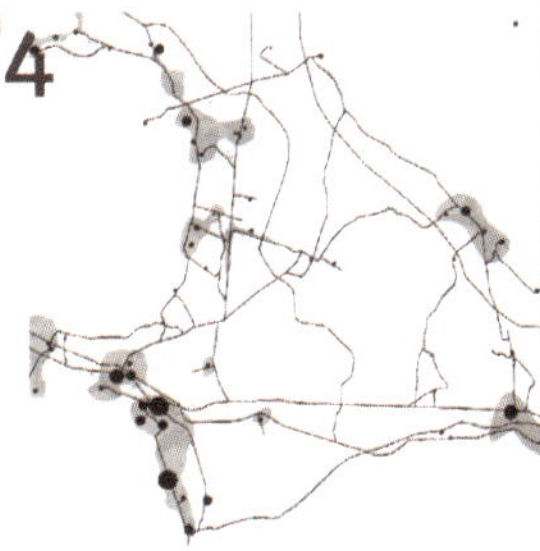

edifici speciali dismessi

T5

cave e discariche

T6

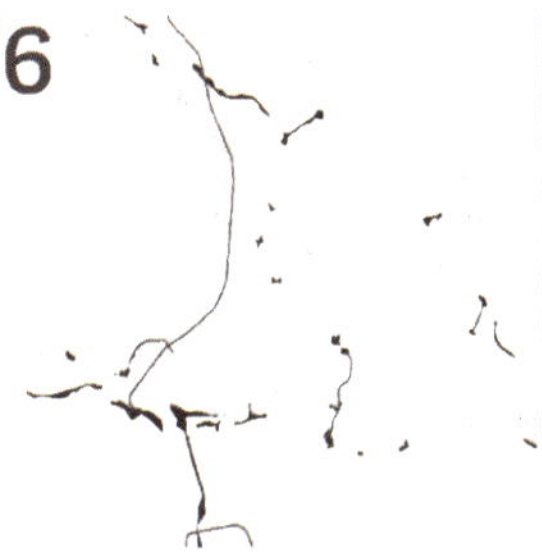

infrastrutture dismesse e aree interstiziali

→ obiettivi di re-cycle

› **Costruire**, per le aree studio in **Campania**, una geografia di materiali e luoghi che **hanno esaurito il loro ciclo di vita** e divengono disponibili all'interno di un processo di **re-cycle** alla costruzione di **nuovi cicli di vita degli stessi**.

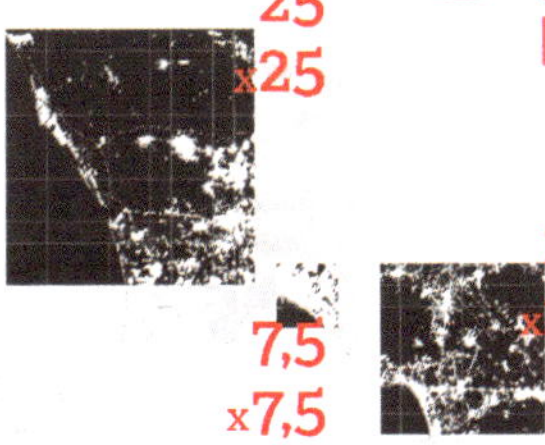

RE-CYCLE ITALY

"E la ***geografia*** *serviva proprio a trasformare la mappa in un* ***discorso****, in una versione alternativa, e perciò potenzialmente critica, rispetto a quella spaziale, e perciò apodittica, della realtà."*

"cit."

disegni di concetti

"E' quel tipo di rappresentazione che si allontana dall'aspetto delle cose; ***disegni che evocano la realtà che rappresentano****."*

"cit."

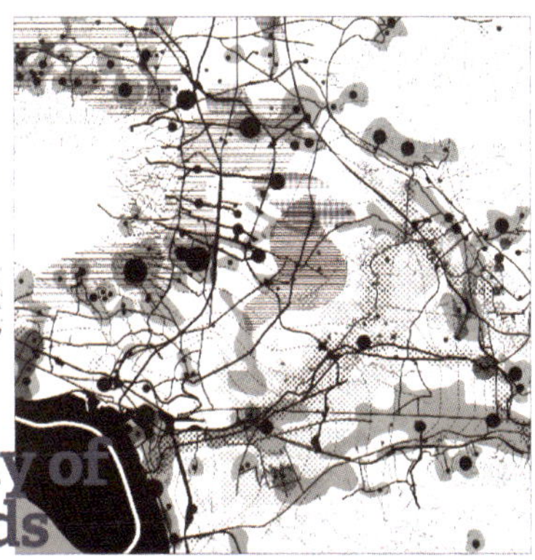

Geography of wastelands

Land use and riverscape: a possible geographical layout for the Sarno river's resilience

Giovanna Ferramosca

Objectives and key-words

The thesis'[79] fil rouge is the river Sarno and the hydrographic system around it, extending to all of the Piana di Sarno, province of Salerno, in the Campania region in Southern Italy. This hydrographic system has through time, not only influenced and defined in many different ways the environmental layout, but also the settlements' organisation, the economic, productive and last, but not least, the cultural structures of the region.

The river Sarno's plain has been dissected in all its essential elements, by activating a "multi-dimensional, multi-scale and multidisciplinary" interpretation and overview, following the French philosopher Edgar Morin's principles of the "epistemology of complexity". Morin asserts that in order to explain a phenomenon it is necessary to go from "the whole to the parts and from the parts to the whole"[80], to have a comprehensive knowledge of the issue being analysed.

The thesis sets three demonstrative goals:
1. To build a basis of knowledge that permits an in-depth examination of the complexity of the relationship between the river and its system and the land's layout;
2. To define a map of the opportunities capable of guiding the choices depending on the issues at stake;
3. To provide a set of qualitative indicators[81] which reflect the current status of the relationship between the hydrographic river system and the use of its soil, and which are structured to be used as "information/rule" for future interventions.

Two are the key-words which have been orienting this research: land use[82] and riverscape, the former in a city planner's view, the latter in the style of an ecologist. These terms make it possible to understand the river landscape and its interaction with the use that is made of the land, and have the aim of acknowledging how the land itself appears when this interaction changes, and, above all, the role that time has played in this process. The soil and water elements represent the basic tools for the examined area's acknowledgement; once these have been investigated, it is possible to rebuild the identity and characteristics of those places and understand their complex dynamics.

79. The research was developed in Giovanna Ferramosca's thesis in Urbanism (UPTA Degree at DiARC, Università Federico II of Naples), and the supervisor was Prof. Arch. Marina Rigillo, with the assistance of Arch. Anna Terracciano.

80. Morin E., Introduction to Complex Thought, Milan, Sperling & Kupfer, 1993. Also in "Le vie della Complessità" by Bocchi G. and Ceruti M, La Sfida della Complessità, Milan, Feltrinelli Editore, 1985.

81. For more details about the difference between qualitative and quantitative indicators, refer to Ittner C.D. and Larcker D., Coming up short on nonfinancial performance measurement, Harvard Business Review, 2003.

82. The methodological reference for landuse is McHarg I., Progettare con la natura, Padua, Franco Muzzio Editore, 2007. For more information about the monitoring of landuse in Europe, visit http://www.eea.europe.eu/themes/landuse

River landscapes, and consequently their ecosystems, are particularly complex and present very unstable balances as a result of being the connection between other two ecosystems, sea and earth. This precariousness surely originates from the rivers' direct contact with the local anthropic activities, such as towns and factories, as well as farms and agricultural lands. Not without reason, in this sense the word "geography"[83] is meant as that science investigating not only the phenomena marking the Earth but also all the social practices within it, that is, the anthropic activities which mark and define the territory[84].

The structure

The thesis is structured in five parts. The first is an assessment phase, in which the analyses of the different relationships between the river's morphology and the evolution in the use of soil, between the river's morphology and the urban developments and, finally, of the whole hydrographic system are made[85].
The second phase is interpretative: several specific dynamics of the territory are analysed, starting from the water's pollution. The issue is first seen at European and national levels, then at the regional one, trying to find common points that could help researchers think and act from an integrated perspective. As regards the river Sarno's pollution[86] and polluted landscapes, the attention goes to the exploitation of the water resource[87] and all the economic and manufacturing supply chains[88] linked to water. The first goal of building a basis of knowledge is met after the first two phases. In this way all the different information which is usually difficult to collect and put in order is given a structure.
The third part analyses the detailed landuse of all the different soils' exploitation along the riverside (keeping in consideration the perimeter marked as hydrogeological risk area by the local Autorità di Bacino, the Basin's Authority[89]) in order to understand how these exploitations work and and are connected to each other via the study of the territory[90].
Thanks to the fourth phase it was possible to reach the third goal: that is, to provide a set of qualitative indicators reflecting the relations between the river landscape and the use of its soil. These were then used to define possible scenarios and

83. Refer to Farinelli F., Geografia. Un'introduzione ai modelli del mondo, Einaudi, 2003.

84. Terracciano A., "Stratificazioni e sovrapposizioni dei territori contemporanei. Il caso dei Campi Flegrei" in Russo M. and Miano P., Città tra terra e acqua. Esplorazioni e progetto nel Dottorato di Ricerca, Naples, CLEAN Edizioni, 2014.

85. The research mainly refers to Misrach R. and Orff K., Petrochemical America, New York, Aperture Foundation, 2012.

86. Manzione R., Parliamentary Committee of Inquiry on the river Sarno's causes of pollution. Final report, Rome, Tipografia del Senato, 2006; Vito M., Siti contaminati in Campania, Naples, ARPAC, 2008; Patti O., Sarno. Ecogeografia dell'acqua dai disastri alla risorsa, Naples, Edizioni Scientifiche Italiane, 1999.

87. Villanova B., La gestione dei mulini e delle acque dei fiumi Sele e Tusciano nel XIX secolo, Salerno, Brunolibri, 2005; Pace L., Pepe V., "Il governo delle risorse idriche", in Quaderni AdB Sarno, n°1/2004.

88. Valentinelli A., "Considerazioni sull'uso e consumo del suolo nella piana del Sarno", in Quaderni AdB Sarno, n°1/2007; Beguinot C., La valle del Sarno. Edilizia minore e sviluppo economico, Naples, Fausto Fiorentino Editore, 1963; Id, La pianificazione urbanistica nella valle del Sarno. Questioni di metodo per la valutazione dell'ambiente, Naples, Regional plan for the Campania region, 1957.

89. In this case, we are talking about the Basin's Authority (ADB) of Campania Centrale, born in 2012 from the union of the ADB of North-West Campania and the ADB Sarno. For more information concerning the Sarno river hydrographic basin refer to the updated version of the Piano Stralcio di Bacino per l'Assetto Idrogeologico (2015). For any other historical or technical detail visit the website http://adbcampaniacentrale2.it/quaderni

90. Amato A., "Caratterizzazione geologica della piana del Sarno da dati di sondaggio", in Quaderni AdB Sarno, Collana di Studi, Documentazione e Ricerca, n°1/2008.

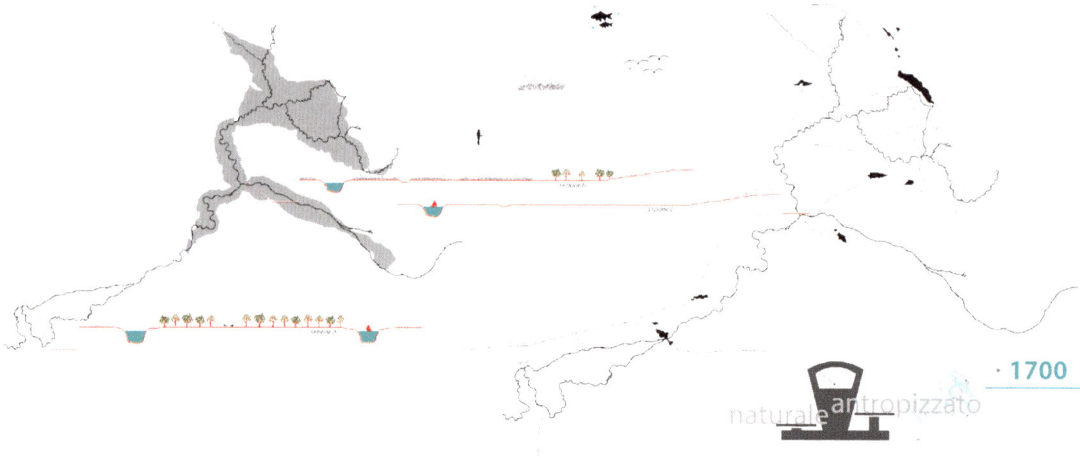

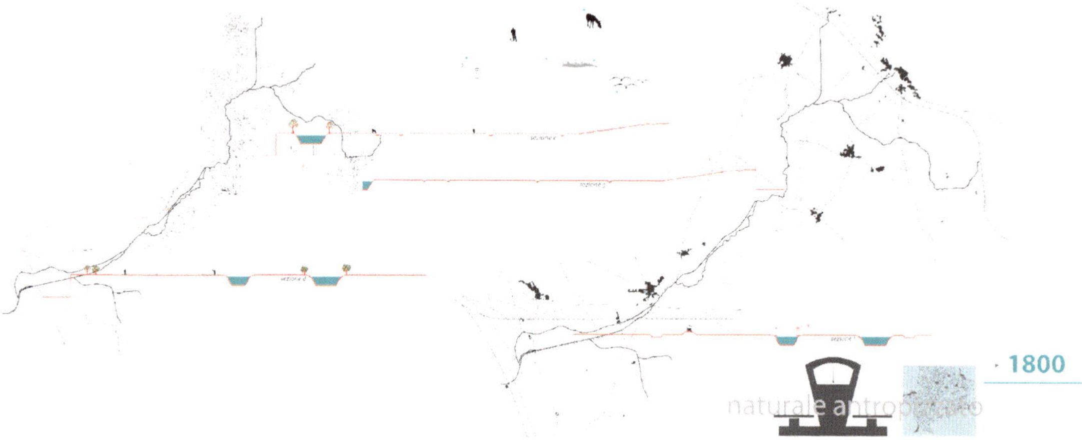

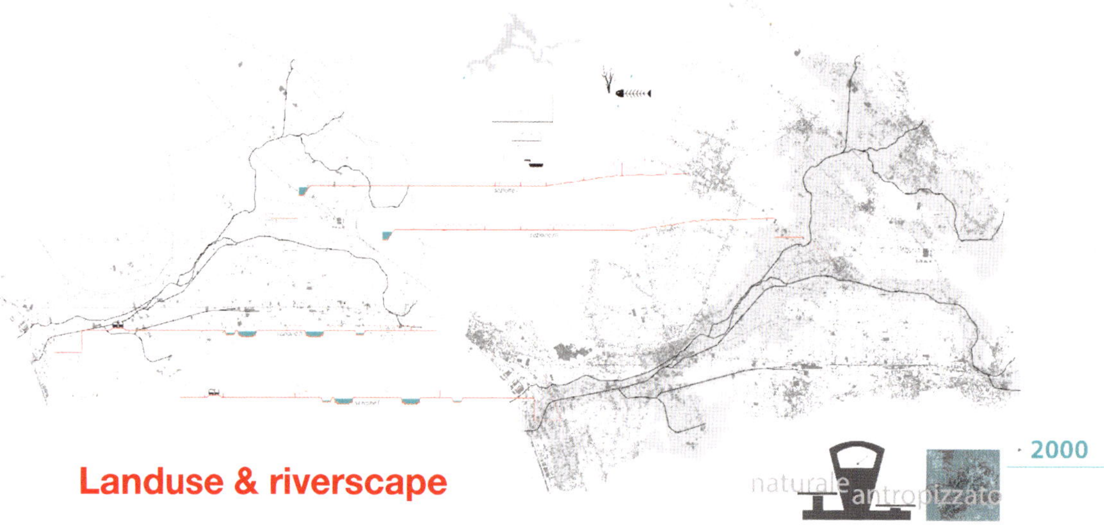

Landuse & riverscape

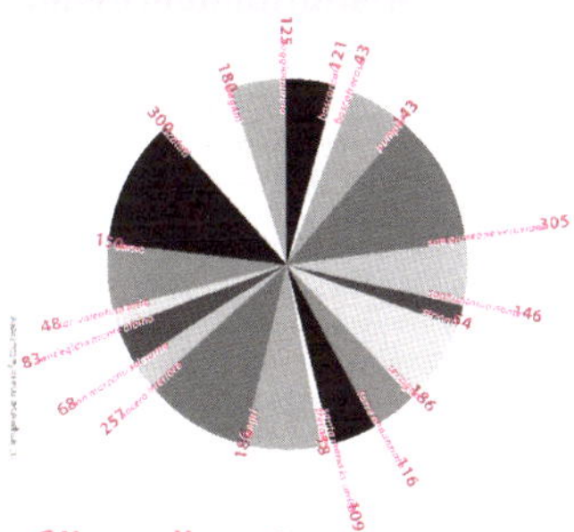

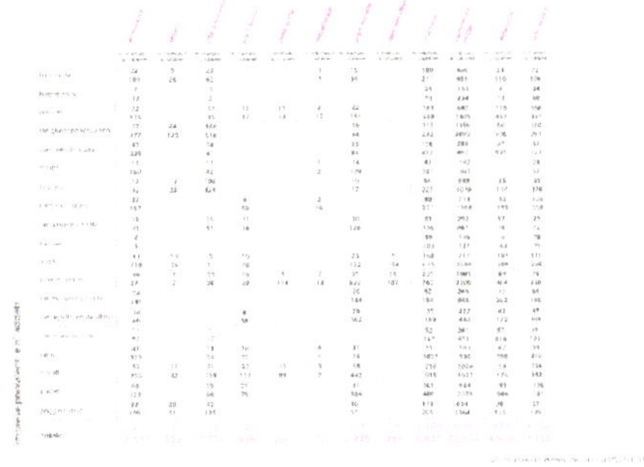

_filiera alimentare

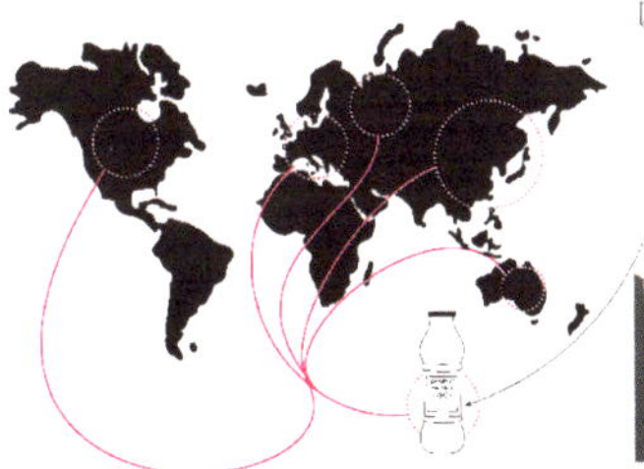

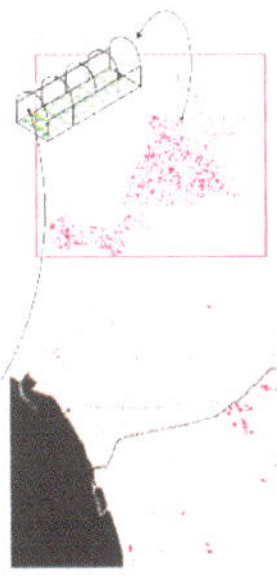

_filiera meccanico-chimica

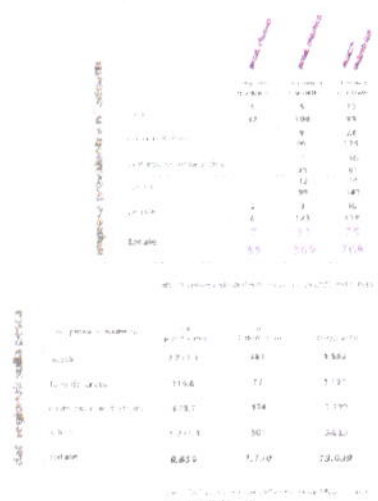

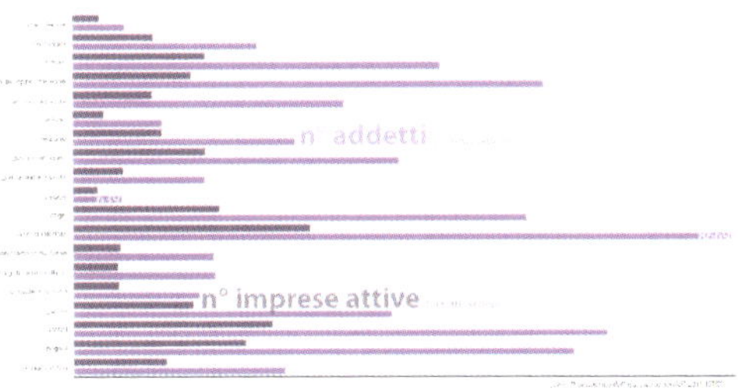

Economies

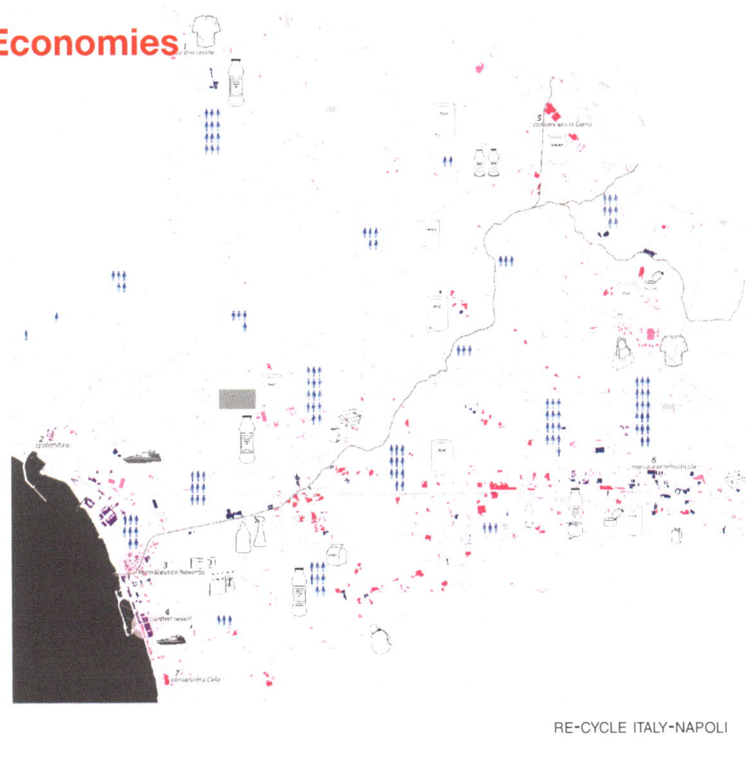

_sistema economico territoriale

Palimpsest of the territory

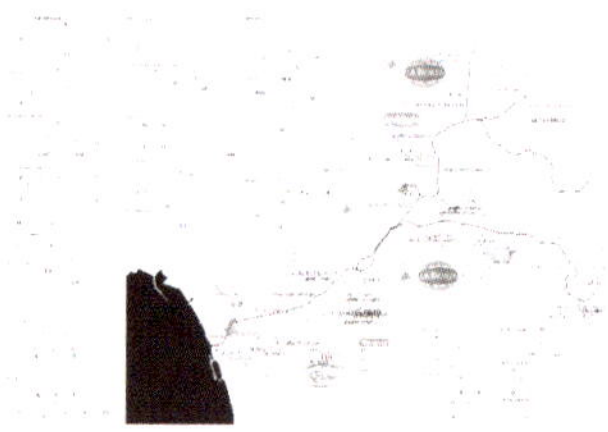

River pollution: Identification of the chemicals produced by industries and causing groundwater and soil contaminations

models to follow. We are not speaking about types but of models because there is a substantial difference between the two terms, since the former is something extremely rigid and can be stereotyped if pushed too far, while the latter is malleable and can adapt to the context[91]. These indicators were identified through the analysis of multiple data collected during the preceding phases and define binding recommendations for the planning phase, which will be monitored in its final results on the territory.
The fifth phase can be defined as proactive: all the collected material is reinterpreted and flows into a more complex strategy in which landuse is defined following a "resilient landscape infrastructure"[92] that meets the territory's needs[93] through a modern interpretation of the historical memory of those places.
With this last phase we are also able to meet the second goal of defining a map of future opportunities, in which all the residual and abandoned areas are recovered and renovated to form a modern river landscape.

Conclusions

As with cities, is it possible to talk about "life-cycles[94]" for a territory or an ecosystem? If so, then it is also possible to suppose that the Sarno river system has gone through a cyclic process of birth-growth-development-exploitation-abandonment to get to today's phases of renovation and rethinking. This process is similar to the concept of resilience, the capacity of a system to rise up and restore a condition of balance and stability following a strong shock.
By studying the system/environment relationship it is possible to analyse the "autopoiesis"[95] and the system's capacity for adaptation that have been activated by the uncertainty of the perturbing event. An autopoietic system is able to re-define itself continuously and to support and reproduce from the inside; this kind of system can be considered as a network of creation, transformation and destruction processes; different components that maintain and regenerate the same system by interacting with each other. This concept of resilience expands further through this renewal capacity.
The union between these two different approaches represents the overview of a city planner, whose goal is to observe and understand the territory in all its aspects. Reality, complex by definition, can indeed be decomposed into different parts, then analysed before being recomposed and understood in its functioning; through the choice guided by our views, the identification, the selection, the interpretation and synthesis, we decide to communicate certain binding values about the territory that will be available and transmitted to those who will be called to work on its planning.

91. De Carlo G., "Note sulla incontinente ascesa della tipologia", in Casabella, n° 509/510, 1985.

92. To have a better idea of landscape infrastructures visit the websites: http://www.scapestudio.com and http://www.swagroup.com and compare two different approaches to landscape architecture.

93. Rigillo M.,"Vulnerabilità e resilienza dell'ambiente urbano", in Ecowebtown, 2015.

94. Marini S., Santangelo V., Nuovi cicli di vita per architetture e infrastrutture delle città e dei paesaggi, Aracne, Roma, 2013.

95. Maturana H., Varela F.G., De maquinas y seres vivos, Santiago, Chile, Editorial Universitaria, 1972.

To retrace the territorial dynamics, Secchi's words are illuminating: "the identification among territories of parts, meaning differences and specificities, represents the exact moment in which our eye becomes active: it recognizes relevant objects and separates them from the background to name them as different. [...] To recognize different parts immediately raises a few questions: those linked to the criteria I use to distinguish them, the relations existing between them, and the ways through which I interpret them."
To conclude, the collection of data from many different subject areas (population, quantity of water pumped from the ground, quantity of waste water treated, economic supply chains, number of employees, polluting chemical agents, contaminated areas, ways of exploitation of the land, etc.), their subsequent elaboration through the use of drawings and representations, and their use not only as a descriptive and methodical tool, but also as a means of divulgation that follows a process of reinterpretation, finally brings us back to the goals set by this research, that is of being an indicative detailed tool of knowledge.

In the previous pages: on the left, restoration of the river-landuse-urbanism relationship on the Sarno plain. Sources: "Pianta topografica per dimostrare tutte le acque che compongono il fiume Sarno", 1769; "Reale Ufficio topografico, carte preparatorie", 1838; Carta topografica e programmatica regionale", 1980.
On the right: Charter of the territorial economic system (productive, food, chemical and mechanical supply chains)

Below, graph of the relations between the river system and the soil's design

In the following pages: on the left, Table of qualitative indicators;
On the right: "Resilient landscape infrastructure" to meet the territory's needs while preserving the natural elements notwithstanding human activity and reclaiming marshlands; restoring the visual contact with water given that physical contact is denied for the "tough" living conditions; and mitigating the relation between the system of production and the river system with the recovery of certain coastal areas.

1 Proper distance between the river system and the settlements

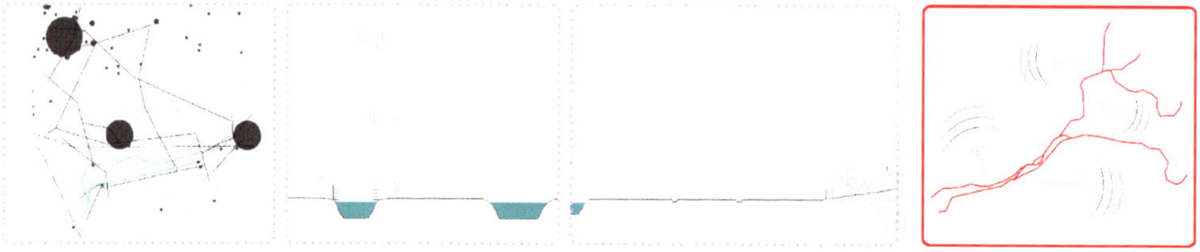

2 Continuity of the river system

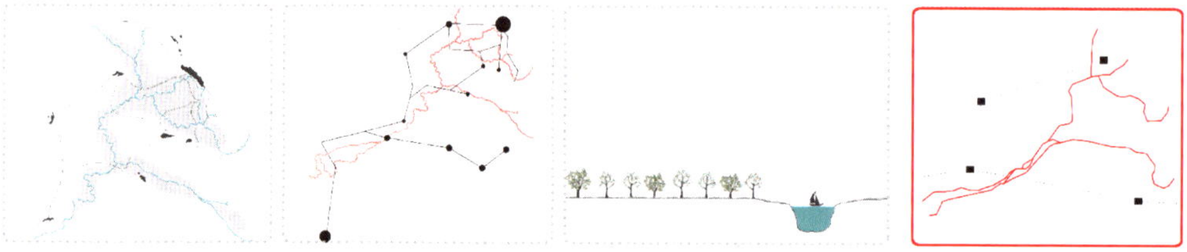

3 Functional relationship between agricultural use of soil and land outline

4 Disorientation and lack of clarity

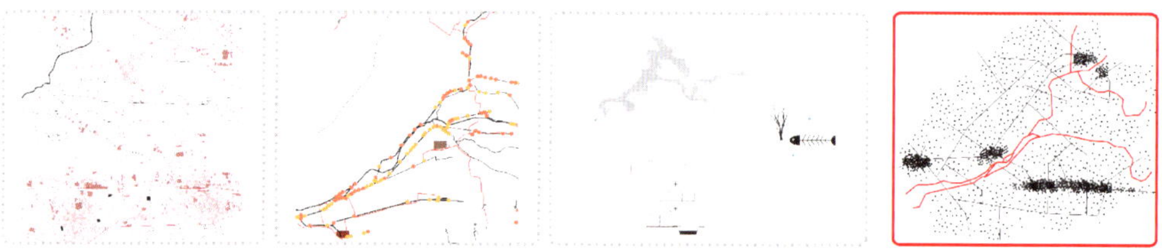

5 Fragmentation and exploitation of resources

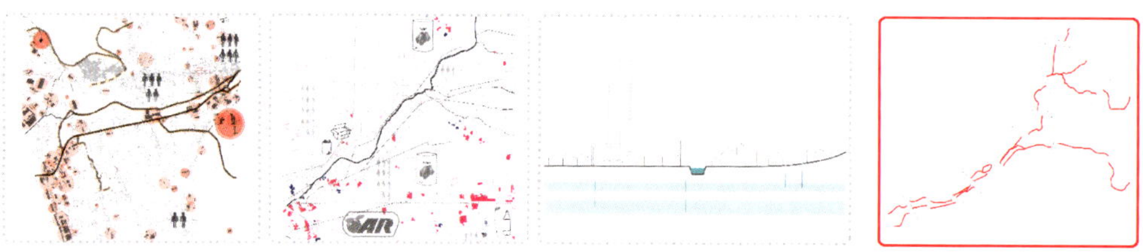

6 Orientation and landmark

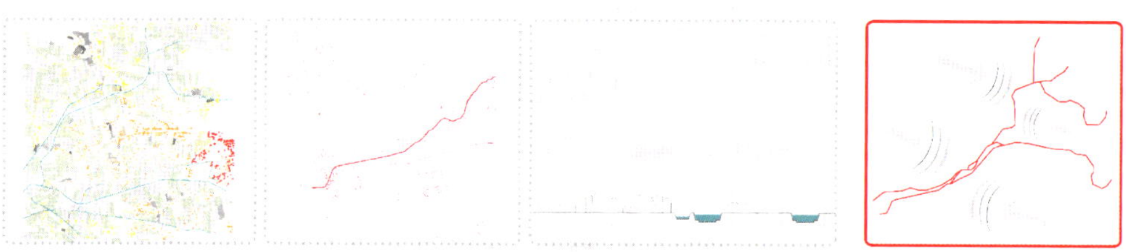

Resilient landscape infrastructure

Evoluzione storica landuse

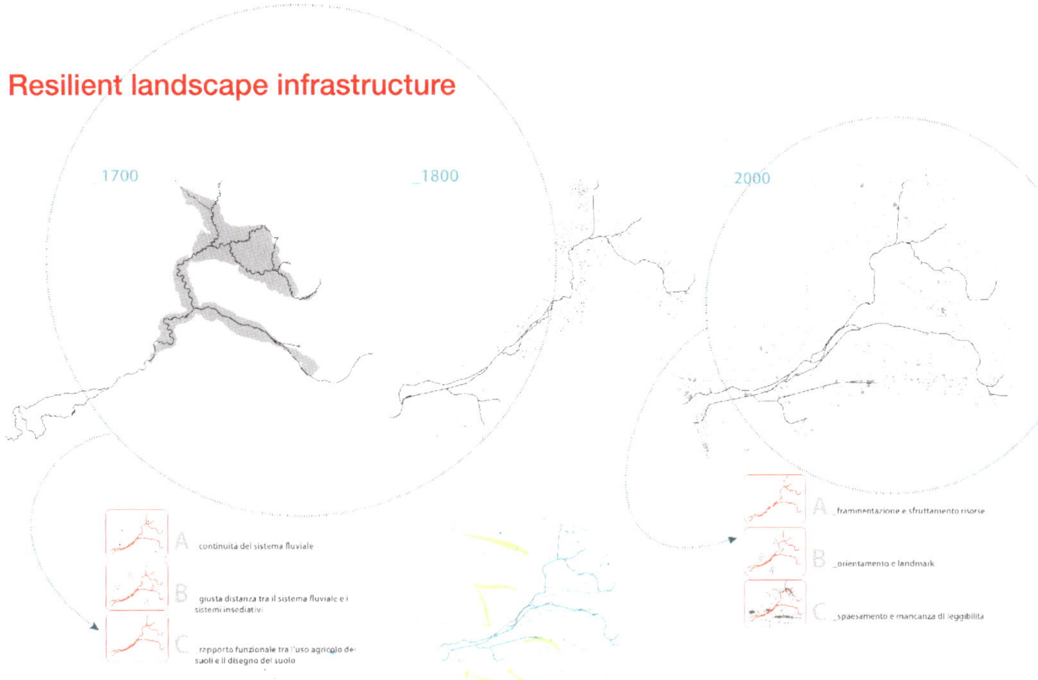

Prospettiva: *INFRASTRUTTURA PAESSAGISTICA*

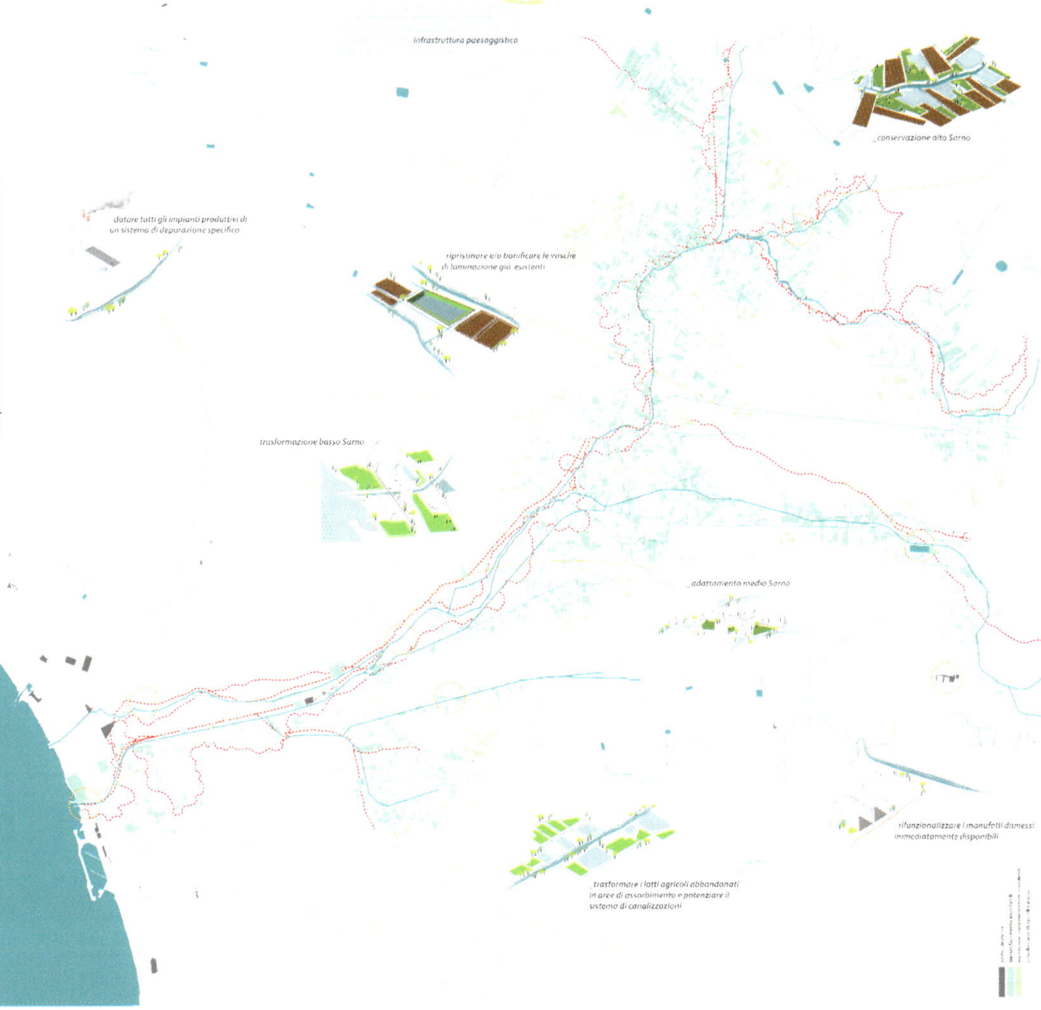

The Sarno river: problems and possible resolutions

Rocco Lafratta

Francesca Montuoro

Territorial framework

The basin of the Sarno river is located between the volcanic complex of Somma-Vesuvio, the Picentini mountain range to the east, the Lattari mountain range and the Tyrrhenian Sea to the west. It covers an area of about 450 km^2. 38 municipalities are located throughout the basin.

The catchment area consists of the Sarno river, 24 km long, fed by the three main sources: Rio Foce, Rio Palazzo and Rio Santa Marina of Lavorate, which crosses the plain bearing the same name, and its two main tributaries which flow into the Alveo Comune Nocerino: the Solofrana and Cavaiola streams. The Solofrana stream was also fed by springs, some of which were lost with the earthquake of 1980, while others were used to supply the city of Naples; the Cavaiola was created as a channel for the collecting and disposing of rainwater in the western part of the plain; today, in fact, it is a sewer and rainwater collector.

Over the past centuries, the entire basin has been subject to strong morphological alterations due to the construction of long artificial stretches, among which is the same Alveo Comune Nocerino, that is, where the waters of the Solofrana and the Cavaiola connect with the Sarno river.

The Sarno river has also undergone several changes, which have - directly or indirectly - modified the course of the river and the supply of water. For example:

- the collection systems at the three main sources have significantly reduced the supply of water to the river;
- the construction of the Scafati crossroads, a barrier to distribute water to the Fienga canal and the Bottaro canal for irrigation purposes;
- the elimination of several meanders, as well as the rectification of the river course in the final stretch from the Scafati dam to the mouth of the Sarno.

Pollution of the Sarno River and its tributaries

The Sarno river has been described by many as the most polluted river in Europe and it is among the ten most polluted in the world. Its pollution is commonly attributed to the tanning industry (184 located on the Solofrana stream), the tomato cannery (108 located on its banks), and the pharmaceutical industry. Nevertheless, the significant dispersion of

small and medium industries on the plain is equally important. Another source of widespread pollution comes from the agricultural sector where many chemicals such as pesticides, fertilizers, fungicides and herbicides are used extensively. During the rains the excesses of these products end up inexorably in the waters of the Sarno river and in the canals of the reclamation. Moreover, as in many municipalities the sewage system is partial or absent, such material is discharged directly into the waters of the river and canals. Even the purification system, where present, is very poor.
The complex environmental situation of the Basin is influenced by strong anthropogenic, agricultural and industrial elements, both for the hydrographic network and for the soils. In this territory made vulnerable by a situation of high hydrogeological risk, such as the rapid flows of mud[96] , there is also a strong and extensively artificialized hydrographic system, which in the last 20 years has not been able to contain the frequent floods, occurring at least 3-4 times a year. To this already fragile environmental picture, we must add a demographic pressure of 2,000 inhabitants per square kilometer, a predominantly intensive agriculture and an industrial activity that today uses the hydrographic grid as drainage infrastructure for its processing waters.
This set of critical issues has caused the Sarno plain to be registered as a Site of National Interest and has led to the declaration of a state of socio-economic-environmental emergency. The Prime Minister's Ordinance No 3270 of 12 March 2003 appointed a commissioner, who was to complete the project for the rehabilitation of the Sarno plain, this being part of the overall framework from the rehabilitation of the entire Gulf of Naples that began in 1973.
The initial project involved the construction of four purification plants in the Medio Sarno (middle Sarno) area (only three of them were eventually built, because of the discovery of the archaeological site of Longola in the municipality of Poggiomarino), two purification plants in the Alto Sarno area and the modernisation of the Basso Sarno purification plant in Castellammare di Stabia.
The three purifiers of the middle Sarno are almost completed but they operate on a very narrow gauge because they still lack sewage networks in some municipalities. What is worse is

96. See in this respect the Sarno flood events of 1998.

that they lack the collection networks from the municipalities' sewage networks to the purifiers.
The Alto Sarno area, which primarily concerns the Municipality of Solofra and the Industrial Leather Processing Centre, includes the municipalities of Montoro and Mercato San Severino where two purification plants have been built. In the municipality of Solofra, in one of the few cases in Italy, the separate sewage systems were built from scratch with three distinct lines: industrial sewers, residential sewers and white water network.
However, even in the Upper Sarno, for some municipalities and land areas there is a lack of sewage network coverage and its related collection to the purifiers.
The lower Sarno area foresees the completion of the adaptation of the Sarno estuary purification plant at Castellammare di Stabia, which will change from chemical-physical to biological. The construction work is well underway, but here, too, there is a lack of connection between the sewage networks of some municipalities and the purifier itself. Therefore, after a further 13 years of intervention, many municipalities continue to pour directly into the Sarno river and/or its tributaries, and/or into the network of drainage channels.
In addition to the socio-economic-environmental rehabilitation project[97], the Sarno plain has been involved in recent years in another project called "Grande Progetto Sarno" ('Great Sarno Project'). This project, which was launched in 1998 following the landslides at Sarno, aims at reducing the frequent flooding along the watercourses of the entire Sarno basin. It provides for the construction of flood lamination tanks and a controlled flooding area, as well as the rebuilding and/or widening of the embankments, spread over all the watercourses. Along the Solofrana torrent there will be nine tanks while the upper part of the Sarno river will have a controlled lamination area. These works will occupy about 100 hectares of highly fertile and productive land.
As the "Great Sarno Project" was conceived, the tanks have the sole purpose of mitigating, with very high soil consumption, the hydraulic risk. They have no function within compensatory and integrative ecosystems to which they could at least contribute as wetlands.
The latter have the capacity to provide specific ecosystem services, such as the reduction of the diffuse pollution load, the increase of typical floro-vegetational settlements, the increase of the stable presence of amphibians and reptiles, odonata, other specific fauna and avifauna, the recharge of the stratum, the regulation of the local microclimate and the development of biodiversity, both within the area and also in the adjacent areas.
Today, because of its intrinsic socio-economic benefits, this project should not be neglected. However, it is necessary to

On the opposite page, springs of the Upper Sarno

97. See former Prime Minister's Office 3270/2003.

rethink a different design configuration with multiple goals, with reference to ecosystem services and new European indications and regulations. The economic value of ecosystem services, estimated per hectare, is between 36,000 and 55,000 euros per year[98].

As proof of how to conceive a type of design of wider boundaries, a wet area has been developed and implemented within a flood lamination tank, already built in the municipality of Mercato S. Severino (SA), in Acigliano-Pandola, already during the implementation of the "Great Sarno Project", completed in 2010. The wetland was designed and built in 2012, inside the rolling basin fed by the waters of the Rio Laura, a tributary of the Solofrana stream. It occupies a total area, between the flooded area and the adjacent area, of about 2 hectares. Both the lamination tank and the wetland were kept under observation to assess their evolution. Analyses were also carried out on the inlet and outlet waters of the wetland, from ARPA Campania, in the Salerno headquarters. The analyses gave excellent results, especially on the reduction of suspended solids, nitrogen and Escherichia coli, in line with the efficiency results of other national and European wetlands. The area has been naturally repopulated with hygrophilous and aquatic vegetation, amphibians, reptiles, odonata and with the presence of avifauna.

The Campania Regional Agency for Soil Protection (ARCADIS), which was in charge of the development of the "Great Sarno Project", did not take into consideration the importance of monitoring the Mercato San Severino basin and the adjacent wetland area. This would have made it possible to remodel the design of the other tanks thus reducing the strong opposition received from the interested bodies and local stakeholders during the presentation of the projects.

Water quality

The degree of water pollution[99] is only the most obvious factor in a systemic problem which affects all elements of the environmental matrices. In fact, according to the WHO (2013), exposure to environmental pollution remains the greatest source of risk to human health. These environmental risk factors account for one third of global health.

It should be noted that many elements (and their concentrations) considered harmful to man in our areas are of natural origin, as they are contained in the various formations and loose materials. These materials derive mainly from volcanic eruptions, both of the volcanic complex of the Phlegraean Fields and of the Sum-Vesuvius.

As a result of recent studies on the state of pollution of the superficial aquifers, from which water is drawn for home use through private wells, it is becoming increasingly clear that

On the opposite page, Pianillo basin, San Giuseppe Vesuviano (Naples)

98. See ISPRA Report, July 2016

99. See in this respect the following publications: S. Albanese, P. Iavazzo, P. Adamo, A. Lima, B. De Vivo, "Assessment of the environmental conditions of the Sarno river basin (south Italy): a stream sediment approach", 2013; D. Chicchella, L. Giaccio, A. Lima, S. Albanese, A. Cosenza, D. Civitillo, B. De Vivo, "Assessment of the top soul heavy metals pollution in the Sarno river basin, south Italy", 2014; M. Arienzo, P. Adamo, M.R. Bianco, P. Violante, "Impact of land use and urban runoff in the Sarno river basin in southwestern Italy", 2001; D. Cicchella, J. Hoogewerff, S. Albanese, P. Adamo, A. Lima, M. V. E. Taiani, B. De Vivo, "Distribution of toxic elements and transfer from the environment to humans traced by using lead isotope. A case of study in the Sarno river basin, south Italy", 2016; P. Adamo, P. Iavazzo, S. Albanese, D. Agrelli, B. De Vivo, A. Lima, "Bioavailability and soil-to-plant transfer factors as indicators of potentially toxic elements contamination in agricultural soils", 2015; Legambiente Onlus - Goletta Verde del Fiume Sarno, 2014, 2015, 2016.

the water should not be used for washing and cooking, in some cases not even for drinking, without receiving any pre-treatment depurative.
In the Piana del Sarno the state of health of its inhabitants is strongly compromised by the direct and indirect use of its polluted waters and by the widespread presence of toxic elements in the soil and in the air. These elements are eroded and transported to river beds and surrounding areas by rain-water leaching.
The studies, reported in the bibliography, highlight what has already been said: the surface waters, the first aquifers and even the soils are polluted, with reference to the parameters of Legislative Decree 152/2006 and subsequent amendments and integrations. However, some are polluted due to their basic chemical composition of volcanic origin.
The same studies show that there is a connection between the pollutants and the health of the inhabitants of the Sarno plain. However, they also hypothesize that such a connection is not made through fruit and vegetable products. In fact, it is very likely that it occurs through the use of water, as well as air pollution and also dust, which, as deposited on fruit and vegetable products, can enter the food chain.
It is therefore necessary to inform citizens well about the data and sources of the real dangers and to train and educate them to consume water and food resources in a responsible way.

On the opposite page, Rovigliano beach, Castellammare di Stabia (Naples).

Below, pollution of the Sarno river

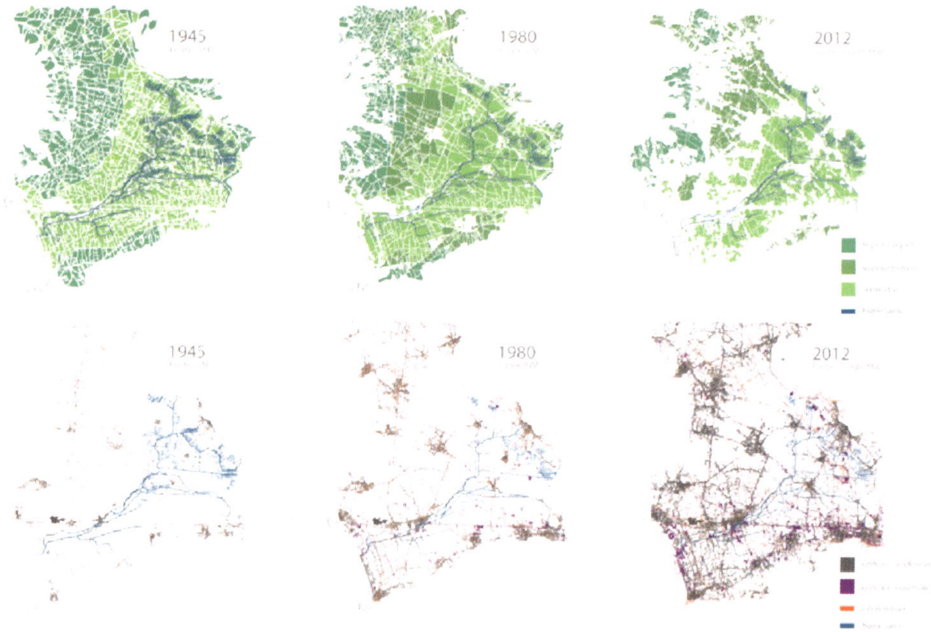

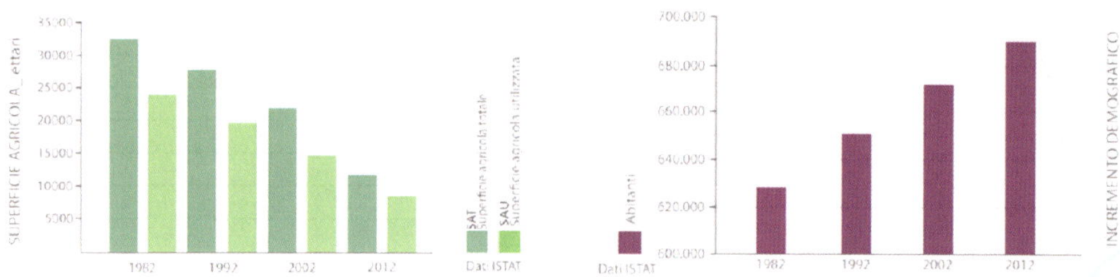

Morphological evolution and use of soil. Source: Consorzio di bonifica integrale comprensorio Sarno, MUDIF, Museo Didattico della Fotografia, Nocera (Salerno).

FIUME SARNO

13.540 Km² area del Bacino Idrografico

3 Provincie attraversate Napoli, Salerno Avellino

39 Comuni

24 Km percorso complessivo dell'asta principale

5-6 Mc/s portata media d'acqua

1630 Km lineari di affluenti secondari

1252 Km di alvei naturali

180 Km di alvei trasformati in strade

98 Km di alvei tombati

100 Km di canali artificiali

3 Sorgenti:
Santa Maria a Foce
Mercato Palazzo
Santa Marina di Lavorate

2500-3400 l/s acqua emunta dall'Acquedotto Campano dalle sorgenti

7 Corpi idrici sotterranei principali

14 Corpi idrici sotterranei secondari

57 Milioni di mc apporto annuo di acqua dalla falda

The figures in these two pages are by Francesca Montuoro and can be found in her degree thesis in Architecture (CdL Architettura 5UE, DiARC – Università degli Studi di Napoli "Federico II"), supervised by Prof. Arch. Marella Santangelo, co-rapporteur Prof. Arch. Marina Rigillo.

STRATEGIC CONTEXTS

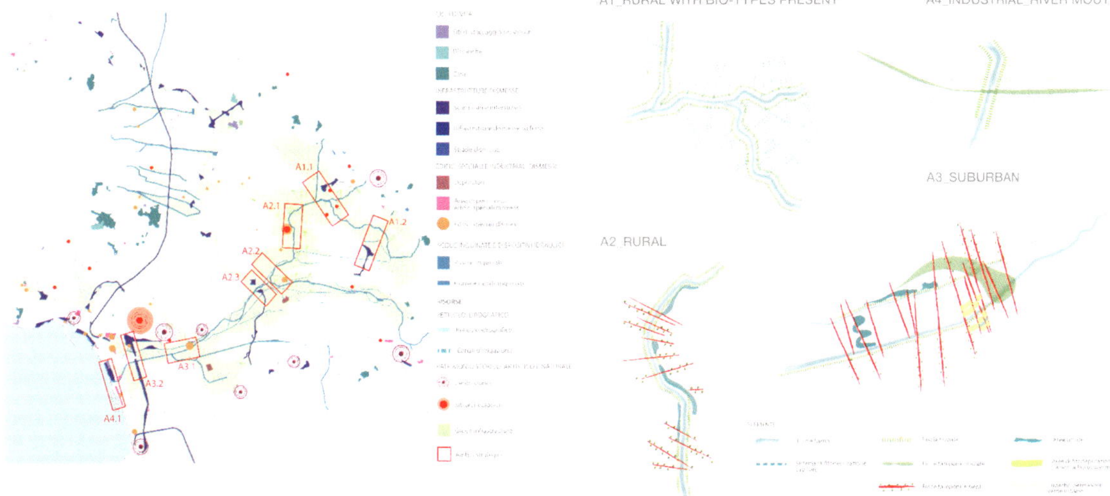

Living machine

Concept and masterplan of the suburban "Ambito 13" in Scafati (Salerno)

Resilient approaches and recycling of compromised areas in the Sarno plain

Francesco Sammarco

Ciro Sepe

Danilo Vinaccia

This dissertation's[100] ambition is to explore and define new "life cycles" for those spaces and elements of the city and the territory which have lost today their meaning use or importance as a result of urban scattering, the fragmentation of the agricultural landscapes' mosaic and the infrastructural's critical aspects in general.

It is only by operating between the contemporary's city crazy metabolism and its waste that the proposed reflections can find the relationships between the confused fragments of the urban explosion, both through the construction of a new territorial frame - consisting mainly of the concatenation of open spaces - and the exploitation of the existing "green and blue infrastructures" and the reuse of the territory's abandoned, degraded or underused areas.

In this way, we proceed to the construction of a "project/process" that contains, manipulates and produces concepts and formulates hypotheses of actions in space, re-interpreting and re-adjusting the compromised areas of the Sarno plain, analysing the different levels of the "waste landscapes" and their reciprocal connections. This work resumes and interprets, on an appropriate scale for the Plain's context, the concept of "drosscape" as introduced by Alan Berger (Co-Director at the Center for Advanced Urbanism - MIT, Boston) in the sense, repeated several times in this volume, that considers waste as a natural component of every city; which, above all, develops dynamically, thus becoming an authentic "indicator" of urban health.

These "drosscapes", however, are not only artefacts or areas that are no longer part of the virtuous urban metabolism; in fact, they also generate critical scenarios which alter everyday life, affecting climate change and the quality of the environment, as well as producing degradation, waste, consumption of energy, materials and soil. The latter are tangible risk factors for the population's health and survival. Moreover, they also represent, from a morphological point of view, the components of a true "reverse city", which cannot be ignored in the establishment of a "resilient/adaptive" recycling model for the future of these territories. The waste areas become, in this way, the central component and an enormous resource for the ecological regeneration and spatial reconfiguration strategy of the contemporary city.

100. The work presented was developed by Francesco Sammarco, Ciro Sepe and Danilo Vinaccia in the scope of the graduate thesis in Urban Planning (CdL PTUPA, DiARC - University of Naples "Federico II"). Supervisor: prof. arch. Carlo Gasparrini. Co-supervisor: architect. Anna Terracciano.

In addition, the extreme complexity of the phenomena involved implies the use of tools capable of collecting a huge variety of interacting data and requires a rethinking of the rendering methods and tools. For this reason, the knowledge in the use and programming of territorial information systems (Sistemi Informativi Territoriali - SIT) has been essential for putting to system and comparing a variety of information with the purpose of reconstructing the territorial programming in this area. Concepts, means and objectives relate and find their form in graphic communication; the territory and the city are questioned with new mechanisms of analysis, representing an unprecedented and comprehensive "multi-layered" scenario including all the Sarno plain's characteristics, amidst existing and imaginable landscapes, with the hope of producing knowledge and increasing awareness in the local population. Different tools and languages are used so that the territory is represented in an easily understandable way not only to experts but to a broader audience as well.
The territory is told through new analytical - "multi-scalar" and "multi-focal" - views in order to identify the entirety of the landscape emergencies, anthropic and natural, whilst devoting particular attention to the dynamics affecting the hydraulic, infrastructural and agricultural systems.
These fields of investigation in fact are privileged, since they are able to produce three different kinds of landscapes in an always changing sequence all along the banks of the river Sarno. These landscapes are "palimpsests in the palimpsest", complex, continuously changing organisms on which the inhabitants constantly write and erase the signs of their actions. In some places, these overlap and generate extremely evocative images, such as those in the infrastructural ruptures, the areas of abandonment, waste, degradation and industrial archaeology. Images that make up a dramatic mosaic within an equally tormented landscape, whose identity and beauty are contaminated and altered.
The attempt to create ambitious industrialisation programmes together with a rapid and disorganised growth of the settlement pattern, mainly during the second half of the last century, has in fact led to a reckless consumption of land. Alarmingly, all this has produced an enormous increase in the impermeable surface, causing an intensive and widespread urbanization

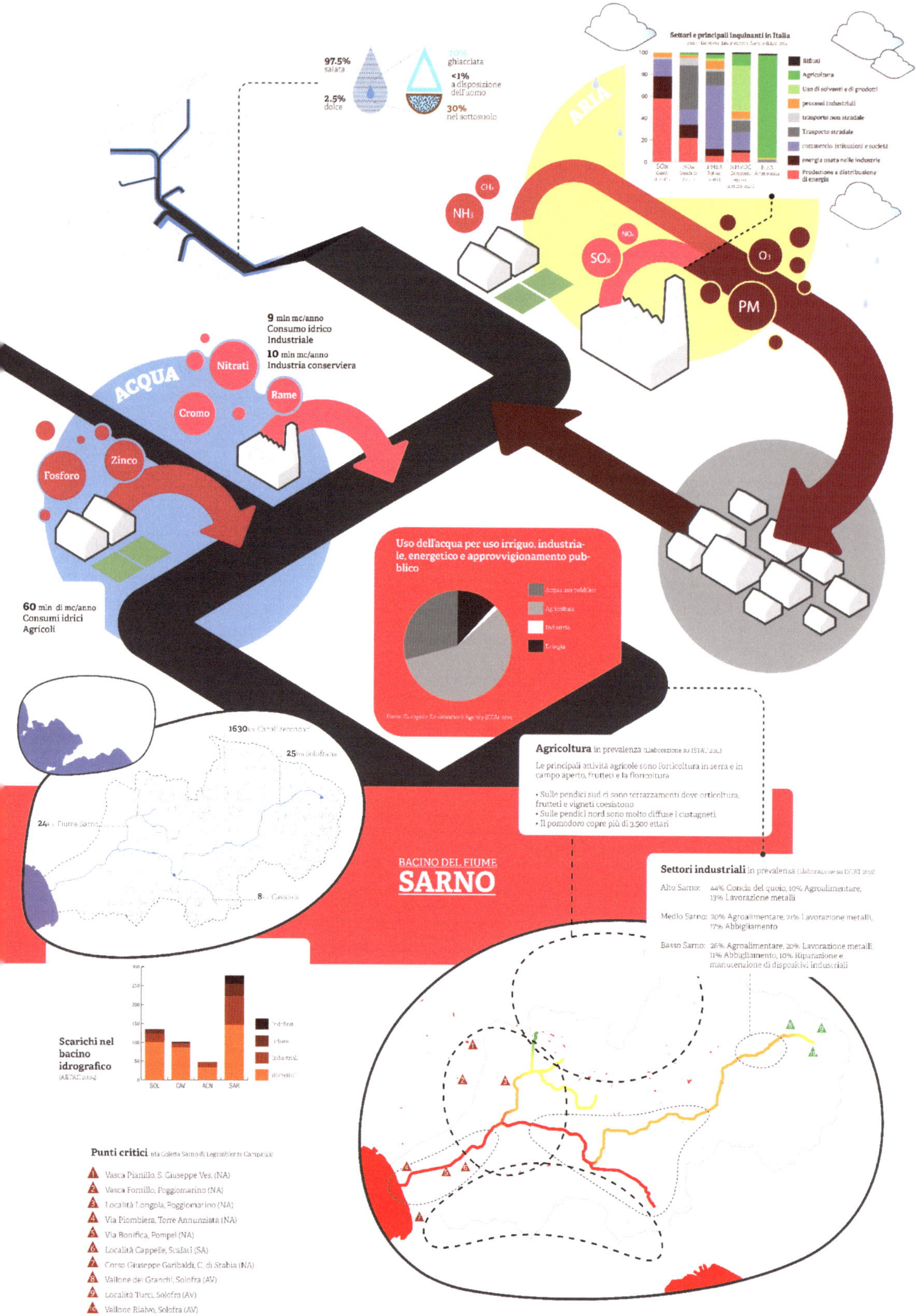
97.5% salata
2.5% dolce
70% ghiacciata
<1% a disposizione dell'uomo
30% nel sottosuolo
ARIA
Settori e principali inquinanti in Italia
SOx
NOx
PM2.5
NMVOC
NH3
Rifiuti
Agricoltura
Uso di solventi e di prodotti
processi industriali
trasporto non stradale
Trasporto stradale
commercio, istituzioni e società
energia usata nelle industrie
Produzione e distribuzione di energia
CH4
NH3
NOx
SOx
O3
PM
ACQUA
Nitrati
Cromo
Rame
Zinco
Fosforo
9 mln mc/anno Consumo idrico Industriale
10 mln mc/anno Industria conserviera
60 mln di mc/anno Consumi idrici Agricoli
Uso dell'acqua per uso irriguo, industriale, energetico e approvvigionamento pubblico
Acqua uso pubblico
Agricoltura
Industria
Energia
1630 km Canali secondari
25 km Solofrana
24 km Fiume Sarno
8 km Cavaiola
BACINO DEL FIUME
SARNO
Agricoltura in prevalenza (Elaborazione su ISTAT 2011)
Le principali attività agricole sono l'orticoltura in serra e in campo aperto, frutteti e la floricoltura
• Sulle pendici sud ci sono terrazzamenti dove orticoltura, frutteti e vigneti coesistono
• Sulle pendici nord sono molto diffuse i castagneti
• Il pomodoro copre più di 3.500 ettari
Settori industriali in prevalenza (Elaborazione su ISTAT 2011)
Alto Sarno: 44% Concia del quoio, 10% Agroalimentare, 13% Lavorazione metalli
Medio Sarno: 20% Agroalimentare, 21% Lavorazione metalli, 17% Abbigliamento
Basso Sarno: 26% Agroalimentare, 20% Lavorazione metalli, 11% Abbigliamento, 10% Riparazione e manutenzione di dispositivi industriali
Scarichi nel bacino idrografico
(ARPAC 2004)
SOL
CAV
ACN
SAR
domestici
Punti critici (da Goletta Sarno di Legambiente Campania)
1 Vasca Pianillo, S. Giuseppe Ves. (NA)
2 Vasca Fornillo, Poggiomarino (NA)
3 Località Longola, Poggiomarino (NA)
4 Via Piombiera, Torre Annunziata (NA)
5 Via Bonifica, Pompei (NA)
6 Località Cappelle, Scafati (SA)
7 Corso Giuseppe Garibaldi, C. di Stabia (NA)
8 Vallone dei Granchi, Solofra (AV)
9 Località Turci, Solofra (AV)
10 Vallone Rialvo, Solofra (AV)

which has compromised the continuity of the agricultural soil, thus damaging the area's main and historical vocation. Uncontrolled urban growth also has repercussions on soil's stability, increasing its fragility and exposing the building stock to considerable risk. Within this research, the aim is to try to give answers and show some possible trajectories, through the construction of some selective but also synergic "visions", which can identify the objectives and strategies of an ecologically oriented regeneration of the aforementioned issues.
The importance of elaborating of images, through the so-called "visioning", is a constant and an essential phase of the large-scale project analysis, precisely because "images are the link between the physical and conceptual territories of the project"[101], and not just a vehicle, although necessary, to communicate the possible choices and scenarios for our cities. The planning approach outlines a complex and organic strategy of integrated interventions and activities that can contribute to risk mitigation only as a whole. The objective of this phase is therefore to build a new "multi-scalar" territorial armour, developed through three spatio-temporal frames.
The three "visions" identify the strategic and structuring objectives to be achieved in the long term on large parts of the territory, not to mention the specific interventions consisting of actions and micro-project actions, which are meant to be implemented in the short and medium term, on the smallest scale of the urban - but also natural - fabric. It is from the overlapping and interaction between these thematic layers that the overall "vision" takes shape and form. All the spaces and materials available for the recycling process play a role, in order to find new configurations and new life cycles.
This is the perspective in which the actions connected to four prevailing interventions are placed.
The first is linked to the recycling of the disused railway line Torre Annunziata-Cancello and the waste areas generated by it, to be achieved through the creation of a network of facilities and collective spaces supporting a "linear territorial park for cycling and walking".
The second is linked to reconsidering the whole system regulating the water coming from the Vesuvius through a sort of "Water Freeway", which not only involves a re-functionalisation of the canals and collection tanks, but also assumes they are catalysing elements for the green networks and the new ecological functions linked to the absorption and filtering of rainwater.
The third objective concerns the "recycling of the Conte di Sarno canal", rethought as an intelligent pipeline and infrastructure to enhance sustainability on an urban scale, for the storage or controlled release of water, supported by devices on a small scale for the collection and reuse of rainwater.

On the opposite page: Pollution, causes and effects on the territory

In the following pages: on the left, Criticalities of the infrastructural system; On the right, incomplete stretch of the SS268 and decommissioning of the "Circumvesuviana" railway in Boscoreale (Naples).

101. P. Viganò, "I territori dell'urbanistica. Il progetto come produttore di conoscenza", Officina, 2010

→ Network of flows

DNA DEI TRASPORTI
Quantità degli spostamenti che attraversano nell' unità di tempo una data superficie.

Matrice **origine-destinazione** elaborata sulla base degli spostamenti per motivi di lavoro o di studio nella piana del Sarno.

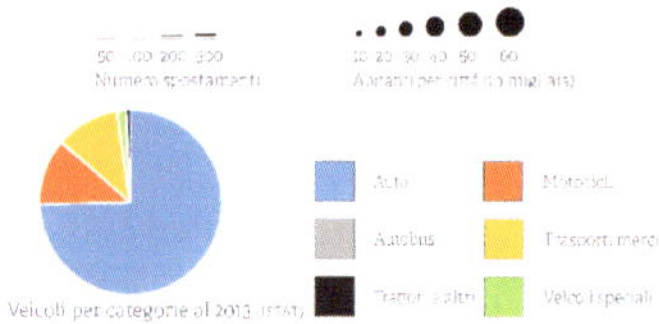

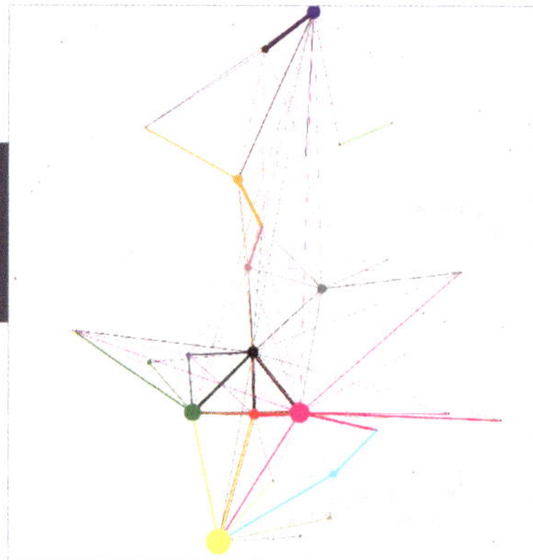

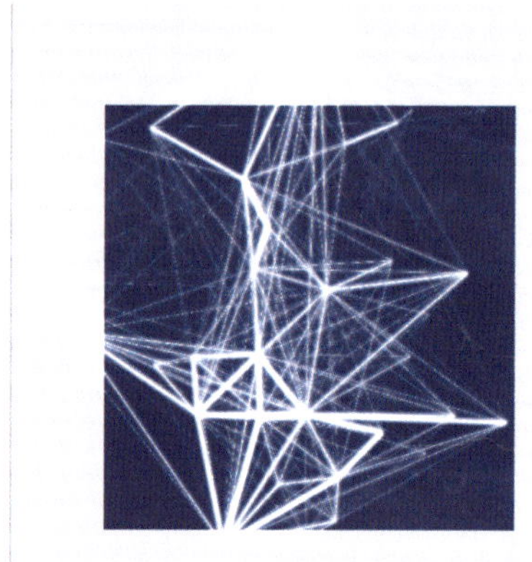

→ Real connections

IMPRONTA DIGITALE DELLA MOBILITA' URBANA
L'idea è quella di calcolare l'area circostante un sito che può essere coperta con un viaggio in un determinato lasso di tempo.
Tutti i percorsi quindi creano "impronta digitale della mobilità urbana".

L'algoritmo utilizzato determina l'isocrona valutando: limiti di velocità, pendenze, incroci e altri fattori determinanti
(da web map di mooviLab team)

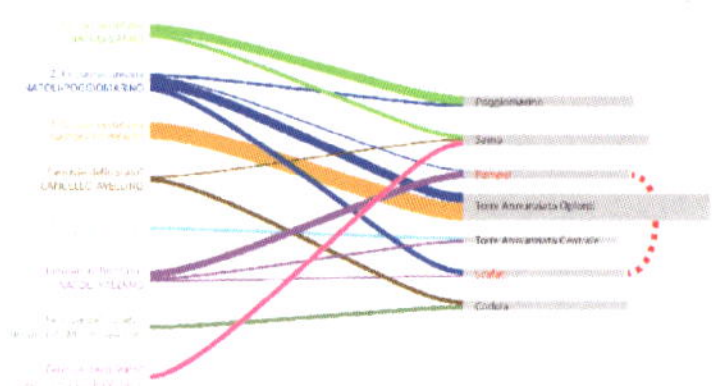

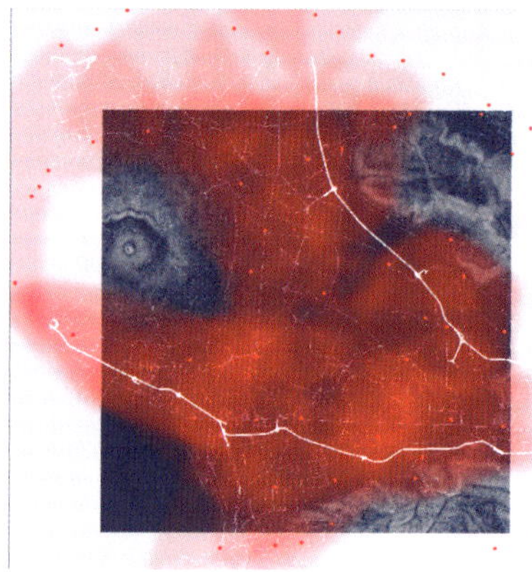

↘ Train stations and areas of influence

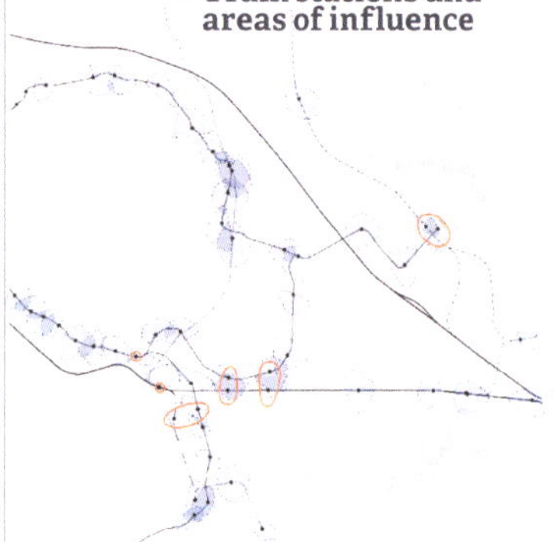

↘ Overview of waste in the Network

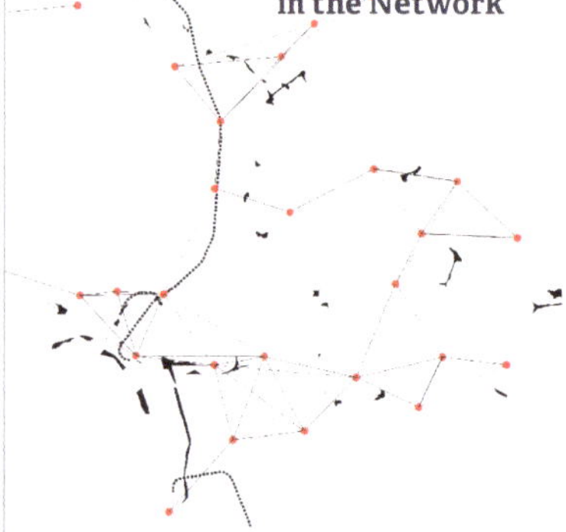

→ Overview and challenges of the infrastructural system

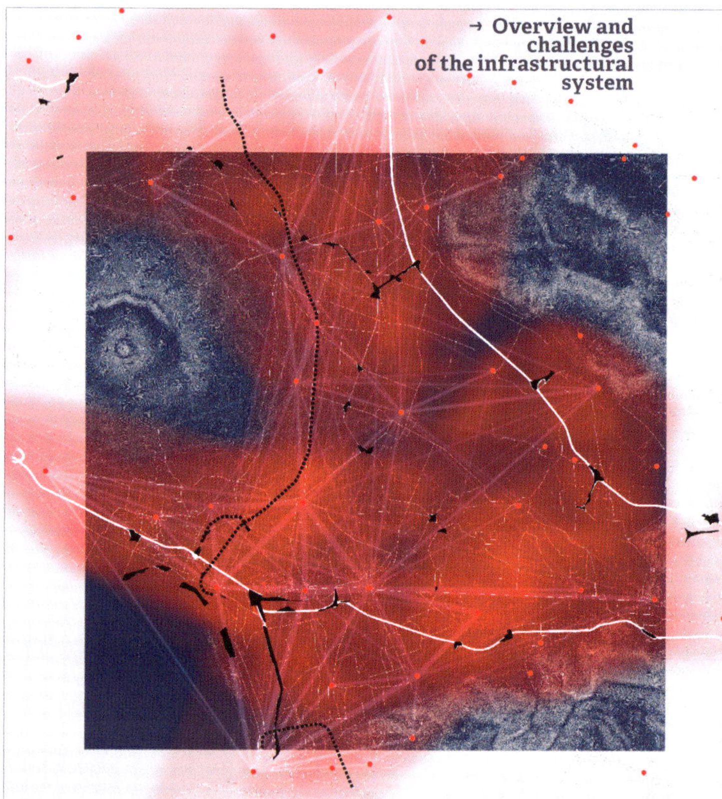

CIRCUMVESUVIANA
BOSCOREALE

In these pages, overall strategic plan, strategic objectives, thematical layers and conceptual schemes.

In the following pages, planning explorations for the "Sarno Archaeo-fluvial park" and for the "Recycling of the Conte di Sarno canal" in an urban context.

The fourth and last objective concerns the reclamation of the relationship with the Sarno river through a new design of the river landscape, its functions and land use within an "archaeo-fluvial park". Putting the areas wetted by the river back in play not only means providing new methods of use linked to accessibility, the quality of spaces and to networking the huge archaeological resources present in the area, but also to give back space to the river and the surrounding natural area.
New, sustainable and innovative ways of cultivation, which are, most importantly, compatible with the flooding areas and the river dynamics contribute to the outline of a new landscape and a renewed and respectful relationship between man and water. Recycling territories therefore means meeting a very often "latent" demand for reactivation, in which the waste produced is not only seen as the problem but also as the opportunity for a new starting point for the territory's regeneration which shall benefit the towns and local communities.

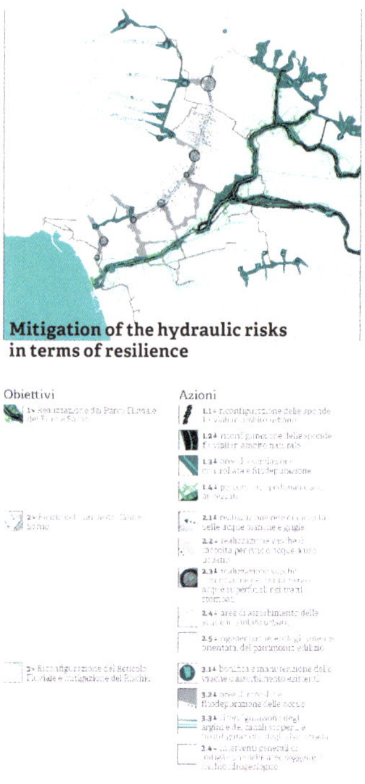

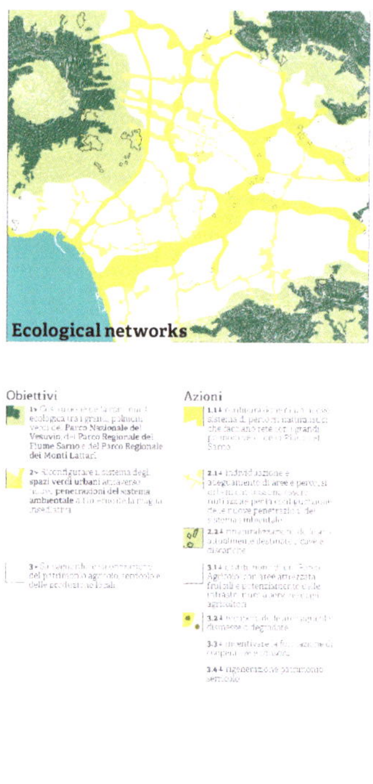

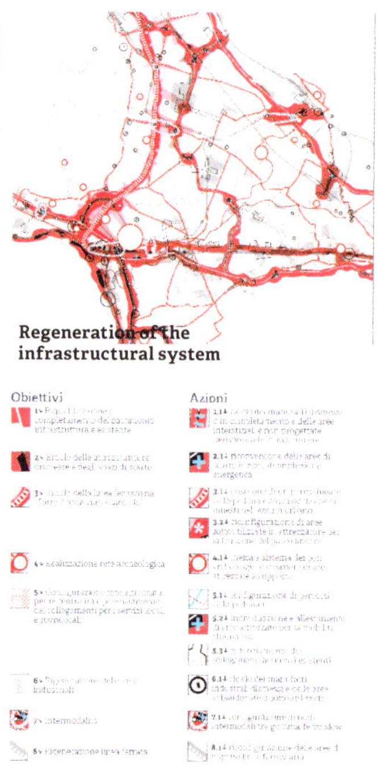

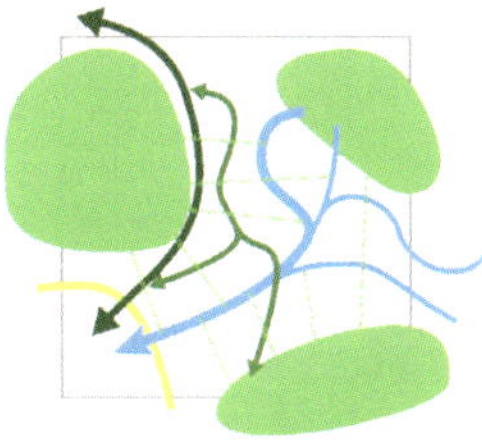

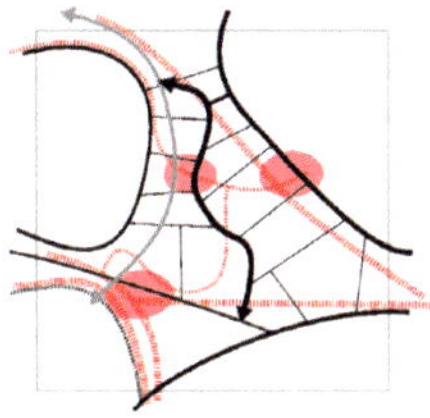

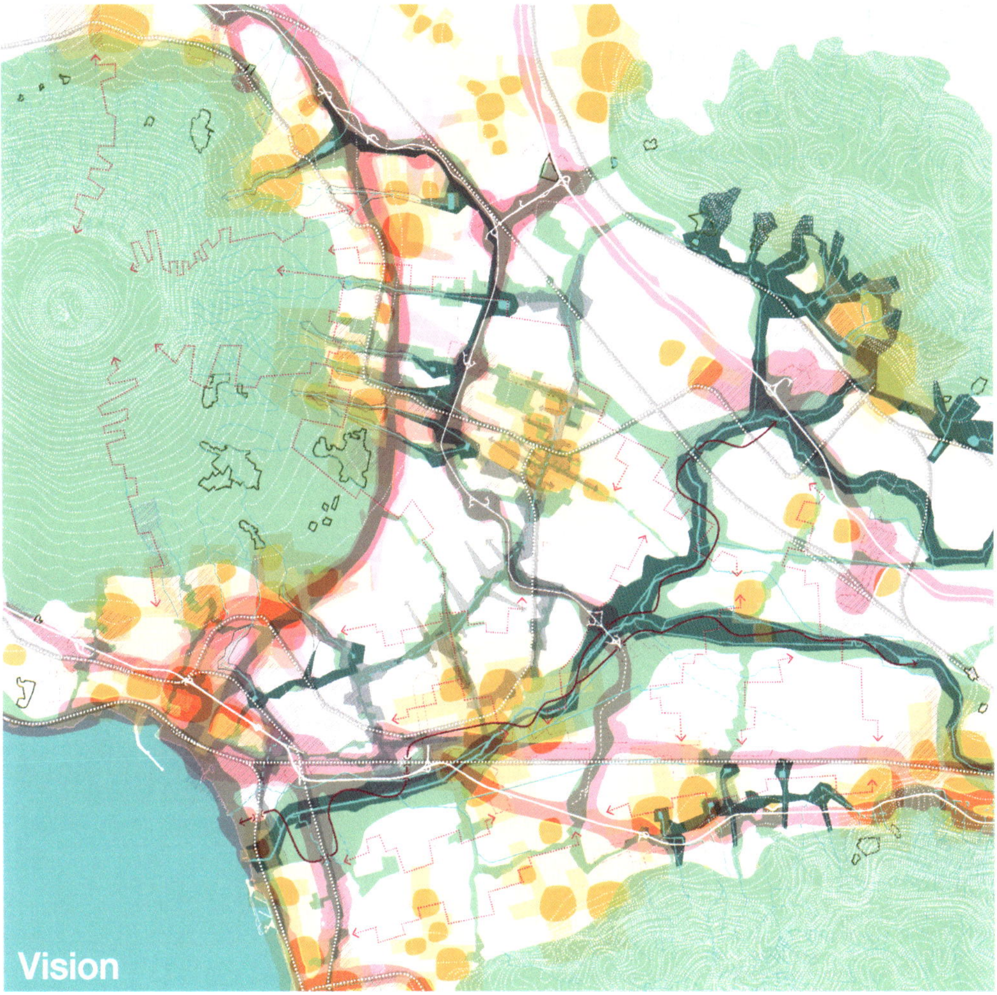

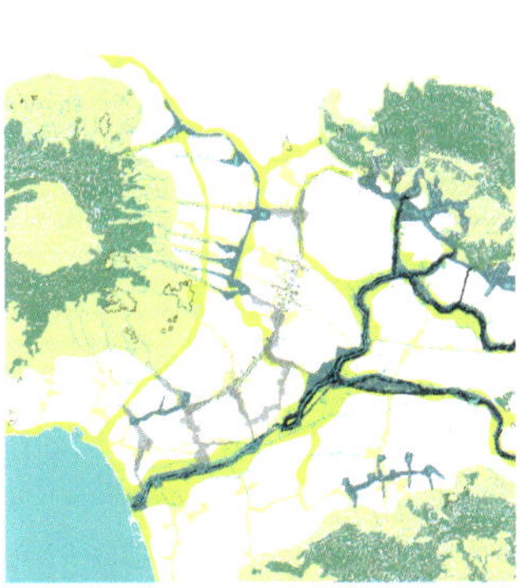

Blue and green infrastructural network

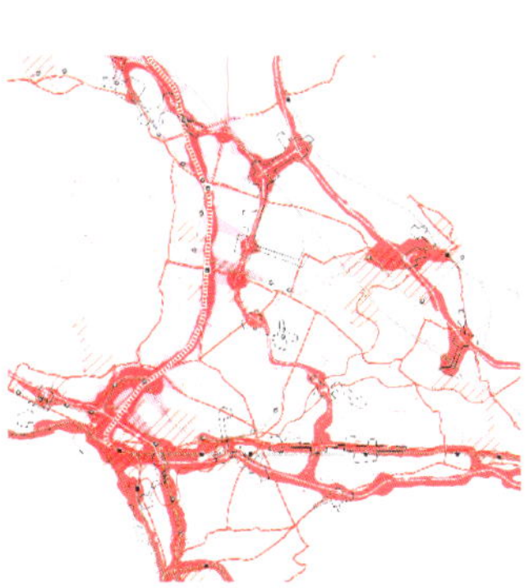

Integrated mobility network

Metropolitan archipelago

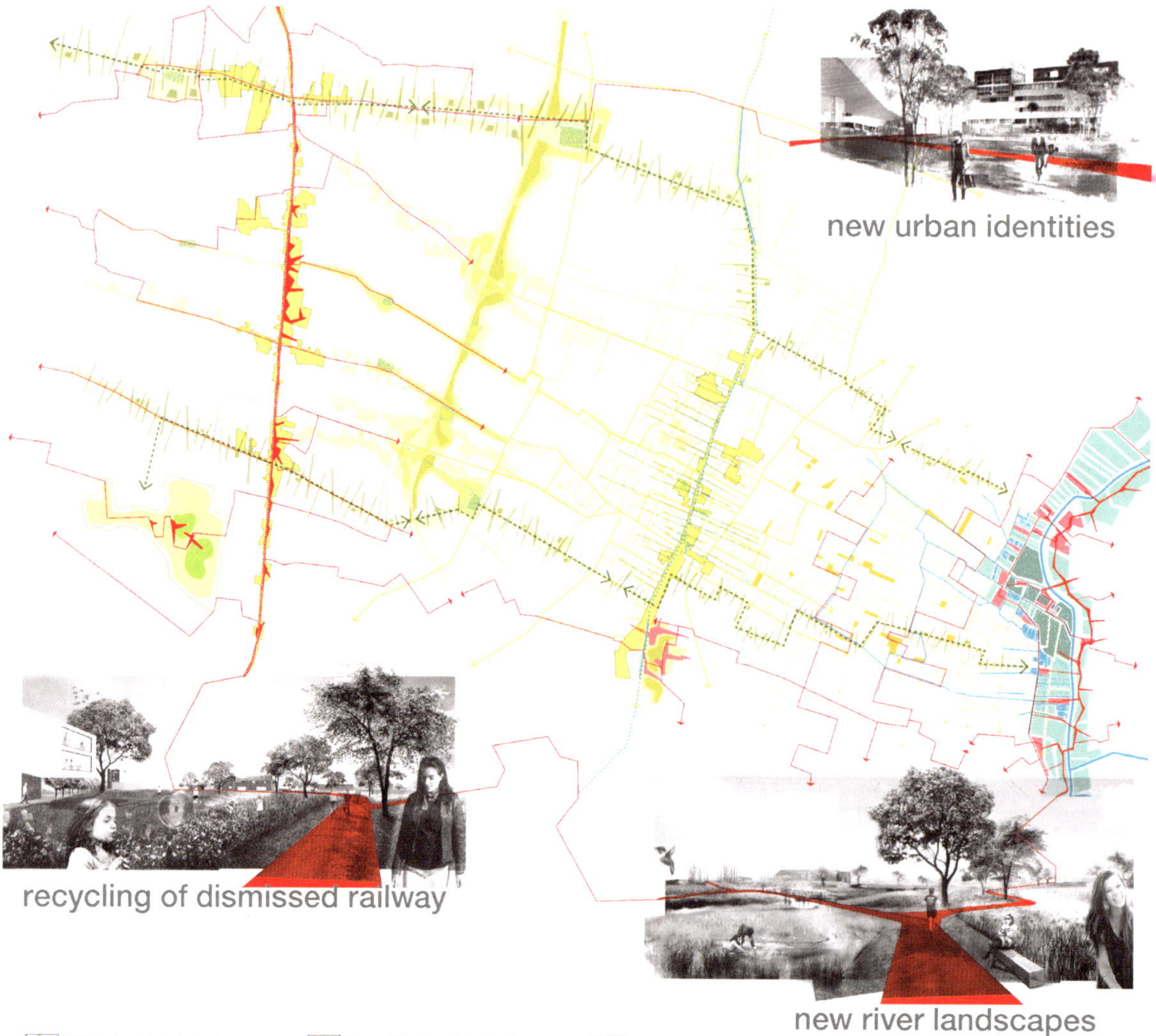

Greenway and ecological reconnections

Existing urban layout to be redeveloped

Pedestrian and cycle paths

Linear park former Torre Annunziata-Cancello railway

Setup of the carriageway axis on the Conte di Sarno canal

Re-qualification of the farming tracks

Agricultural park

Reconfiguration of agricultural land in equipped and rest areas

Ecologically oriented conversion of the greenhouse system

Reconfiguration of greenhouses for activities and equipment

Equipped areas of the river park

Reconfiguration of cultures in wetlands

Archaeological river park of Longola

Reconfiguration of the irrigation canals network

Supervised flooding areas of the Sarno river

Urban green areas

Rolling tanks

Filter zone of the "water beltway"

Ecologically oriented conversion of the built heritage

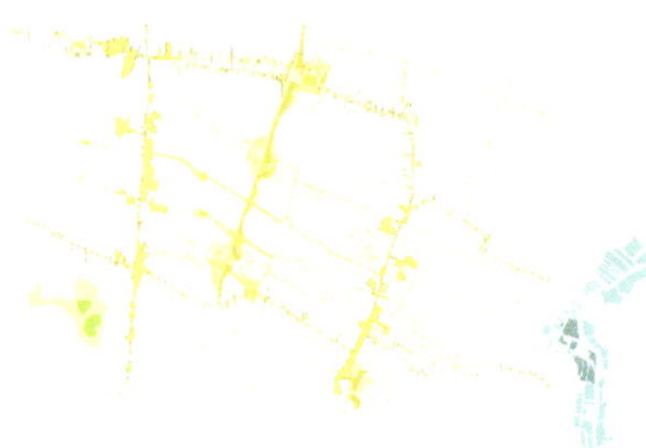

Open spaces network

Network of paths

Network of facilities

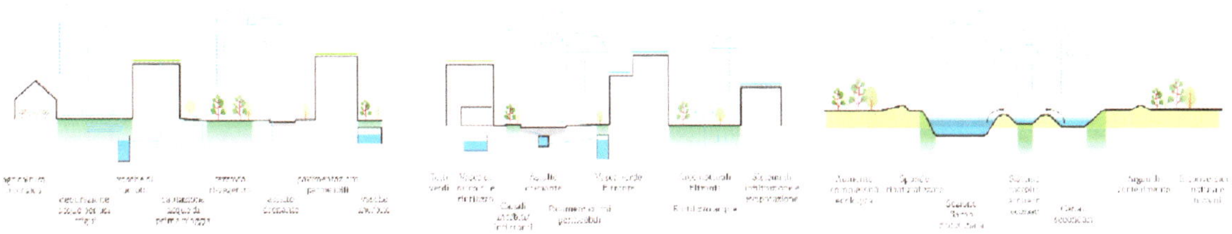

section of the dismissed railway recycling

section in urban context

section of the canal network

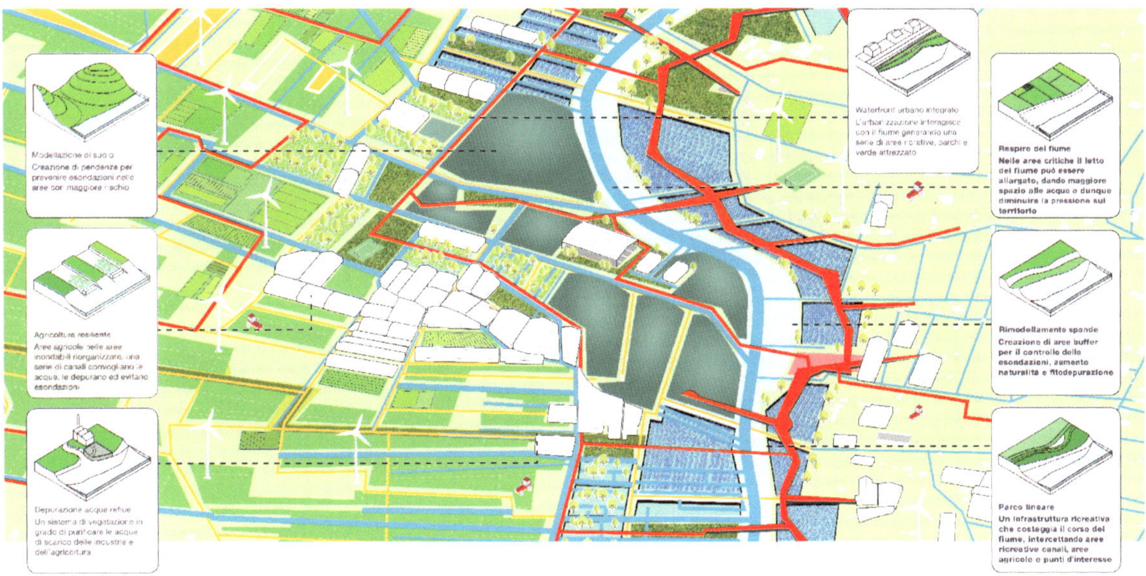

Borderline. Reclamation of the FS (State Railways) Torre Annunziata - Cancello

Alessia Franzese

Over recent decades, the issue of abandoned railways has become a real problem in terms of safety and maintenance, outlining a geography of wounds in both urban and agricultural fabrics and arousing growing interest both from institutions[102] and from communities brought together in the form of cooperatives. Examples such as the Spanish national program "Vias Verdes", but above all the cases of "Promenade Plantèe" and "High Line", have represented a design opportunity to bring city fragments back together, but have also become powerful attractors of economic-cultural interests, as well as a driving force for urban regeneration.

The research subject of the dissertation[103] focuses on the railway line Torre Annunziata - Cancello that runs through the territories of the urbanized countryside of the Vesuvian area for 31 km. From the coastal area to the plain of Nola, up to the slopes of the regional park of the Partenio, cutting through towns or touching extensive cultivated fields, thus crossing areas that are typologically different but have in common a lack of identity, this being a result of the intense phenomena of illegal building and in the total absence of urban and territorial planning.

The study of the railway section recycling as reclamation of a "socio-environmental infrastructure of slow mobility" aims at prefiguring a strategic vision of the possible urban relations that can be established with the (tourist, agricultural, naturalistic, and productive) resources that are still unexpressed in the territory, yet are also capable of triggering processes of resilience in the Neapolitan hinterland. This study is based on a critical re-reading - throughout the administrative territory crossed by this section - of the morphological aspects, the naturalistic structure, the anthropic and environmental risks, and the forecasts on the strategic development of the regulatory system on a supra-municipal scale (large but never implemented), followed by an analysis of the spatial relations that exist between the line and the context. Such a study needs to be carried out in a precise manner, in order to recognize and to theme the common and repeated modes of relation.

The analytical apparatus merges into a "strategic concept" of territorial development that is connected as much to the Vesuvius National Park (P.P.N.V.) as it is to the archaeological areas scattered over the territory - and is divided into 4 thematic

102. Agenzia del Demanio and RFI are promoting processes to enhance the value of disused railway assets.

103. The work presented was developed by Alessia Franzese, as part of her degree thesis in Architecture (CdL Architettura 5UE, DiARC - Univesità degli Studi di Napoli "Federico II"). Supervisor: prof. arch. Michelangelo Russo. Co-supervisor: architect Enrico Formato.

scenarios that identify as many possible configurations of grafting. This possible intervention strategy was then based on the in-depth design of three crucial nodes which represent models that can be proposed at different points along the linear dimension of the section.

TO-CA - A development opportunity for the Vesuvian suburbs

The Torre annunziata - Cancello line runs for about 31 kilometres through the suburban area behind the Monte Somma-Vesuvio system, as far as the Nola - Acerrana plain. The context is mainly one of urbanized countryside where social structures are lacking - the only places of aggregation are often the parishes - and is connected to the main urban centres by an infrastructure that is undersized compared to the population density of the area. The line was inaugurated in 1895 to offer a connection service between the inland areas, industrial areas and shipyards on the coast south of Naples.

After a phase of expansion and modernization in the 50s, its decline was marked by the industrial crisis and especially by the co-location of the Circumvesuviana: a railway line that crosses, for many sections, the same municipalities. The railway service was suspended in 2005 and the actual dismissal took place by means of Ministerial Decree no. 420 in 2014. Although its current state of abandonment contributes to the degradation of these suburbs, it also represents a fragment of "third landscape"[104], a shelter for diversity, an ecological corridor that preserves the heritage of spontaneous vegetation typical of the Mediterranean. Its path develops in varied contexts since it runs through built-up areas and open countryside, agricultural production fields and industrial areas, bridges or trenches. It touches not only the Vesuvius National Park but also the municipalities included in the "Red Zone " of volcanic risk (a factor that has stopped urban development and, as a result, provoked a strong deregulation in the morphological structure of the built environment), becoming from time to time an element of interruption "in between" or else a privileged point of view to start "visual dialogues" within the context. Sometimes, it even ends up dissolving "into the wild" due to invasive vegetation.

104. Clement G., Manifeste du Tiers Paysage ("Il manifesto del terzo paesaggio", Italian translation by De Pieri F.), Quodlibet, 2005.

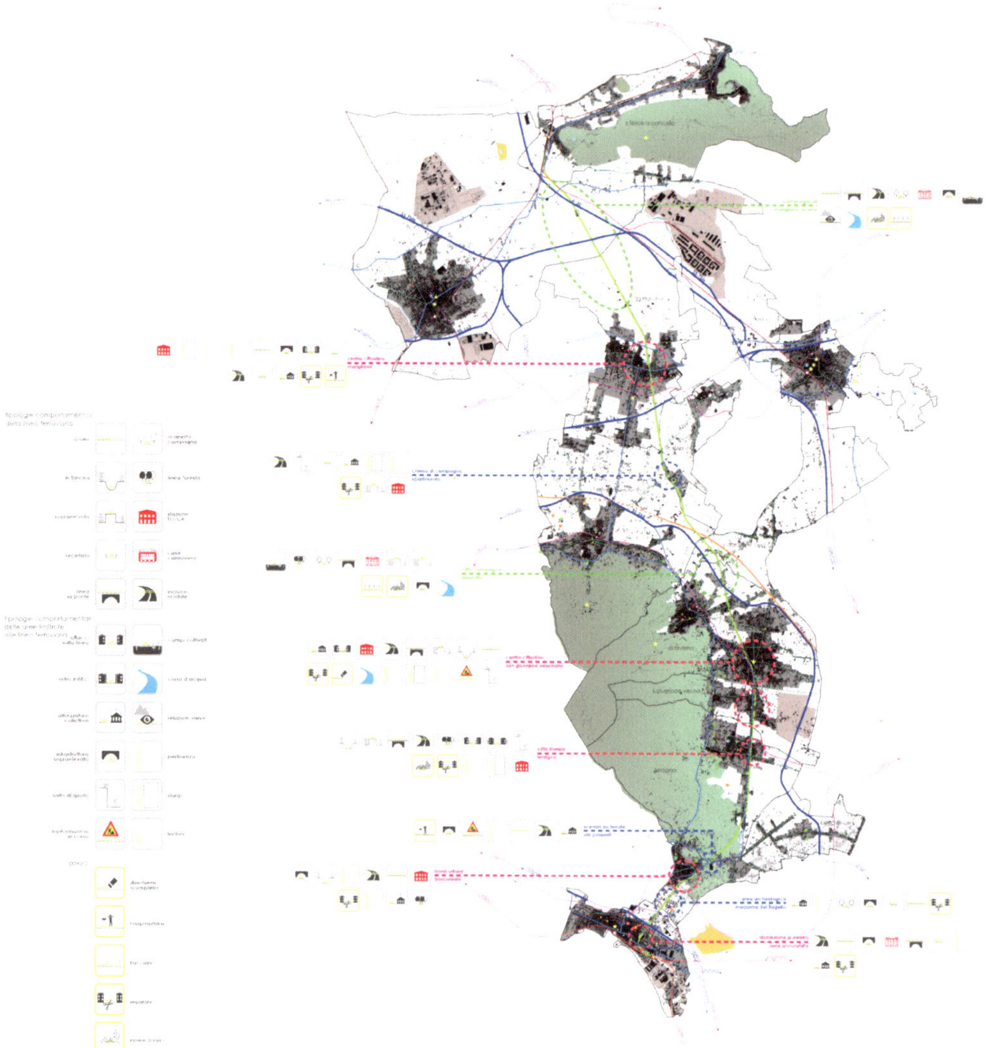

The analytical reading of the context has revealed many critical points but, at the same time, unexpressed resources.
The settlement system, typical of urban sprawl, is lacking social facilities but has urban centers and evidence of high historical and artistic value, as well as a constellation of unappreciated archaeological sites and artifacts, despite their proximity to the UNESCO site of Pompeii. The mobility system is mainly organised by road, with a weak interconnection system both towards the large cities such as Naples or Nola, and towards the neighbouring municipalities, in the total absence of bike lanes or "slow" mobility infrastructures in general.
From an environmental point of view, the composition of a "soil mosaic" restores the richness of the vast green areas of landscape and agronomic importance - such as the Vesuvius National Park - which are very different from the large and fragmented agricultural expanses of the Nola countryside, where you can find the cultivation of many local and traditional products (PDO, DOC, PGI).
The concern for seismic and volcanic risks makes this territory fragile and complex. This is also due to the fact that these threats are compounded by the risks deriving from hydrogeological instability (also as a consequence of the strong presence of buried canals) and the hazards deriving from multiple sources of pollution (such as the presence of many quarries, landfills, contaminated sites, etc.). These problems have been tackled over time in a sector-specific and specialist manner, translating and producing a strongly binding apparatus that only partially attempts to preserve the biodiversity of the site.
The presence of such challenges, which are added to an abusive urbanization over a long period of time and, consequently, a strong anthropic pressure, ends up delegitimising entirely a territory in which there are significant elements of value such as archaeological sites, architectural artifacts of great value, folklore traditions and agri-foodstuffs that are strongly rooted in popular culture.

In addition, the current regulatory apparatus at municipal level - if any - lacks implementation, as well as the urban planning tools (P.P.N..V., .PT.P. in Vesuvian municipalities, P.T.C.), which provide, among other measures, strategic guidelines for the creation of an ecological Vesuvian network capable of connecting the urban centers that are to be enhanced through pedestrian and cycle paths, strategic projects to access the Vesuvius, and urban agricultural parks for the enhancement of the Vesuvian riverbeds.
In this configuration, the TO-CA line would either take on the role of an ecological corridor for slow mobility (Naples P.T.C.P.) or it would be part of the projects for the ascent to the Vesuvius cone (Vesuvius National Park Plan). By making a leap forward in scale and approaching the morphological dimension of the area of study, we have analyzed the "ways" through which the abandoned line relates to the context, and vice versa. Three "macro-systems" thus allow us to find the relationships with the urban context, the agricultural context and the equipment, with the aim of recognizing areas with similar characteristics that are repeated along the way, thus establishing in the next phase the possible strategies that can be replicated in similar nodes. The construction of a "behavioural abacus" has, in this way, made it possible to describe three different ways of interacting: the way in which the line crosses urban and agricultural territories, the way in which the areas directly facing the infrastructure relate to it, and the actions that derive from its presence.

Recycling strategies for the landscape of the Vesuvian hinterland

The critical-cognitive analysis has made it possible to collect an enormous amount of data dispersed over the territory and has brought out resources and potential that need to be systematized, through a network, in order to be exploited.

The proposed regeneration strategy is thus developed through scenarios capable of indicating the dynamics of transformation and recycling according to the open logic of the "project/ process". The themes identified - "tourism and leisure", "culture and heritage", "agriculture and economy", "nature and landscape" - establish paths for the development of the internal Vesuvian identity, merging into that strategic "concept" previously identified, which considers the disused railway line as a "porous and infiltrating" infrastructure that penetrates the surrounding territory through the residual and waste spaces. A motorway for slow mobility transforms the railway line into a "botanical park/bike path", thus constituting the "hard part" of the process, linking and enhancing the new polarities through connections, temporary uses, and pop-up spaces, which - precisely because they vary over time - constitute the "soft part" of the process itself. The scenarios thus have a revealing effect.
The first theme - "tourism and free time" - is linked to the routes of ascent to Mount Vesuvius: the interchange nodes on rubber and iron that intersect with the TO-CA line become strategic. The second theme - "culture and historical heritage" - enhances the archaeological areas scattered over the territory: the line becomes the crucial axis from which the connecting ramifications branch out. The third and fourth themes - "agriculture and economy" together with "nature and landscape" - propose an alternative way of crossing the territory by developing an equipped park that enhances the native and spontaneous tree species and promotes local food and wine production through the reuse of the stations and the track maintenance houses as social containers of aggregation and dissemination of local culture.
The proposed project exploration foresees some variations of scale, even if very wide, in which we switch from the urban dimension to the detailed one, also involving a continuous redefinition of the road sections. "Designing relationships" has thus meant working on even the smallest parts, through the definition of elements of space qualification and soil design, such as flooring, urban décor, botanical species.

In the three nodes, the "masterplans" highlight the "contextual interferences": road crossings, height jumps, overcoming barriers, inclusion of "friches" as elements of landscape construction as well as the recovery of stations, rest areas and interchange of travel modes.
The Terzigno node has been themed on the enhancement of typical products. Possible interventions include "botanical construction" with themed gardens and educational gardens, the reuse of abandoned buildings as info stations and hubs for local gastronomic culture, temporary spaces dedicated to markets for local products and sports facilities to support existing meeting areas with staircases that act as terraces, a system of ramps to overcome differences in level and create new entry points to the linear park.
In the node of San Giuseppe Vesuviano, right in the city centre, the line becomes a green infrastructure for intermodal connection and exchange, overcoming a jump in altitude and breaking down physical limits. Also, the proximity to the Circumvesuviana station allows it to be turned into a strategic element for slow mobility and an alternative access to the Vesuvius park, aimed at specialized tourism.
The Spartimento node, on the other hand, relates the green infrastructure to the recovery of the riverbeds that descend from Monte Somma, acting as a connector between transversal routes (leading to Vesuvius on one side and to the archaeological sites of Nola on the other) and focusing on the problem of hydrogeological pressure.
In short, we can say that the construction of regeneration scenarios for the Vesuvian area through the recycling of the railway line needs a "cultural" activation. This to be achieved by involving citizens in the education of the shared space so as to make them protagonists of the process and not only users of an urban product.

Walking along the river. The Royal Bourbon Powder Factory of Scafati, a link between the historic centre and the Archaeological Excavations of Pompeii

Valentina Vittiglio

Lorena Santamaria

Extending over an area of about 11 hectares, the former Royal Bourbon Factory of Scafati is now an important resource for the city and an opportunity for urban transformation that, while preserving the historical value of the site, is being reused through the introduction of new functions in order to develop and increase the activities in the territory of Scafati.

Scafati, located in the Sarno Valley's focal point, plays a significant role in the provincial settlement system, especially in relation to its proximity to an archaeological area of undisputed importance: the Excavations of Pompeii.

The historical and urban evolution of the area has been strongly influenced by the presence of the Sarno river system, considered a natural element which generates all those components that, at various scales, intertwine in the territory. In this context, the centurial system, recognizable in a conspicuous series of linear recurrences on the modern territory, is of considerable importance for its coherent arrangement with respect to the ancient artefacts located throughout the area and in the course of the"centuratio" itself, which appears faithful to a large part of the modern road network.

Considering some territorial limits, such as poor connections, insufficient public spaces, unexploited pre-existing structures and unusable river axis, the intervention strategy identifies a number of potentialities as the project's cornerstones, starting from the negativity of the area. The planning concept is expressed by the presence of the cycle-pedestrian itinerary, accessible in both directions, which reaches the archaeological excavations of Pompeii from the historical centre, a "grand boulevard"which enables the perception of space and can be used as a tool for learning about the urban system; a walk along the river and through the places that have marked the origins of the territory to observe, assimilate and remember today what is memory of a past time.

Given the geographical proximity between Pompeii and Scafati, the project's aim is to create a network between the two municipalities and a new centre for tourism, in synergy with the Excavations. The intervention also outlines[105] an alternative to traditional mobility by conceiving the new routes as green infrastructure and by inserting them harmoniously into

105. The work was developed by Lorena Santamaria and Valentina Vittiglio, within the degree thesis in Architecture (CdL Architettura 5UE, DiARC - University of Naples "Federico II"). Supervisor: prof. arch. Michelangelo Russo. Co-supervisors: prof. arch. Bianca Gioia Marino, architect. Enrico Formato and prof. geol. Rocco Lafratta.

the urban fabric, without altering the existing road network. Given the strong state of abandonment and degradation in which the historical centre and the municipal villa of Scafati find themselves, far from the vitality and ferment you ideally find in Pompeii and completely disconnected from the flow of tourists, the design strategy is dual: on one side to rediscover an area of great historical and natural value and on the other to reconnect it both to the historical centre and to a tourist pole of primary importance, restoring a no longer fragmented general context. The "linear park" is developed starting from the municipal villa, which is perhaps the most picturesque area of the historical center, in which the river's "blue infrastructure" is not perceived as an environmental critical factor but as a potential component in the urban landscape's re-qualification. The existing external car parks have been rearranged and the villa's green areas redesigned. The idea was to give new functions to an abandoned building, now transformed into a museum, and to enhance the thirteenth cenury wooden drawbridge, now made of stone, and the last existing mill in the Mills' district, which is still a ruin. A number of pedestrian and cycle paths start from the villa, surrounded by new green areas. New spaces for rest and relaxation lead to a new exhibition area at the service of citizens. The entire area is made by alternating paved and green areas reaching beyond the villa's perimeter in order to connect it to the Polverificio's park while ensuring the different fabrics' continuity. The cycle-pedestrian itineraries extend to the Porta del Polverificio, crossing the entire park and connecting it to the fabric of the historical city. Finally, from the secondary entrance, a "boulevard" connects the city park and the Polverificio to the Pompeii Excavations, reconnecting the fabrics and sewing these areas together in a single heterogeneous and complete system. Access to the Polverificio's park will only be allowed from the Bourbonic Portal which marks the entrance placing itself as the landscape's identification element since the complex's earliest days of activity, preserving its original charm and character still today. The central passage was created as a carriage door, while the side ones were used by pedestrians.
Originally, the gate was incorporated into the city wall which enclosed the entire area of the Polverificio and, near the

Parco del Polverificio, natural reserve

Parco del Polverificio, Bourbonic portal

Parco del Polverificio, parks and orchards

openings, had rooms used for traffic control and management of goods. To date, the situation is significantly different. There is no trace of the surrounding wall and the gate has lost its fundamental character of transition; not far from the other buildings, but ideally isolated and "defragmented" from the system of which it used to be the focal point. There is a lack of connection with the surrounding fabric, together with a lack of both driveway and pedestrian viability within it. As a result, the surrounding square improperly became a parking lot for the employees of the experimental research institute, before the area was finally closed down. The project involves the revalorisation of the door and of its original symbolic vocation as an element of transition. The project proposal takes concrete form in the definition of a water boundary which imposes the passing through the Portal to access to the park. The wall of water, thin and impenetrable, consists of a glass facade on which a battery of nozzles distributed by a pipe placed at the top emits a continuous flow of water, enough to permanently veil the transparency. At the base of the glass, a linear collection tank with a suitable slope collects the water and then conveys it to a pump that finally puts it back into circulation. Finally, considering the hydrogeological risk to which the area is subject, its original agricultural vocation and the presence of historical irrigation channels, it was decided to design a natural area to modify and contain the environmental needs with the inclusion of a wetlands system, an extra-alveous surface flow in this case.

The opening of the "Traversa di Scafati" (Scafati's Cross) in 1600, whose function was to supply power to the mills, was a prerequisite for the design of the wetland inside the park. It had, in fact, the function of feeding the mills in the area of Bottaro through an artificial channel which took its name. In 1850, when the Polverificio Borbonico was founded, all the waterfalls of the Traversa were closed except for the first one, intended to supply power to the waters of the Bourbon plant by feeding water to the Sarno river's main course, which bordered the southern side of the Polverificio. Following the closure of the Traversa, the bed of the Bottaro Canal now appears to have dried up completely and it is therefore essential, for the purposes of hydrogeological and environmental balance, to restore its course by making it the primary vector transporting water into the tank system. The project's ultimate aim is to highlight the importance of rethinking the role of natural and infrastructural networks which, by reconnecting the historical fabrics and recovering and strengthening the ancient routes, establish themselves in the pre-existing network, thus defining new synergic relationships between anthropic and natural components and allowing the public space to restore the city's power of narration and discovery of those places which, although currently forgotten, emphasize its historical importance.

N
Scavi Archeologici di
Pompei
Centro storico
di Scafati
Real Polverificio
Borbonico di Scafati

Sustainable water management. Recycling strategies in the PUC* for Poggiomarino

Anna Terracciano

Emanuela De Marco

Recycling potential: Social marginality and geographical importance

The scarce supply of services at the urban and territorial scales, the strong residential specialization, the fragmentation of the open spaces system, the discontinuity of the infrastructural network and a widespread condition of decadence highlight the persistence of the Municipality's strong condition of remoteness - despite its position, an imaginary hinge between the Vesuvian area and Nocera Inferiore. One good example of this condition is the old city, where it is possible to find scenarios of abandonment and decrepitude. In addition, the progressive alteration of the morphogenetic and architectural characteristics results in the replacement of ancient artifacts with new buildings that are strongly incoherent with the historical fabrics in terms of form and size. This situation clearly indicates a lack of awareness of the town's history and identity, still strongly marking the territory's interactions with the anthropized environmental heritage. All this translates into a poor consideration for existing resources and spaces as they are not recognized or perceived to belong to the territory's deepest identity.

The total lack of enhancement of the archaeological site of Longola is an even more significant example in terms of historical and environmental resources. Its potential value as an attractor within the circuits of the Vesuvian area and its archaeological sites is huge. The degradation of the historical farmhouses, an integral part of the agricultural territory, is also evident. They could instead play a central role in seeeking a multifunctional contemporary agricultural landscape. The transformations of the urban form over the last few decades have thus led to a progressive weakening of the existing links between natural and anthropic components.

The complexities and the importance of environmental and infrastructural networks in the territory's planning and in rethinking the urban structure in this particular historical phase - in which the ecological crisis linked to the major effects of climate change on the quality of water and soil is closely intertwined with the effects related to volcanic and hydrogeological risks[106], the chemicals used in agriculture and

*Piano Urbanistico Comunale – Municipal Development Plan

106. Fabian L., Viganò P., "Extreme City. Climate change and the transformation of the waterscape", Lorenzo Fabian, Paola Viganò eds. Published by IUAV University of Venice, 2010

the anthropic pressure - make of the municipality of Poggiomarino (Naples) an area of particular importance for research and experimentation of recycling strategies.
The P.U.C.[107] experience highlights the need to make choices within a multiscalar territorial framework of relations and tools for planning and programming on a supra-local scale in relation to a wider territory than the municipal administrative border. The nature and the very dimension of some problems linked to connecting infrastructures, in fact, make it essential to coordinate the objectives and urban policies within wider interpretative, planning and decision-making scenarios[108].

Resistance and resilience

The restoration of signs and traces in the territory's hydraulic networks reveals a series of "environmental invariants". Such invariants are compromised in different ways but they all represent potential elements for regeneration as well. The interruption of the virtuous cycle of water, as it was originally conceived, has resulted in fact in the deposit and sedimentation, on the surface as well as through the various soil's stratifications, of a multiplicity of materials that have exhausted their life cycle. These considerations necessarily encourage a reflection towards the construction of innovative planning tools to stimulate new processes in the reactivation of these drosscapes[109].
We are therefore trying to reflect on the precise and systemic value of these materials for which recycling is not just an isolated gesture of planning regeneration; it rather means activating a process of reclamation through incremental procedures capable of promoting a complex and "adaptive" project for the territory. The goal is to try to convey the water network signs towards the definition of a "territorial hydraulic system" able to establish a new balance between urban settlements and the environment.
It is a matter of contextualizing these signs, spaces and practices within a perspective that recognizes which parts of the territory are elements of "resistance" and which instead can be adjusted to an "evolutionary and adaptive model", the so-called "resilience" of our times. The aim is to resume

107. The activities set in the PUC of Poggiomarino (Naples) refer to the Convention signed on 29/10/2012 between the Municipality of Poggiomarino (Naples) and the University of Naples "Federico II", DiARC - Department of Architecture; for the activities of support and technical-scientific assistance to the City Planning Sector aimed at drafting the PUC, the SEA (Statistic Environmental Evaluation), and the RUEC under the regional law n. 16 of 22/12/2004 and subsequent amendments and additions. The DiARC group is composed as follows: Scientific director: Prof. Arch. Carlo Gasparrini; Specialist consultants for SEA: Arch. Maria Cerreta and Pasquale De Toro with Arch. Giuliano Poli; Specialist consultant for mobility: Prof. Ing. Claudio Troisi; Specialist consultant for the RUEC: Prof. Arch. Valeria D'Ambrosio with the Arch. Eduardo Bassolino; Operational coordination: Arch. Anna Terracciano; Consultant for GIS and territorial information systems: Arch. Marco Facchini; Arch. Emanuela De Marco, Francesco Stefano Sammarco, Ciro Sepe, Danilo Vinaccia.

108. Gasparrini C., "Cities to recognize and eco-landscape networks", in "PPC" n. 25, Listlab, 2011.

109. Berger A., "Drosscape, Wasting land in urban America", Princeton Architectural Press, New York, 2006.

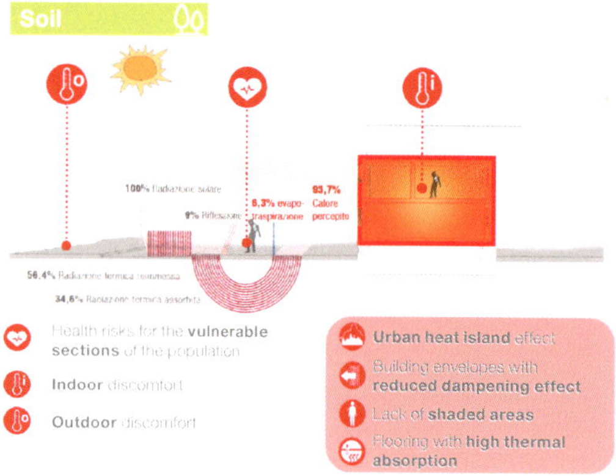

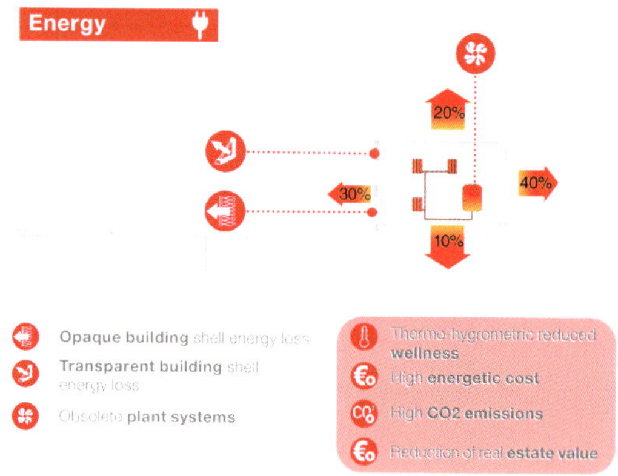

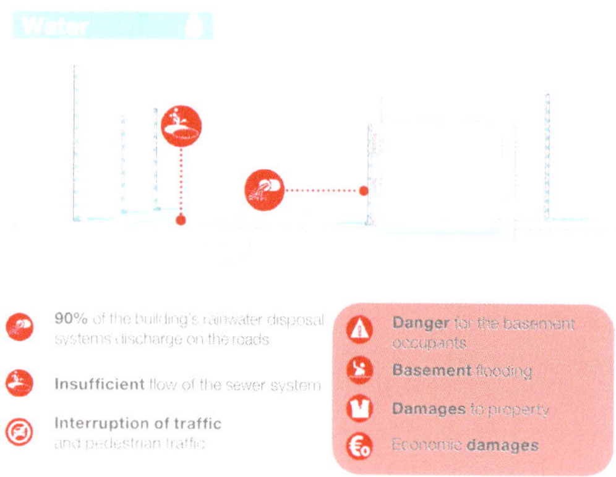

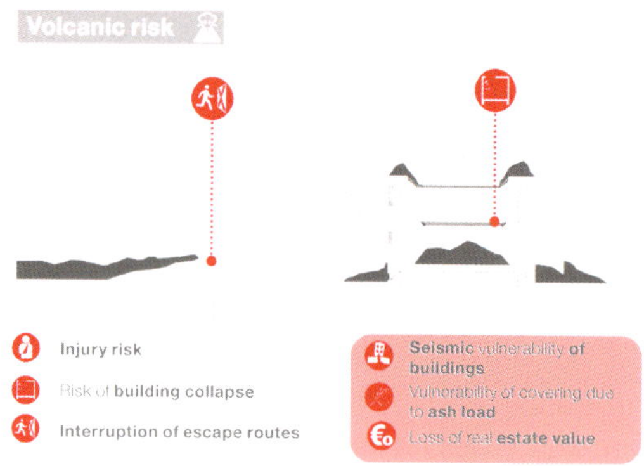

On the left, critical issue concerning the buildings and the system of open spaces. The pictures are by Valeria D'Ambrosio and Eduardo Basolino; On the right, Current water cycle and recycling proposal for a Sustainable Water Management in the specific and territorial dimensions. The images are by Anna Terracciano.

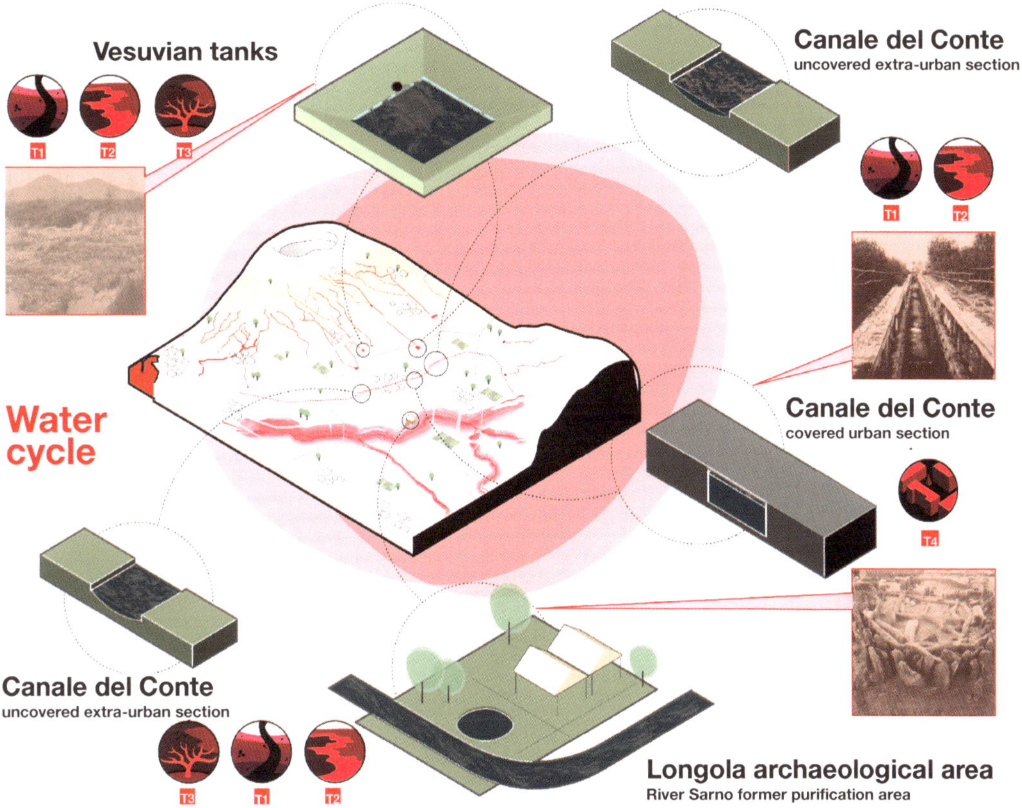
Vesuvian tanks
T1
T2
T3
Canale del Conte
uncovered extra-urban section
T1
T2
Water
cycle
Canale del Conte
covered urban section
T4
Canale del Conte
uncovered extra-urban section
T3
T1
T2
Longola archaeological area
River Sarno former purification area

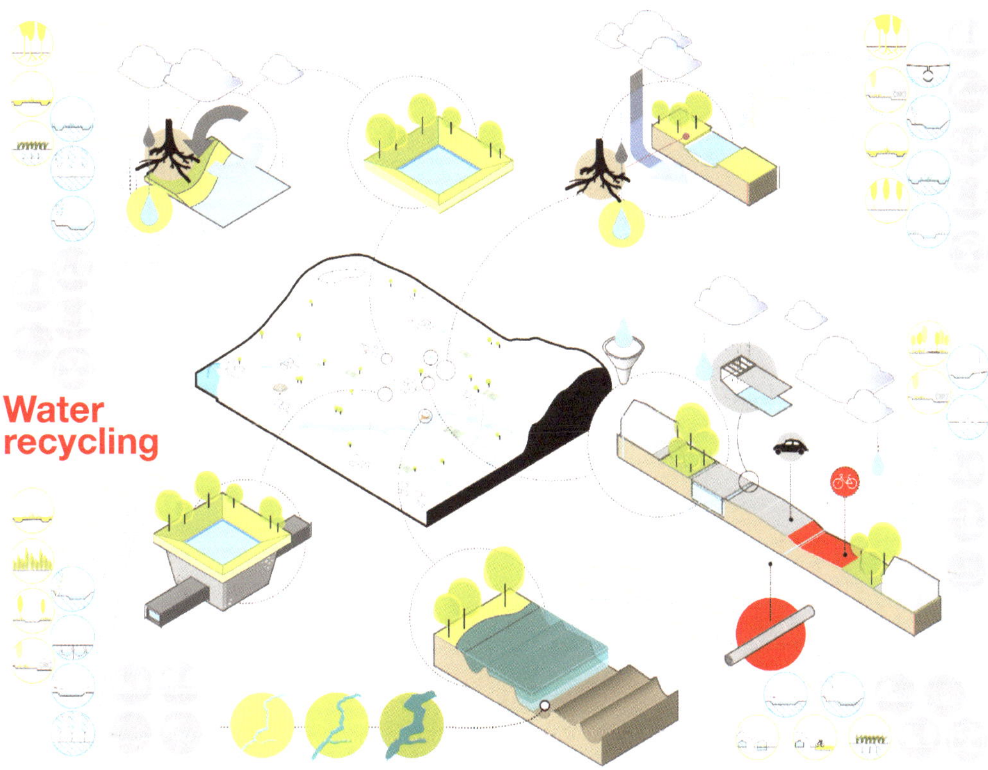
Water
recycling

and update a narrative of places that has been interrupted for almost a century, within a project of modification not only capable of protecting but also of defining new landscapes on the territory while working on the "networks" to recreate the conditions for its use and sustainability. This is why the suggested planning action must necessarily address the specificity of the context, while ensuring the right balance among "resistance and resilience" elements in order to reinterpret in a contemporary perspective the ancient attitude of preserving the territory. In other words, it is necessary to aim towards a "sustainable water management" capable at the same time of restoring a virtuous cycle of surface and deep waters and of guaranteeing water and soil reclamation. New urban and peri-urban landscapes progressively emerge from the expression of different forms of coexistence between decontamination, environmental transformation and social re-appropriation, within a more complex project of innovative open public spaces that is capable of meeting the local population's needs, largely neglected so far.

Vision 1 "Poggiomarino eco-city with a strong identity"; general objectives: 1. To consolidate the ecological network and rethink the water system; 2. To safeguard the natural-historical resources and the local traditions; 3. To develop the agricultural landscapes.

On the opposite page: Strategies for a networked territory

Strategies for a networked territory

It is therefore a question of defining strategic actions capable of implementing, in the long term, the continuity of the public dimension and collective living. New policies and, more generally, a new vision for the contemporary territory are needed. A frame in which multiple temporary practices and design actions, spread throughout the territory, are able to identify the interstitial times of a "project/process "[110] that, unfolding in an incremental way, recycles and returns those "drosscape" geographies thanks to new methods compatible with the reclamation and realization phases of the large-scale territorial requalification projects, delivering new and infinite configurations.

A new territorial framework emerges, articulated along three "landscape networks" and inspired by the diffusion, capillarity and pervasiveness of the "drosscape" territories - a viable and desirable strategy to give spatial and organizational coherence to the territory's recycling process. We are not only talking about searching new configurations to interpret the chaotic settlement that urban sprawl has produced in these areas, or the new relational densities that that same condition entails, but also about expressing new perspectives and new models for the management of urban areas[111].

The "blue network" is the system of spaces intercepting the superficial and deep hydrographic network and, more generally, the environmental continuities. Along these networks, the hydraulic reclamation systems become tools to design the territory, placing the hydrogeological risk as a founding principle in the reconsideration of the settlements' geography,

110. Corner J., "Recovering Landscape. Essays in Contemporary Landscape Theory", Princeton Architectural Press, New York, 1999.

111. C. Gasparrini, "Multiscalar and multiscapes visions to tell Naples", in M. Gausa, M. Ricci, "Med.net. rep.01", Listlab, 2014

Green network

Blue network

Quarries, landfills and storage sites to reclaim and re-cycle

Dismantled industrial areas to reclaim and re-cycle

Archaeological sites

Archaeological areas and fruition tools

Pedestrian and cycle paths

Green way network

Actions

Blue Network
[e water management]

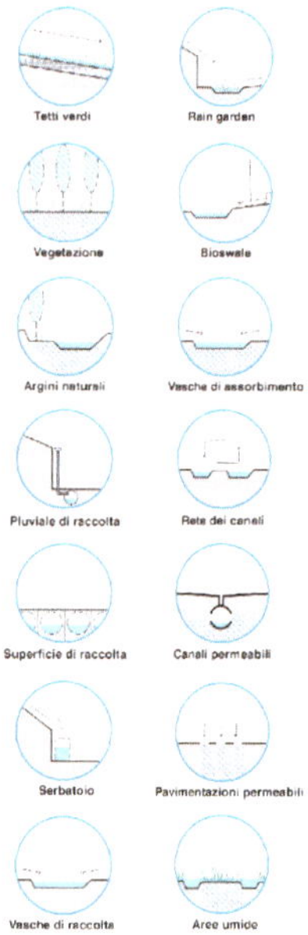

Green Network
[tutela della biodiversità]

New urban and suburban landscapes

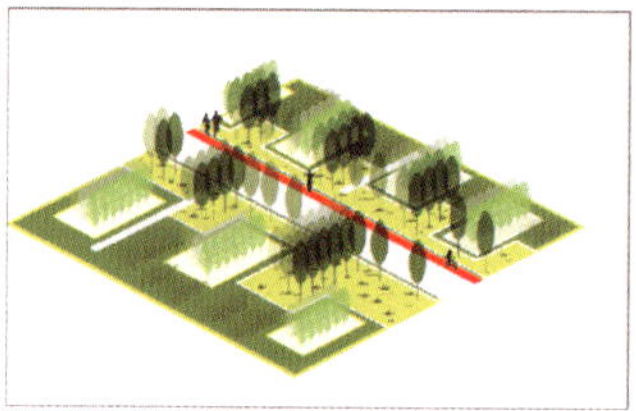

Fasce filtro

Aree di mediazione tra lo spazio della strada e le aree agricole e per la fitodepurazione delle acque piovane

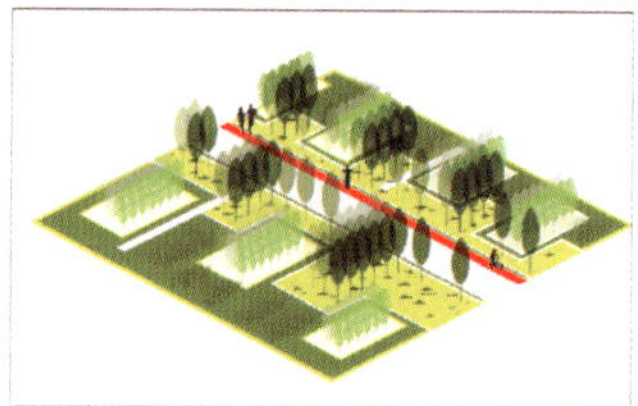

Fasce filtro

Aree di mediazione tra lo spazio della strada e le aree agricole e per la fitodepurazione delle acque piovane

Spazi aperti attrezzati

Aree per la sosta, per il tempo libero e per lo sport

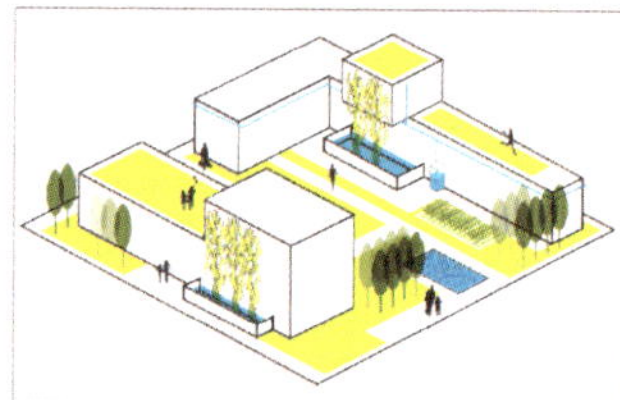

Green-Blue residential

Aree per il water management in ambiente urbano e riequilibrio tra il costruito e le superfici permeabili

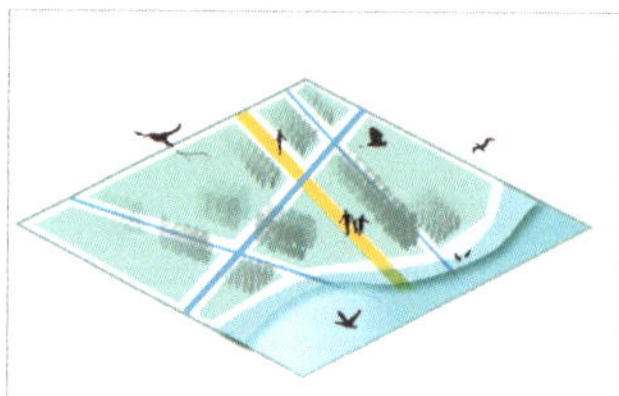

Aree umide innovative

Aree per la riconversione delle colture nelle aree umide, per la sperimentazione di colture legate all'acqua

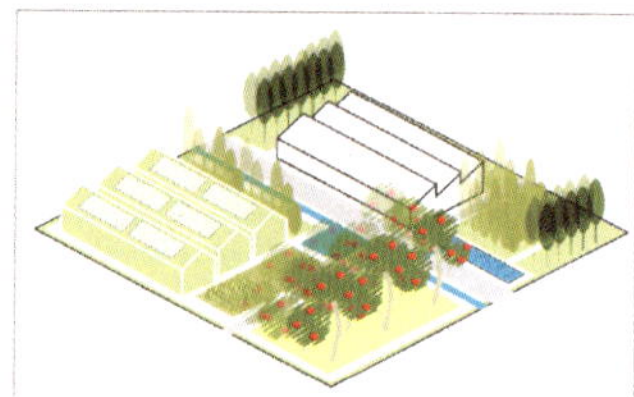

Innovative green productive

Aree per la sperimentazione di un rapporto innovativo tra produzione agricola e nuove filiere produttive

Aree di esondazione controllata

Aree della rinaturalizzazione fluviale, della biodiversità e della fruizione controllata

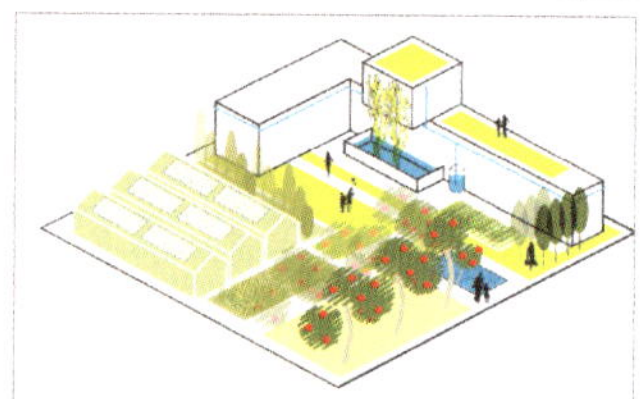

Mixed innovative residential-productive

Aree di commistione virtuosa tra gli spazi dell'abitare e gli spazi produttivi

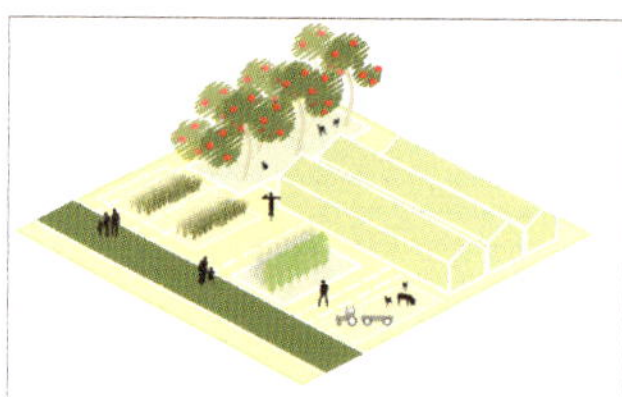

Paesaggi produttivi

Aree per l'agricoltura di qualità, per la trasformazione sostenibile dei prodotti e la vendita a km0

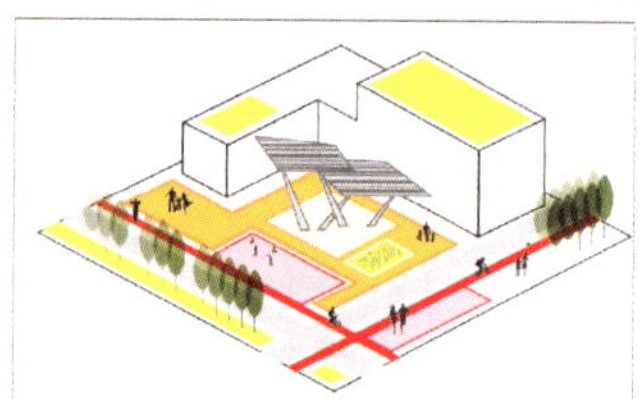

Urban open space

Aree per il leisure e la convivialità nell'ambiente urbano

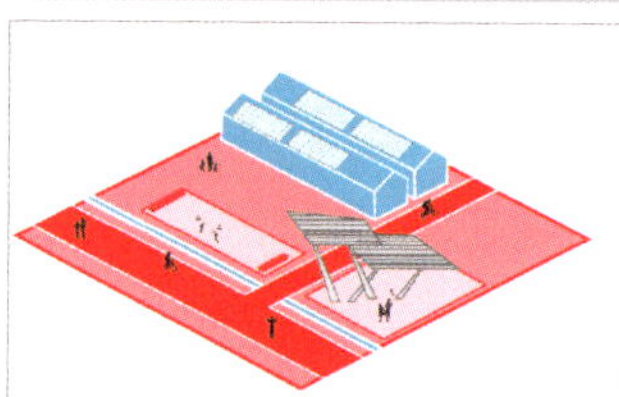

Spazi aperti innovativi

Aree per la sperimentazione, la didattica e le energie alternative

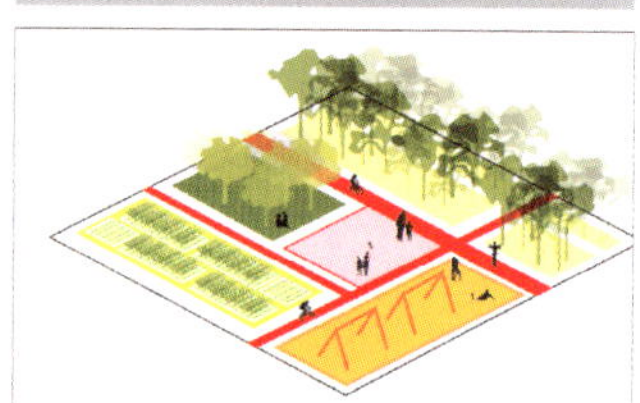

Parco urbano

Aree per il contatto con la naturalità, per il leisure e la riappropriazione degli spazi anche attraverso pratiche spontanee

326

thus transforming the territory's fragilities and the population's vulnerability into an entirely modern design opportunity.[112]
The "green network" is the system of urban green spaces and nature that aims to the construction of a continuous, complex and productive, mosaic in the economic, energetic and social terms related to open spaces. The design issues addressed are many and often closely related to those of water. They range from overcoming fragmentation to the de-waterproofing of soils, which represent one of the main components of the settlement system's porosity. This can also be achieved through strategies and micro-actions for the recovery of grey and white waters, even at the urban level, in line with the multi-scalar vision of sustainable water management on a broader territorial scale[113].
The "innovative use network", instead, is the range of possible uses, including temporary and informal, of public open spaces . In this way a new potential network is being defined, which involves the weak and accidental spaces of the urban experience and the waste areas created by a planning model that has not been very careful to creating those connections in the first place. The aim is to recycle these categories of public space into a network of new spaces and tools capable of making the "functional and landscape mix" denser and more vital and to propose an all-round fruition of the urban space.

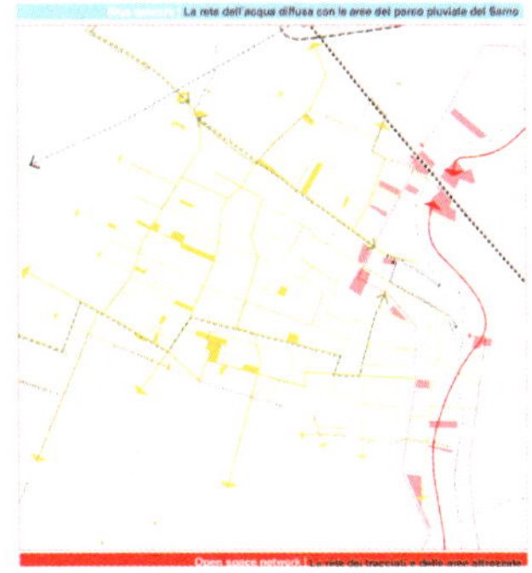

A few planning topics

Along the ridge of the SS 268 State Road it is possible to see the "water beltway", through various land projects[114], intended as a filter strip capable of intercepting the waters coming from the Vesuvius and the overflow of its tanks. The aim is to address, within a "resilient" and "adaptive" logic, the issues of hydraulic risk in the exposed areas[115] and those close to the urban centers.
The same logic is applied, although with different results, to the exposed areas along the river Sarno, transforming them into a sequence of green areas and parks in which the prevailing vegetation is able to re-create the wet landscape eroded by intensive agricultural us but also to allow, through mechanisms of "phytoremediation", the depollution of both the surface water table and the river's water. These areas, which also fall within the perimeter of the Parco Regionale del Fiume Sarno (Regional Park of the River Sarno, or P.R.F.S.), reveal the "Sarno River-park" plan, ideally capable of restoring its vocation as the Plain's main ecological and landscape connection as well as to encourage the connection, on a broader scale, between the two large natural reservoirs of the region: the Parco Nazionale del Vesuvio (Vesuvius National Park, or P.P.N.V.) and the Parco dei Monti Lattari (Lattari mountains Park). The project also foresees the recovery of the dense network of irrigation canals, most

112. H. Bulkeley, "Cities and Climate Change", Routledge Critical Introductions to Urbanism and the City, Routledge, 2013; K. Shannon, B. De Meulder, "Water urbanisms east", Park Books, Zurich, 2013; P. Viganò, G. Giannotti, (ed. by), "Our Common Risk. Scenarios for the diffuse city", in et al/ED, 2012; D. Watson, M. Adams, "Design for flooding: architecture, landscape, and urban design for resilience to climate change", John Wiley and Sons, 2011.

113. AAVV, "Urban ecology. An International Perspective on the Interaction Between humans and nature", Springer New York, USA, 2008.

114. Secchi B., "Progetto di Suolo", in "Casabella", n. 520/521, 1986.

115. See PSAI AdB of Campania Centrale 2015.

Soil

- Rainwater seepage into the ground
- Use of open joint pavements with SRI > 29
- Integration of green areas

- Increased soil permeability
- Increased "evaporating-transpiring" processes
- Increased shaded areas

Volcanic Risk

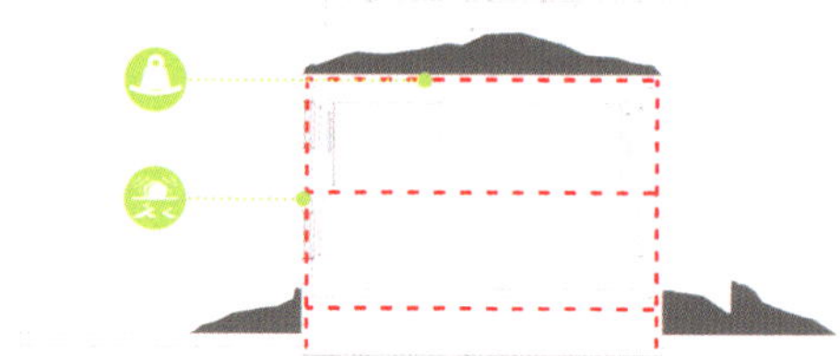

- adjustment of covering to volcanic ash load
- Structural reinforcement / seismic improvement

- Safe construction / functionality of escape routes
- Preservation of buildings and economic revaluation of real estate

Water

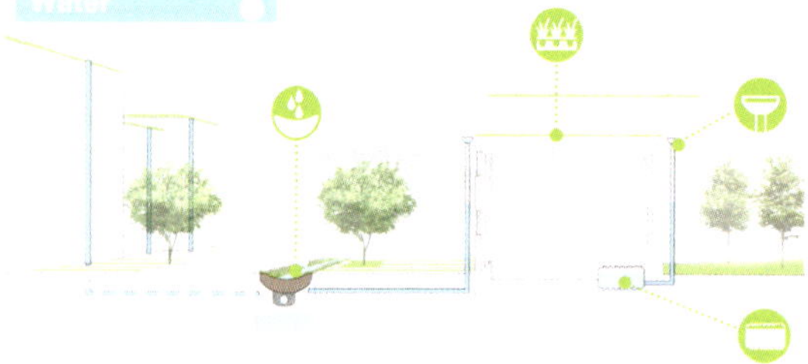

- Rainwater collection from covering and re-use
- Treatment of rainwater with vegetated systems
- Reallocation of basements as technical rooms
- Integration of roof gardens

- 30% water load reduction in the sewers
- 20% reduction of water consumption

Energy

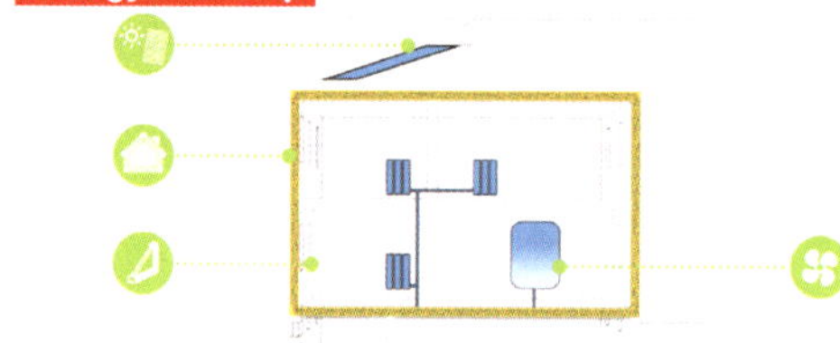

- Thermal insulation of the opaque building shell
- Improvement of transparent building shell
- Improvement of efficiency of plant systems
- Free contribution from renewable sources

- Improvement of indoor comfort
- Reduction of energy costs
- Reduction of CO2 emissions
- Appreciation of real estate value

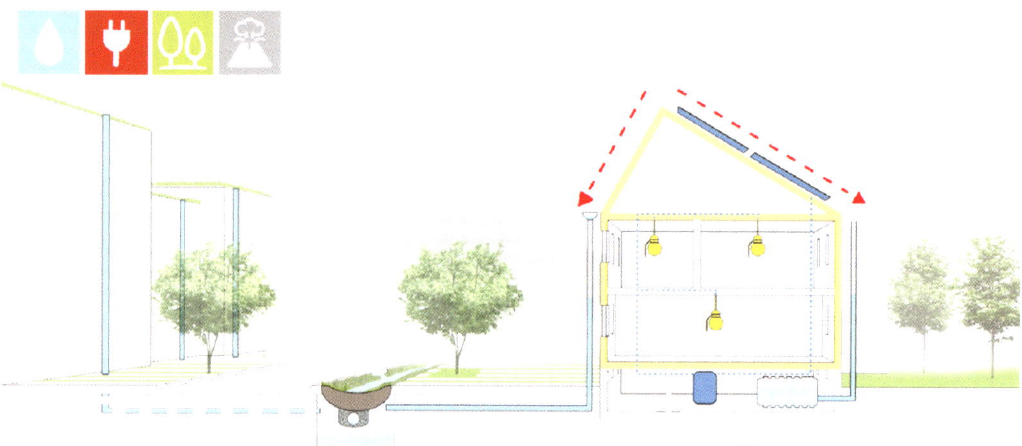

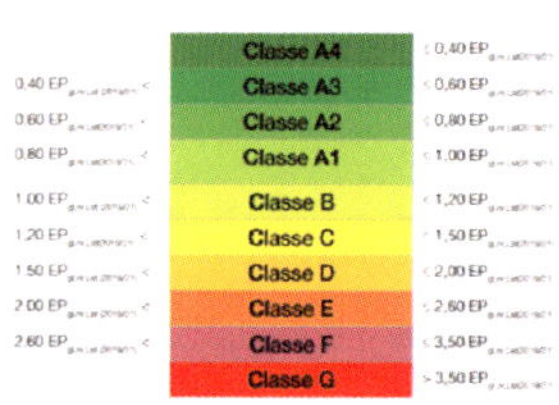

+

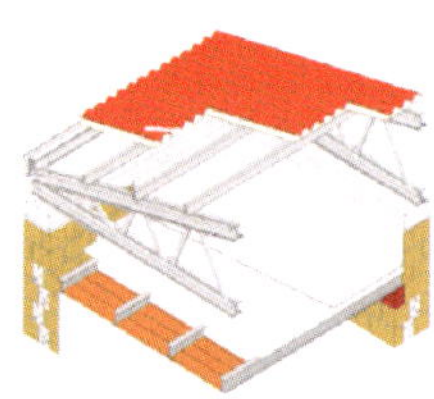

+

of which still recognizable, to ensure a èervasive water system managing the flow in the event of severe flooding.
In addition, the redevelopment of agricultural tracks and the design of new paths for "slow mobility", identifies and involves a variety of uncultivated and degraded areas that can be dedicated to create innovative and equipped open spaces.
In urban contexts, instead, it is hoped that a collection network will be created for the recycling of grey and white waters, running in parallel to the collection network of black water and through all the settlement fabric, and that drains these waters into the Count of Sarno Canal. The Canal, after being reclaimed, cleaned and preserved, shall thus assume the role of main collection conduct through all the municipalities it crosses.

On the opposite page: planning actions for an integrated recycling and an ecologically oriented regeneration of the buildings and open spaces' critical heritage.
The images are by Valeria D'Ambrosio and Eduardo Bassolino

Below: Planning actions of densification to heal the urban form. Images by Emanuela De Marco

In the following pages: Global policy framework; incremental scenarios; illustrations of different recycling proposals for the Canale del Conte di Sarno and proposals for hydraulic mitigation in terms of resilience. Images by Anna Terracciano.

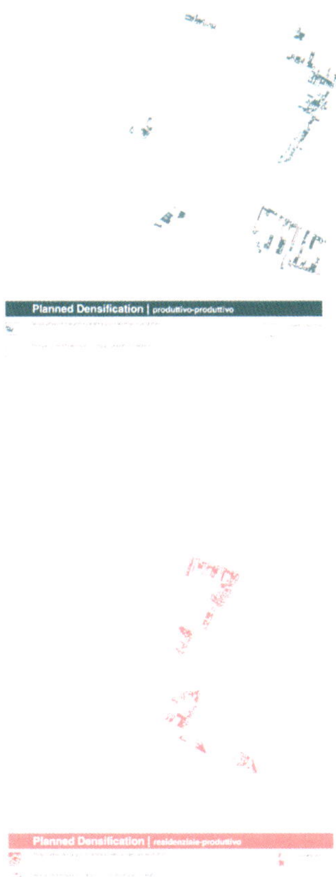

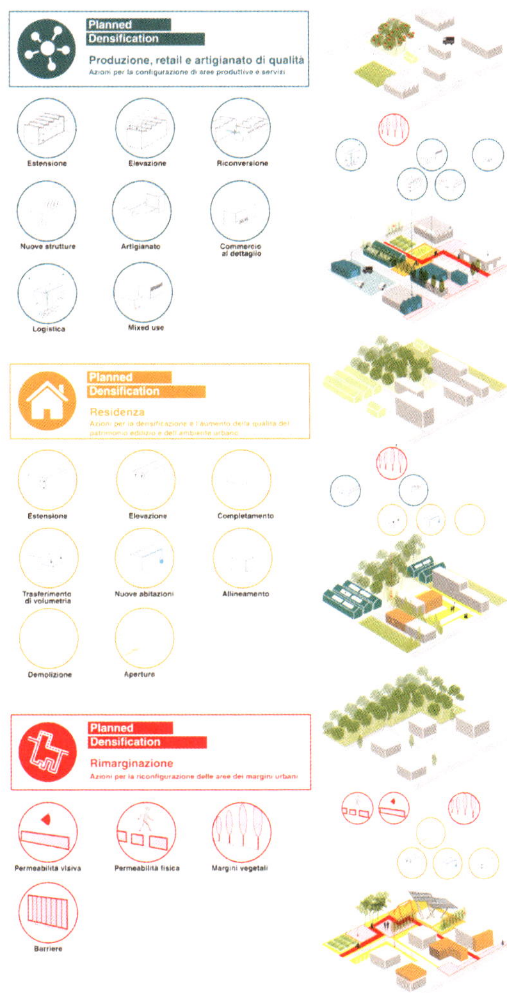

Strategic Plan

Process

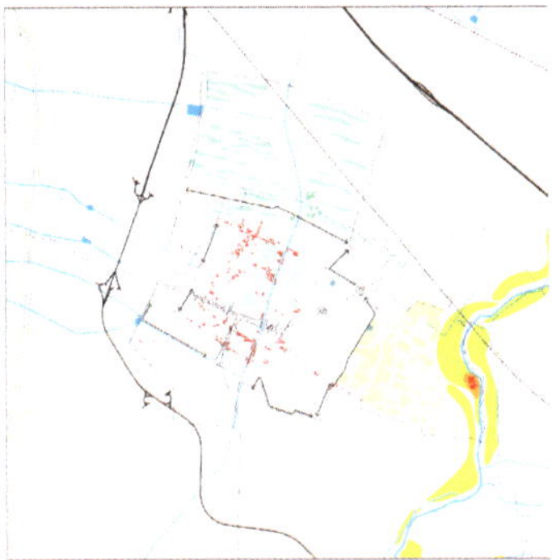

Scenario di breve periodo

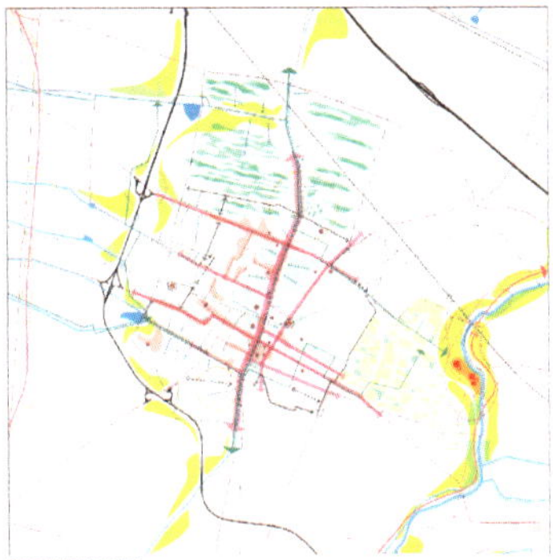

Scenario di medio periodo

Scenario di lungo periodo

Planning actions

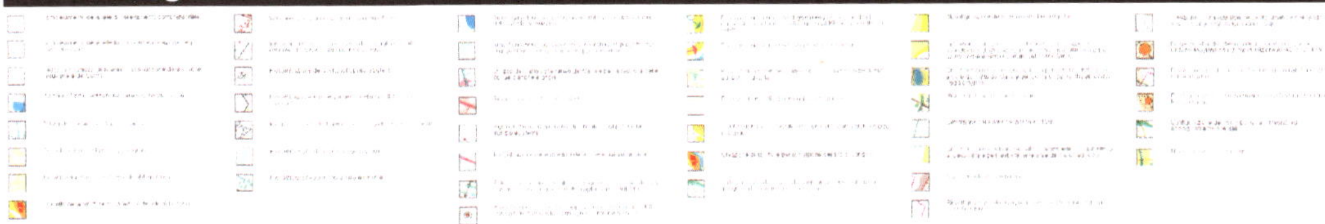

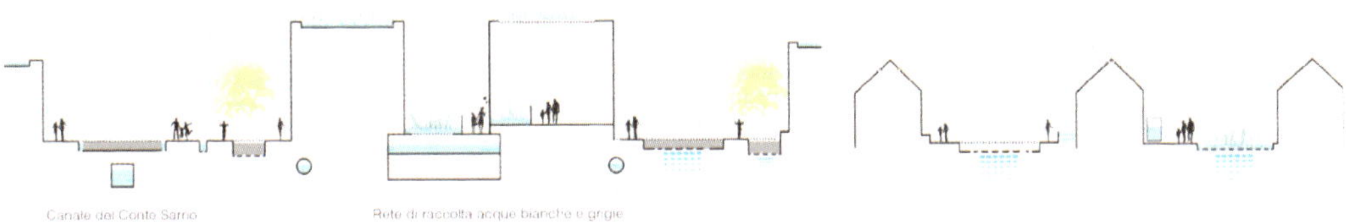

Sezione tipo sulle aree di esondazione controllata

Sezione tipo sulla rete di canali

Project/Process

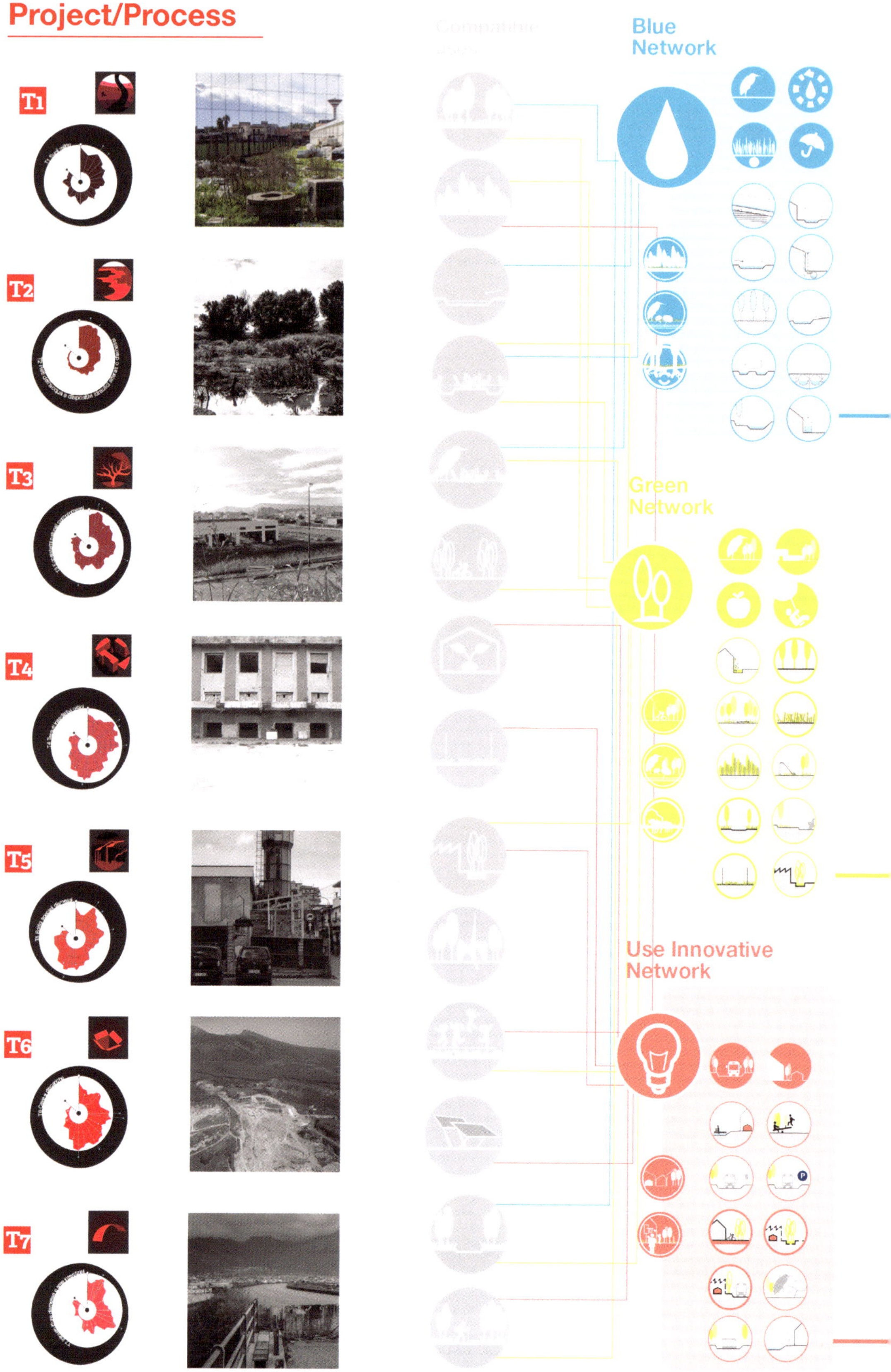

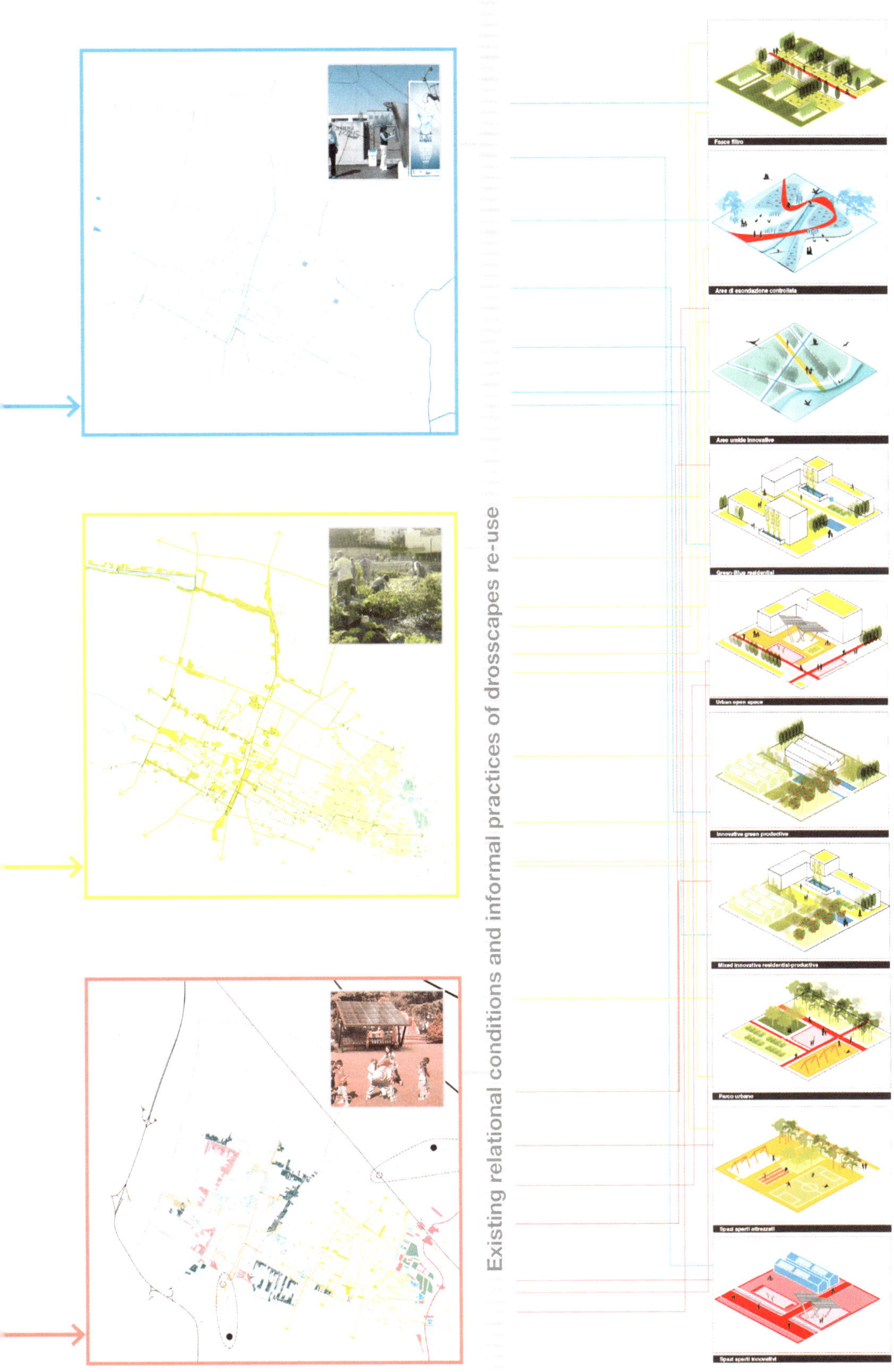
Existing relational conditions and informal practices of drosscapes re-use
Fasce filtro
Aree di esondazione controllata
Aree umide innovative
Green-Blue residential
Urban open space
Innovative green productive
Mixed innovative residential-productive
Parco urbano
Spazi aperti attrezzati
Spazi aperti innovativi